Coarse-Grained Molecular Dynamics

Coarse-Grained Molecular Dynamics

Edited by

Maria Emilova Velinova

Delve
PUBLISHING

www.delvepublishing.com

Coarse-Grained Molecular Dynamics

Maria Emilova Velinova

Delve Publishing
2010 Winston Park Drive,
2nd Floor
Oakville, ON L6H 5R7
Canada
www.delvepublishing.com
Tel: 001-289-291-7705
001-905-616-2116
Fax: 001-289-291-7601
Email: orders@arclereducation.com

ISBN: 978-1-77361-085-6 (Hardcover)

DECLARATION

Some content or chapters in this book are open access copyright free published research work, which is published under Creative Commons License and are indicated with the citation. We are thankful to the publishers and authors of the content and chapters as without them this book wouldn't have been possible.

ABOUT THE EDITOR

Maria Emilova Velinova

Maria Velinova is Ph.D. holder in Quantum chemistry at the University of Sofia since April 2012. Her major research experience is in the field of Computational Chemistry, especially in statistical mechanics methods applied to different sorts of biomolecules. Member of the Laboratory of Quantum and Computational Chemistry at the University of Sofia.

TABLE OF CONTENTS

PART III: COARSE-GRAINED MODELS FOR DNA AND RNA

CONTRIBUTORS

Maria Velinova

Ken Takahashi
Department of Cardiovascular Physiology, Graduate School of Medicine, Dentistry and Pharmaceutical Sciences, Okayama University, Okayama, Japan

Takayuki Oda
School of Medicine, Okayama University, Okayama, Japan

Keiji Naruse
Department of Cardiovascular Physiology, Graduate School of Medicine, Dentistry and Pharmaceutical Sciences, Okayama University, Okayama, Japan

Junfeng Gu
State Key Laboratory of Structural Analysis for Industrial Equipment, Department of Engineering Mechanics, Dalian University of Technology, Dalian 116023, China

Fang Bai
Faculty of Chemical, Environmental and Biological Science and Technology, Dalian University of Technology, Dalian 116023, China

Honglin Li
School of Pharmacy, East China University of Science and Technology, Shanghai 200237, China

Xicheng Wang
State Key Laboratory of Structural Analysis for Industrial Equipment, Department of Engineering Mechanics, Dalian University of Technology, Dalian 116023, China

Ronald D. Hills Jr.
Department of Molecular Biology and Kellogg School of Science and Technology, The Scripps Research Institute, 10550 N. Torrey Pines Rd. TPC6 La Jolla, CA 92037, USA

Charles L. Brooks III
Department of Molecular Biology and Kellogg School of Science and Technology, The Scripps Research Institute, 10550 N. Torrey Pines Rd. TPC6 La Jolla, CA 92037, USA
Department of Chemistry and Biophysics Program, University of Michigan, 930 N. University Ave, Ann Arbor, MI 48109, USA

Jacek Wabik
Laboratory of Theory of Biopolymers, Faculty of Chemistry, University of Warsaw, Pasteura 1, 02-093 Warsaw, Poland

Mateusz Kurcinski
Laboratory of Theory of Biopolymers, Faculty of Chemistry, University of Warsaw, Pasteura 1, 02-093 Warsaw, Poland

Andrzej Kolinski
Laboratory of Theory of Biopolymers, Faculty of Chemistry, University of Warsaw, Pasteura 1, 02-093 Warsaw, Poland

Rino Ragno
Laboratory of Theory of Biopolymers, Faculty of Chemistry, University of Warsaw, Pasteura 1, 02-093 Warsaw, Poland

Peter Spijker
Department of Biomedical Engineering, Eindhoven University of Technology, 5600 MB, Eindhoven, The Netherlands

Bram van Hoof
Department of Biomedical Engineering, Eindhoven University of Technology, 5600 MB, Eindhoven, The Netherlands
Division of Immunology, Beckman Research Institute of the City of Hope, 1500 E. Duarte Road, Duarte, CA-91010, USA

Michel Debertrand
Department of Biomedical Engineering, Eindhoven University of Technology, 5600 MB, Eindhoven, The Netherlands
Division of Immunology, Beckman Research Institute of the City of Hope, 1500 E. Duarte Road, Duarte, CA-91010, USA

Albert J. Markvoort
Department of Biomedical Engineering, Eindhoven University of Technology, 5600 MB, Eindhoven, The Netherlands

Nagarajan Vaidehi
Division of Immunology, Beckman Research Institute of the City of Hope, 1500 E. Duarte Road, Duarte, CA-91010, USA

Peter A.J. Hilbers
Department of Biomedical Engineering, Eindhoven University of Technology, 5600 MB, Eindhoven, The Netherlands

Benjamin A. Hall
Oxford Centre for Integrative Systems Biology, Department of Biochemistry, University of Oxford, Oxford, United Kingdom

Judith P. Armitage
Oxford Centre for Integrative Systems Biology, Department of Biochemistry, University of Oxford, Oxford, United Kingdom

Mark S. P. Sansom
Oxford Centre for Integrative Systems Biology, Department of Biochemistry, University of Oxford, Oxford, United Kingdom

Nikolay Korolev
School of Biological Sciences, Nanyang Technological University, 639798, Singapore

Di Luo
School of Biological Sciences, Nanyang Technological University, 639798, Singapore

Alexander P. Lyubartsev
Division of Physical Chemistry, Arrhenius Laboratory, Stockholm University, Stockholm, S 10691 Sweden

Lars Nordenskiöld
School of Biological Sciences, Nanyang Technological University, 639798, Singapore

Yichen He
State Key Laboratory of Chemical Engineering and Department of Chemistry, East China University of Science and Technology, Shanghai 200237, China

Yazhuo Shang
State Key Laboratory of Chemical Engineering and Department of Chemistry, East China University of Science and Technology, Shanghai 200237, China

Yu Liu
State Key Laboratory of Chemical Engineering, East China University of Science and Technology, Shanghai 200237, China

Shuangliang Zhao
State Key Laboratory of Chemical Engineering, East China University of Science and Technology, Shanghai 200237, China

Honglai Liu
State Key Laboratory of Chemical Engineering and Department of Chemistry, East China University of Science and Technology, Shanghai 200237, China

Aarón Ayllón Benítez
Departamento de Química Física, Universidad de Murcia, 30100 Murcia, Spain

José Ginés Hernández Cifre
Departamento de Química Física, Universidad de Murcia, 30100 Murcia, Spain

Francisco Guillermo Díaz Baños
Departamento de Química Física, Universidad de Murcia, 30100 Murcia, Spain

José García de la Torre
Departamento de Química Física, Universidad de Murcia, 30100 Murcia, Spain

Liang Liu
Department of Physics and Department of Biochemistry, University of Missouri, Columbia, Missouri, United States of America

Shi-Jie Chen
Department of Physics and Department of Biochemistry, University of Missouri, Columbia, Missouri, United States of America

William Hoiles
Department of Electrical and Computer Engineering, The University of British Columbia, Vancouver, British Columbia, Canada

Rini Gupta
Department of Electrical and Computer Engineering, The University of British Columbia, Vancouver, British Columbia, Canada

Bruce Cornell
Director of Science and Technology, Surgical Diagnostics Pty Ltd., Unit 6 30-32 Barcoo Street, Roseville, New South Wales, 2069, Australia

Charles Cranfield
School of Life Sciences, University of Technology Sydney, Broadway, New South Wales, Australia

Vikram Krishnamurthy
Electrical and Computer Engineering, Cornell University, New York, New York, United States of America

Artturi Koivuniemi
Department of Physics, Tampere University of Technology, Tampere, Finland

Timo Vuorela
Department of Physics, Tampere University of Technology, Tampere, Finland

Petri T. Kovanen
Wihuri Research Institute, Helsinki, Finland

Ilpo Vattulainen
Department of Physics, Tampere University of Technology, Tampere, Finland
Department of Applied Physics, Aalto University School of Science and

Technology, Espoo, Finland
MEMPHYS – Centre for Biomembrane Physics, University of Southern Denmark, Odense, Denmark

Marja T. Hyvo¨nen
Department of Applied Physics, Aalto University School of Science and Technology, Espoo, Finland
Department of Physics, University of Oulu, Oulu, Finland

PREFACE

The book entitled Coarse-Grained Molecular Dynamics Simulations provides an overview of the most used coarse-grained (CG) techniques and their applications on different systems. The CG methods offer results for larger systems over a long period of time that are not obtainable with the all-atom description of these systems. The thermodynamics-based, structural-based and force matching methods for constructing potentials, used in the building of effective CG models, are discussed in the first part of the book. In this part also is highlighted some significant application of the coarse-grained molecular dynamics simulations in biochemistry and medicine. The rest of the book gives a detailed description of different CG models, constructed for specific molecules and systems.

Proteins and peptides, topic of part 2 of the book, are scientifically interesting objects, as well as, for their medical importance as potential drugs. Most of the current docking procedures are focused on all-atoms conformational adjustments and fail to reproduce large-scale protein motion. CABS-dock, which employs a coarse-grained representation of proteins, is one of the very few tools that allow unrestrained docking with large conformational freedom of the receptor. Another CG technique used for a description of protein folding and dynamics is the Gō-like model in which each amino acid residue is represented as a single unified atom, and either an attractive or a repulsive interaction is defined between any two residues. In this part of the book are also shown interesting investigations of interactions of proteins with lipid bilayers on CG level.

Computer modelling of very large bimolecular systems, such as long DNA and RNA cannot reach the all-atom resolution in a foreseeable future and this necessitates the development of CG approximations. Part 3 of the book present different CG models used for prediction of structures, solution properties and dynamics of DNAs and RNAs.

The last part of the book, Coarse-Grained Models for Lipids, collect a big selection of researches in the area. The fundamental building blocks of all cell membranes are phospholipids. Phospholipid molecules can demonstrate more than one micellar structures, which play important role in drug delivery systems as well as in biological systems. The use of CG models in exploring a variety of structural and dynamic properties of large molecular systems built from lipid molecules is discussed in details. The CG-MD technique can provide a good approximation based on the selection of the interaction parameters, as well as filling the gap between theoretical and experimentally determined results. Although CG-MD has provided valuable information for understanding the different phenomenon, there is still limits in its application, but the continuous theoretical developments guarantee the fast progress of this area.

Editor

INTRODUCTION TO COARSE-GRAINED MOLECULAR DYNAMICS SIMULATIONS

By Maria Velinova

Molecular dynamics (MD) is the study for chasing the motion of particles in a system by computer simulation. The particles are allowed to interact with each other for a specific time and their trajectories are calculated based on Newton's laws of motion. The forces between atoms and molecules are defined by a force field, usually based on experimental data.

MD simulations are an ideal method for modelling the structure and function of biomolecules. The most common technique used in MD is to track explicitly each individual atom in the system – also known as an all-atom description of the system. Along-with high-throughput computing and parallel MD codes, the all-atom description can permit the simulation of larger systems over a long period of time. Nevertheless, atomic level MD simulations cannot access every spatial and time scale. Therefore, an additional level of simplification with no loss of atomic detail is required by adopting a continuum model of the system. The 'coarse-grained' (CG) models are such kinds of assumption, which can simulate biomolecular systems and processes over long time scales.

Coarse graining refers to the elimination of non-essential degrees of freedom from a system with the all-atom level of description. The goal, therefore, is to develop fast simulation techniques, which can access long length, and time scales without relying exclusively on the increase in computing power set out by Moore's law. Coarse-graining methods have been successfully used to estimate various biological systems previously considered too big to study using existing all-atom MD techniques: a recent example being the shape and the stability of an entire HIV viral capsid [1].

AN OVERVIEW OF COARSE-GRAINING METHODS

The bottleneck – in terms of time and costs – of any MD simulation is the evaluation of the forces which takes maximum simulation time respect to the total [2]. The aim of an effective CG method will therefore be to speed up this part of simulation. There are two main techniques to achieve this:

- coarse-graining in time
- coarse-graining in space

The coarse-graining in time is an approach, which artificially increases the time step used for solving Newton's equations of motion. This result in fewer iterations of the force evaluation loop in the MD code needed to simulate the same total time, thus reducing the computing costs. The only limitation of this approach is that it requires a time step of minimum duration to ensure that the dynamics of interest are captured accurately by the numerical integration scheme. For example, since atomic vibrational frequencies typically range from 10^{12} to 10^{14} Hz, time steps on the order of 1 femtosecond are traditionally used to capture this motion. Another method for coarse-graining in time contains speed up the simulation to focus on points of interest [3]. In sampling the free energy landscape of a system, for example, this can be achieved by progressively adding a sequence of Gaussian functions to the potential, which does not allow the system of returning to a previously visited state.

Speeding up the simulation by coarse-graining in space occurs through either a reduction in the number of particles, which need to be simulated, or a simpler functional form for the inter-particle force field, or a combination of both. A number of particles can be reduced by grouping the individual particles into larger CG units and treating each CG unit as an individual (Figure 1). All above, reduce the number of degrees of freedom obligatory to keep track of in the simulation. However, there is no consensus to determine the minimum number of atoms that can be grouped into one coarse-grained particle, or where the CG bead should be placed: centre-of-mass or centre-of-geometry. Several studies have shown that smaller degrees of coarse-graining lead to better potential transferability to different temperatures [4, 5, 6] or chain lengths [7, 8]. Nevertheless, it is usually a compromise between capturing the details of the system of interest and computational convenience, and is best decided by the developer.

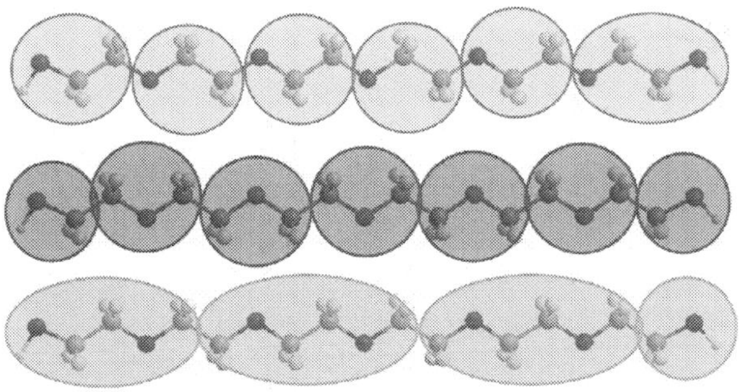

Figure 1: A section from a polyethylene glycol (PEG). On the top, schematic CG representation of the molecule is shown in blue, with six CG units. In the middle, a different CG representation is shown in red with 7 CG units. And down, it is shown in green, CG representation with 4 CG units.

A varied coarse-graining method have been described in the literature. All of them start with a detailed atomistic model or experimental data and then proceed to certain common steps such as CG mapping of atoms and molecules to define a coarse-grain model and constructing an effective potential (usually pairwise and additive) to describe the statistical mechanics of these CG sites. To derive the interaction potentials, three main properties are commonly used as targets:

- thermodynamics;
- structure;
- force.

To determine which of the above needs to be reproduced during the simulation, it is to know which one of them plays the decisive role in the system of interest.

Thermodynamics-based coarse-graining

Thermodynamics-based coarse-graining relies on the following target properties that are experimentally measured – critical points [9], interfacial tension [10] and hydration free energy [11], among others. This type of coarse graining is often described as a "top-down" bridging approach since microscopic CG representations are derived from macroscopic properties. The success of the thermodynamic approach is best illustrated in the popular MARTINI model [12] where partitioning free energies between polar and apolar phases were used to optimize coarse-grain interaction parameters. The MARTINI CG force field is especially effective for modelling membranes and membrane-related phenomena because these are essentially driven by amphipathic assembly forces (Figure 2). There are some drawbacks to the MARTINI system as well: since it is targeted at reproducing energies, it cannot guarantee the reproduction of the structure of the system and structures of peptides and proteins must be fixed by an elastic network while implementing this method. Furthermore, although MARTINI has additional features such as CG water particles, which represent four water molecules, there appears to be no real way to derive such a CG mapping.

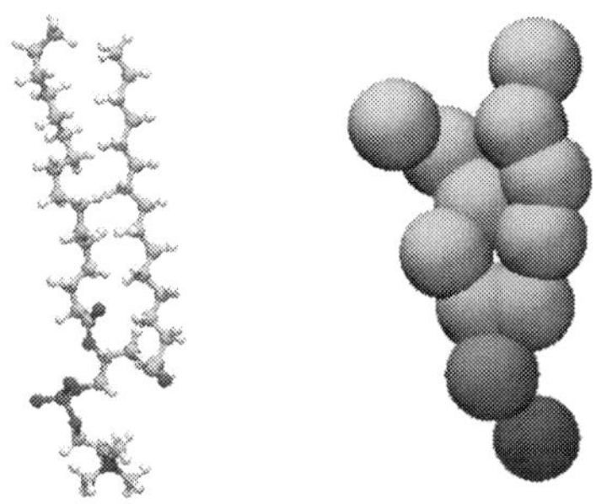

Figure 2: Atomistic (AT) vs. coarse-grained (CG) representations of phosphatidyl choline (PC), namely DMPC (atomistic) and DPPC (CG).

Structural-based coarse-graining

Structural-based coarse-graining targets to find CG interactions that reproduce predefined target structures. Since this method obtains distribution functions from more detailed simulations, it is described as a "bottom-up" approach. The pair correlation function g(r) is mainly used for this aim, although theoretically, any other distribution function such as angular or dihedral can be used.

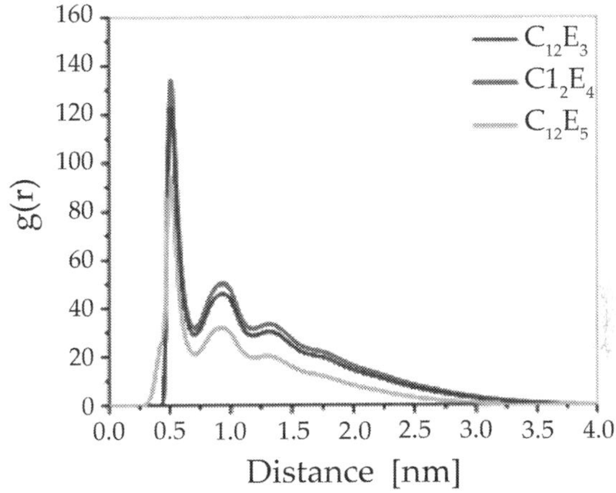

Figure 3: Pair correlation function g(r) in the solutions of poly(ethylene oxide) surfactants

Structural-based coarse-graining does not promise reproduction of thermodynamic properties of the system unless one applies a post-simulation correction for this property of interest [13, 14, 15]. There are two techniques often used to derive structural-based CG potentials: (1) iterative Boltzmann inversion (IBI) [16] and (2) the inverse Monte Carlo method (IMC) [17].

IBI is assured to generate a CG potential, which accurately reproduces the input distribution function when it is used to perform a CG simulation. The IBI method consists of three main steps:

(1) Performing a simulation on all-atom level and to generate the particle trajectories from which target function(s) can be extracted later;

(2) Mapping the atoms onto the CG level and extracting the coarse-grained topologies and target functions;

(3) Carrying out the actual iteration by repeatedly simulating the

system in its mesoscale representation and calculating the target function(s).

An effective CG potential that replicates the exact structural features from the corresponding atomistic simulations is achieved when the trial and target distribution functions are equal or almost equal to some recommended tolerance value.

There are some restrictions with this method:

- The pair correlation functions created at each iteration are not smooth due to statistical noise which in turn leads to energy conservation problems. Although this can be easily avoided by applying a Gaussian smoothing function or fitting a smooth cubic spline, the latter might cause modifications in the potential function.
- In order to reproduce the structure in g(r) correctly, the pair potentials generated should have more than one minimum. There is no certainty that IBI potentials will forecast any material property other than the input distribution function correctly. This significantly limits their application and effectiveness.

However, the IBI method has been used to simulate numerous different phenomena in systems like polymer melts [18, 19], liquid crystals [20], solutions [21] and low molecular weight species [22, 23].

The inverse Monte Carlo (IMC) approach once more tries to reconstruct effective interaction potentials for coarse-grained systems from a radial distribution function which is calculated from the all-atom description of the system [24]. IMC generally offers faster convergence than IBI since the potential update is rigorously derived using statistical mechanics. Another advantage of the IMC is that it is efficient for multicomponent mixtures as it considers cross-correlations between different particles for every potential update. The drawback of IMC is that it needs much longer sampling to get statistics that are good enough to calculate the potential update to a similar accuracy as IBI.

Force matching method

The force matching (FM) method consist of minimizing the difference between the forces defined by the CG force field and the original atomistic force field, instead of reproducing distributions functions [25]. It has been used successfully in developing CG models for many different systems, in particular biomolecules [26].

The FM technique employs the so-called multiscale coarse-graining method [27] wherein an atomistic MD simulation of a system is executed and snapshots of the forces and positions of each atom are stored at exact time intervals. The individual atoms are grouped into various clusters, which are designated as the CG particles. The location of every CG particle is the centre of mass of the corresponding cluster and the force on each CG particle (f_{CG}) – also called the reference forces - is then simply the sum of the forces acting on the individual atoms of that cluster. The challenge of the FM method is then to find a functional form for the CG force field, which matches, as much as possible with the reference forces.

There are some limitations of the FM approach:

- The system of linear equations that need to be solved for matching forces can be very complex, which may coax the authors to use various approximations to find the optimal as opposed to the exact solution. As an example, small subsets of atomistic simulation snapshots can be used to perform a block averaging procedure.
- It is well known that molecules repel each other at short distances; therefore, a very high number of atomistic snapshots need to be taken in the least squares to properly sample these distances. A possible solution is to calculate the potential at distances as close as is practically possible – using linear, quadratic or exponential functions.

To improve the simulation of the underlying atomistic qualities through the FM method, it has been recommended by scientists to include explicit higher-order interactions [28].

APPLICATIONS OF THE COARSE-GRAINING METHODS

There has been a surge in studies using coarse-grained (CG) simulations for molecular modelling in recent years. We have summarized the studies using CG models for proteins, polymers and nucleic acids modelling.

Coarse-grained models for proteins

In structure-based CG models, there is significant variation in the balance between the native bias and generic physical-chemical interactions, which is a key issue in CG modeling for proteins. In strongly biased structure-based CG models, two kinds of models are widely used: the elastic network models (ENMs) [[29],[30]] and Go models [[31]].

The ENM system consists of a network of beads connected by elastic springs, with usually one bead per amino acid. The easiness of the parameterization is balanced by the equilibrium reference configuration, from which only harmonic fluctuations are possible. Nevertheless, an ENM also includes the topology of the system and is able to replicate the correct pattern of the principal modes, which is usually the most significant to the protein function.

The Go model – although proposed long ago for the specific simulation of protein folding – has been only recently out to use for its innovative applications and modifications. In the original version, the protein was represented as a chain of one-bead amino acids whose structure was biased toward the native configuration through simple attractive or repulsive non-bonded interactions between those beads. This version, although highly simplified, was successful in reproducing several features of the thermodynamics and kinetics of folding; probably because of the fact that the folding process has usually evolved to satisfy the principle of minimal frustration [[32]]. On the other hand, completely unfrustrated Go models are unsuccessful to describe intermediate metastable folding states.

The MARTINI force field [[33]], weakly biased to the native protein structure, models mostly via the secondary structure constraints.

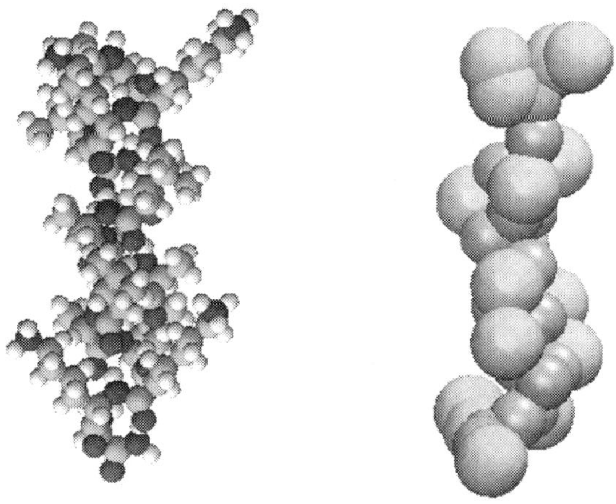

Figure 4: Atomistic (AT) vs. coarse-grained (CG) representations of KALP peptide.

Coarse-grained models for polymers

Different simulations models have described polymers using different length and time scales. An atomistic model of a polymer contains all the atoms which are present in the real polymer, with their individual interactions modelled as realistically as possible. In contrast, a CG model of a polymer consists of super-atoms which may represent groups of atoms – typically 10 real atoms into one super atom, monomers or parts thereof, or even several monomers. The interactions in a CG polymer model is therefore only defined between these super-atoms. Nevertheless, these interactions are effective in that they also incorporate the effects of the neglected degrees of freedom resulting in a reduction of the number of degrees of freedom. Polymers have been modelled using both lattice and off-lattice coarse-grained models which have been successful in reproducing properties not only in the target set, but also for polymer chains much longer than the oligomers used at the parameterization stage [[34]].

Since the CG model is computationally cheaper, longer polymer chains can be simulated in a completely relaxed configuration. This provides the advantage of the reliable calculation of properties like the hydrodynamic radius and the radius of gyration even for chains of 2000 and more monomers – as is the case with real and commercially used polymers.

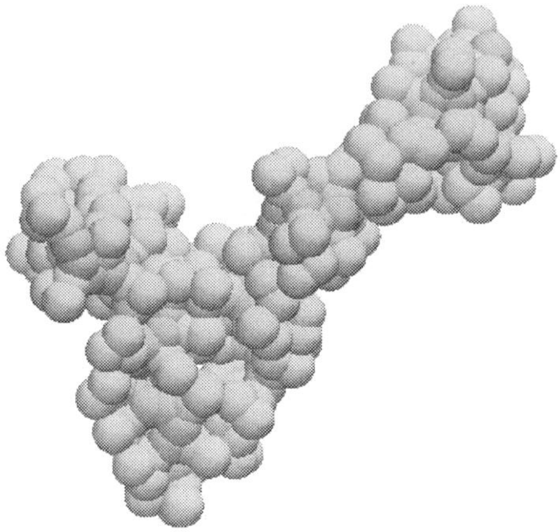

Figure 5: Coarse-grained (CG) representations of Polystyrene chain of 100 monomers [35].

Coarse-grained models for nucleic acids

The CG modeling of RNA and DNA is significantly different due to the difference in their physical states in cells; while DNA strands are much longer and exist either as double helices or flexible single strands once 'melted', the RNA single strands are shorter and often fold into their native structures. Methods similar to protein modelling have been used successfully for RNA [36] whereas different resolutions are needed for the CG modelling of DNA, depending on the length scales of interest. A few CG models for DNA have been recently developed that use a few particles per nucleotide [37-, 38, 39, 40].

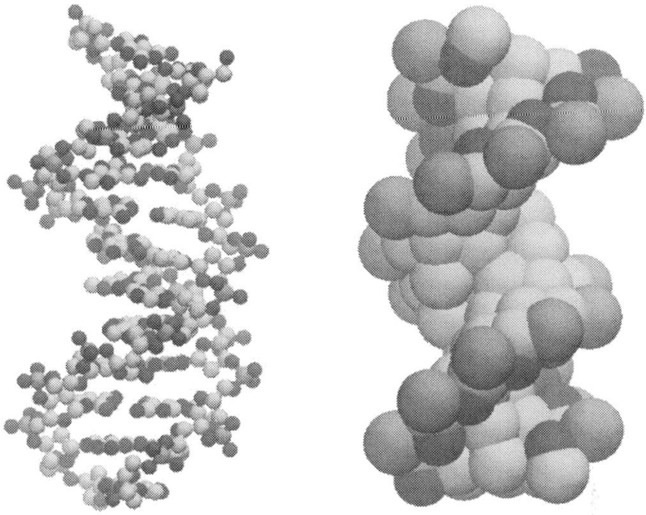

Figure 6: Atomistic (AT) vs. coarse-grained (CG) representations of DNA [41].

FUTURE CHALLENGES

Regardless of its rapid innovations, CG modelling continues to meet a number of challenges. The most common problem with some CG models is that they may have inbuilt bias into them due to researcher's bias regarding the structure and function of a biomolecule. Therefore, clear standards and guidelines based on statistical mechanics need to be adopted for CG modeling for a more accurate prediction. It is also important to continually adapt existing CG models to newer applications and review the failed modellings as well. Nevertheless, there is no question that CG technologies and their applications will relentlessly expand in the future.

As mentioned above, standardized guidelines are required to transfer CG models between various physical systems as well as between different thermodynamic conditions. It is unlikely that any CG model can ever be completely transferable because it is highly simplified reproduction of a complex system, and with this caveat, certain information will always be available in the form of averages only. However, since certain features of a CG model can be completely transferable, the key challenge will be to define those features and reasons for their transferability.

Another important concern is whether CG technologies will be of any real relevance in the future. Despite the driving need of scientists to access biomolecular information that cannot currently be achieved through all-atom MD or MC methods, the latter have also been projected to considerably increase in computing and operating scales in the foreseeable future. In that case, it is reasonable to question if the need for CG modeling and simulation will remain strong in future. However, since the methodologies and problems studied by CG simulation are continually evolving with time, CG modeling is likely to be always valued and needed. To bring simulation closer and closer to the real molecular world, it is vital that the orders of magnitude in length scale and timescale be bridged. Moreover, CG will also remain a vital methodology in the interpretation of the behavior of complex biomolecular systems as it offers a highly simplified picture which can be analyzed with ease; the all-atom description on the other hand contains more detailed information than is needed and offers no specific benefit as an analysis tool. Since this aspect of coarse-graining will likely remain an important and valuable advantage to numerous scientists it will be useful to develop a more rigorous and systematic approach to coarse-graining.

CONCLUSIONS

Coarse graining is an exciting technology in terms of conceptual, theoretical, and algorithmic development in the fields of computer simulation and statistical mechanics. It offers a great step forward in the molecular modeling and simulation of real complex systems. Furthermore, coarse-graining provides the critical connection between the highly detailed world of the atomistic interactions and the highly averaged world of thermodynamics. In other words, coarse-graining and CG modeling "ample" the endeavor of using computer simulation and molecular modeling to understand and predict the behavior of biomolecules and other complex condensed phase systems. Another important feature of CG modeling, especially in the context of biophysics, is to couple the physical processes and forces occurring at each scale for which a system is represented. The multiscale paradigm offered by various CG modelling techniques facilitates this connection between different scales. Taking all the pros and cons together, there is every reason to be optimistic about the future growth and impact of this revolutionary technique in molecular simulation and prediction.

REFERENCES

(Endnotes)

1 V. Krishna, G. S. Ayton, G. A. Voth, Biophys. J. 98, 18 (2010).

2 D. E. Shaw, K. J. Bowers, E. Chow, M. P. Eastwood, D. J. Ierardi, J. L. Klepeis, J. S. Kuskin, R. H. Larson, K. Lindor-Larsen, P. Maragakis, M. A. Moraes, R. O. Dror, S. Piana, Y. Shan, B. Towles, J. K. Salmon, J. P. Grossman, K. M. Mackenzie, J. A. Bank, C. Young, M. M. Denero, B. Batson, "Millisecond-scale molecular dynamics simulations on Anton," in Proc. Conf. High Perform. Comput. Networking, Storage Anal. - SC '09, (New York, New York, USA), p. 1, ACM Press (2009).

3 A. Laio, M. Parrinello, Proc. Natl. Acad. Sci. U. S. A. 99, 12562 (2002).

4 P. Carbone, H. A. K. Varzaneh, X. Y. Chen, F. Muller-Plathe, J. Chem. Phys. 128 (2008).

5 J. Ghosh, R. Faller, Mol. Simul. 33, 759 (2007).

6 T. Vettorel, H. Meyer, J. Chem. Theory Comput. 2, 616 (2006).

7 K. Kamio, K. Moorthi, D. N. Theodorou, Macromolecules 40, 710 (2007).

8 H.-J. Qian, P. Carbone, X. Chen, H. A. Karimi-Varzaneh, C. C. Liew, F. Mueller-Plathet, Macromolecules 41, 9919 (2008).

9 S. Albo, E. A. Muller, J. Phys. Chem. B 107, 1672 (2003).

10 S. O. Nielsen, C. F. Lopez, G. Srinivas, M. L. Klein, J. Chem. Phys. 119, 7043 (2003).

11 W. Shinoda, R. DeVane, M. L. Klein, Soft Matter 4, 2454 (2008).

12 S. J. Marrink, A. H. de Vries, A. E. Mark, J. Phys. Chem. B 108, 750 (2004).

13 D. Reith, M. Putz, F. Muller-Plathe, J. Comput. Chem. 24, 1624 (2003).

14 J. Fischer, D. Paschek, A. Geiger, G. Sadowski, J. Phys. Chem. B 112, 13561 (2008).

15 H. Wang, C. Junghans, K. Kremer, Eur. Phys. J. E 28, 221 (2009).

16 D. Reith, M. Putz, F. Muller-Plathe, J. Comput. Chem. (2003).

17 A. P. Lyubartsev, A. Laaksonen, Phys. Rev. E 52, 3730 (1995).

18 B. Hess, S. Leon, N. van der Vegt, K. Kremer, Soft Matter 2, 409 (2006).

19 K. Kamio, K. Moorthi, D. N. Theodorou, Macromolecules 40, 710 (2007).

20 C. Peter, L. Delle Site, K. Kremer, Soft Matter 4, 859 (2008)

21 G. Milano, S. Goudeau, F. Muller-Plathe, J. Polym. Sci., Part B: Polym. Phys. 43, 871 (2005).

22 J. Ghosh, R. Faller, Mol. Simul. 33, 759 (2007).

23 J. R. Silbermann, S. H. L. Klapp, M. Schoen, N. Chennamsetty, H. Bock, K. E. Gubbins, J. Chem. Phys. 124 (2006).

24 A. P. Lyubartsev, A. Laaksonen, Phys. Rev. E 52, 3730 (1995).

25 S. Izvekov, G. A. Voth, J. Chem. Phys. 123, 134105 (2005).

26 J. Zhou, I. F. Thorpe, S. Izvekov, G. A. Voth, Biophys. J. 92, 4289 (2007).

27 S. Izvekov, G. A. Voth, J. Phys. Chem. B 109, 2469 (2005).

28 L. Larini, L. Lu, G. A. Voth, J. Chem. Phys. 132, 164107 (2010).

29 M. M. Tirion, Phys. Rev. Lett. 77, 1905 (1996).

30 L. W. Yang, A. J. Rader, X. Liu, C. J. Jursa, S. C. Chen, H. A. Karimi, I. Bahar, Nucleic Acids Res. 34, W24 (2006).

31 C. Clementi, H. Nymeyer, J. N. Onuchic, J. Mol. Biol. 298, 937 (2000).

32 D. Baker, Nature 405, 39 (2000).

33 L. Monticelli, S. K. Kandasamy, X. Periole, R.G Larson, D. P. Tieleman, S. J. Marrink, J. Chem. Theor. Comput. 4, 819 (2008).

34 D. Reith, B. Muller, F. Muller-Plathe, S. Wiegand, J. Chem. Phys. 116, 9100 (2002).

35 G. Rossi, L. Monticelli, S. R. Puisto, I. Vattulainen and T. Ala-Nissila Soft Matter, 7, 698-708 (2011).

36 S. S Cho, D. L. Pincus, D. Thirumalai, Proc. Natl. Acad. Sci. USA 106, 17349 (2009).

37 T. A. Knotts, N. Rathore, D. C. Schwartz, J. J. de Pablo, J. Chem. Phys. 126 (2007).

38 A. Morriss-Andrews, J. Rottler, S. S Plotkin, J. Chem. Phys. 132, 035105 (2010).

39 A. Savelyev, G. A. Papoian, Proc. Natl. Acad. Sci. USA 107, 20340 (2010).

40 J.J. Uusitalo, H.I. Ingólfsson, P. Akhshi, D.P. Tieleman, S.J. Marrink, J. Chem. Theory Comput. 11, 3932 (2015).

41 J. Corsi, R. W. Hawtin, O. Ces, C. S. Attard, S. Khalid, Langmuir 26, 12119 (2010).

PART I: COARSE-GRAINED MOLECULAR DYNAMICS SIMULATIONS: THEORY AND APPLICATIONS

CHAPTER 1

COARSE-GRAINED MOLECULAR DYNAMICS SIMULATIONS OF BIOMOLECULES

Ken Takahashi[1], Takayuki Oda[2], Keiji Naruse[1]

[1] Department of Cardiovascular Physiology, Graduate School of Medicine, Dentistry and Pharmaceutical Sciences, Okayama University, Okayama, Japan

[2] School of Medicine, Okayama University, Okayama, Japan

INTRODUCTION

Richard P. Feynman stated, "Everything we know is only some kind of approximation, because we know that we do not know all the laws as yet." In contrast, simulations attempt to reveal nature's secrets by applying approximations of nature's laws. As the name suggests, coarse-grained molecular dynamics (CGMD) simulates the behaviors of atoms and molecules by audacious approximations. Nevertheless, when used effectively, CGMD simulations can explain the physicochemical nature or even predict the behavior of a biological system, which may be impossible experimentally. Because of this advantage, the amount of research that uses CGMD simulations has steadily increased in recent years (Figure 1).

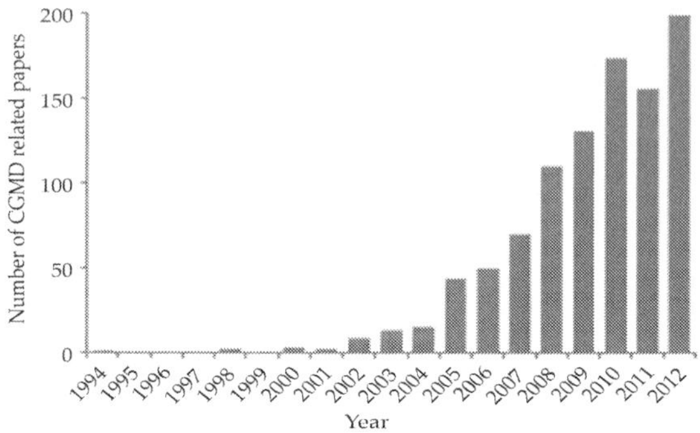

Figure 1: Increase in the number of CGMD-related papers. Pubmed was used to search for the numbers of articles published per year that included "coarse grained" and "molecular dynamics".

Although simulating a molecular system based on quantum chemistry is possible [1], molecular dynamics usually employs classical Newtonian physics to simulate molecular motions. The force between atoms is derived from bonded interactions, including 2-,3-,and 4-body interactions, and non-bonded interactions such as van der Waals and electrostatic interactions. All-atom molecular dynamics (AAMD) calculate the motions of every atom, including those of hydrogen, and thereby simulate the behavior of a system with considerable accuracy.

However, high computational costs limit the spatial and temporal scales that can be used with AAMD simulations. In contrast, CGMD substantially reduces these costs by replacing multiple atoms with a larger, unified particle (coarse-grained atom); thus, the degrees of freedom of a system are limited. This makes the energy function smoother and allows the use of larger time steps (e.g., 30 fs) as compared with that of AAMD (e.g., 2 fs). By appropriately setting up a CGMD system, the number of simulations can be exponentially increased with a limited loss of accuracy.

In this review, we discuss the variations, potentials, and limitations of CGMD simulations. Strategic designing of a molecular system of interest and applying appropriate analyses are the keys to success with CGMD simulations. The validity of CGMD simulations can be supported by comparing the simulation results with experimentally derived evidence. Here, we describe specific examples of CGMD applications and the methods used to correlate the results obtained from simulations with those obtained from experiments.

OVERVIEW OF CGMD

Configuring a CGMD system

In essence, coarse-graining of a molecular system is a process to describe the behaviors of molecules in a simpler way by discarding less essential effects. Historically, a simple model in which amino acid residues were placed in a two-dimensional square lattice, the Gō model,was introduced to investigate protein folding [2]. The Gō model takes into account only those interactions between residues that are present in the native protein structure. This strikingly simple theory was considerably successful in describing the folding mechanisms of real proteins. It inspired many researchers and, after various modifications, is still being applied.

There are infinite ways to construct CGMD models. Variations of these models are derived from the differences in the natures of molecular systems. For example,there is no need to treat solvent molecules,such as water, explicitly when the target molecular event involves DNA self-assembly [3]. In contrast, incorporating solvent molecules, ions, and electrostatic interactions between charged molecules is necessary when the object of interest is the translocation of cationic nanoparticles through a lipid bilayer [4]. When proteins are the molecules of interest, the numbers and types of coarse-grained atoms assigned to represent each amino acid side chain are crucial factors that define a CGMD model [5].

In more complex CGMD models, four to six coarse-grained atoms are allocated to one amino acid residue. Some CGMD models explicitly incorporate hydrogen bonding [6]. In contrast, only one pseudo atom is assigned for an entire subunit of a protein in a huge molecular system, such as a virus capsid.

Some CGMD models are individually tailored for each study, whereas others are generalized for various molecular systems. Parameterization of atomic interactions (forcefield) requires expertise in physics and chemistry; thus, it is particularly difficult for researchers who are not expert in these fields to construct a universal CGMD model. However, there are some generalized CGMD models, including Martini [7, 8] and NAMD-CG builder [9, 10]. Among these,the Martini forcefield has recently gained popularity [4, 11, 12, 13, 14, 15, 16, 17, 18, 19, 20, 21, 22, 23, 24]. Because this CGMD model runs under AAMD software Gromacs [25] with a modified forcefield for coarse-grained atoms, the computational algorithm for CGMD simulation is exactly the same as that used for AAMD.

The Martini forcefield incorporates preset parameters for general phospholipids and proteins. CGMD simulations can basically be run for lipid membranes and/or proteins with known three dimensional structures. However, there are barriers during actual simulations because some knowledge of several processes is required, including the construction of a molecular system and potential energy minimization. Applications of the Martini forcefield have increased after additional coarse-grained parameters were proposed for lipids, including cardiolipin [26] and glycolipids [22], carbohydrates such as monosaccharides, disaccharides, and oligosaccharides [27], DNA [23], and artificial carbon nanotubes [19]. Among CGMD models that have to be reparameterized each time and whose targets are restricted to specific types of molecules (e.g., peptides), Martini's ready-to-use and highly applicable system is very attractive.

In addition, the Martini forcefield can simulate the dielectric responses of solvent molecules. Coarse-grained water atoms in the Martini forcefield were originally apolar, similar to several CGMD models. A recently proposed polarized water model enabled simulations of realistic phenomena, such as electroporation of a lipid bilayer with voltage applied across a bilayer membrane [28].

What can be done with CGMD simulations

During a CGMD simulation, analysis methods and perturbations similar to those used with AAMD can be used. Basic quantities used in a CGMD analysis include indicators of structural alterations such as root mean square deviation (RMSD) and root mean square fluctuation (RMSF), tilt angles of lipids and proteins, area per lipid within a lipid membrane, and diffusion constants for molecules.

Local pressure profiles [29], binding free energy, and potential of mean force are analyzed. In addition, voltage can be locally analyzed [4].

These energy profiles are important for large molecular systems, such as for protein-protein interactions.

Regarding perturbations, the positions of every atom can be controlled during a CGMD simulation. Position restraints and force applications are used during steered molecular dynamics. These features are used for protein and ligand docking. Replacing amino acid residues (mutation) is possible for proteins, which is useful for comparing the results of simulations with those of experiments. In addition to temperature and pressure, the surface tension of a lipid bilayer is controllable during a CGMD simulation. Applying a stretch force to membrane proteins is achieved by incorporating surface tension.

Membrane potential, the difference in electric potential between the inside and outside of a cell, is important for certain cellular functions. This voltage difference can be mimicked during a CGMD simulation by creating two compartments by modeling two parallel planes of lipid bilayers inside a simulation box, and then adding the appropriate types and numbers of ions into each compartment.

In addition, pH effects can be mimicked during a CGMD simulation. The method used to alter the pH in a molecular system by modifying the protonation state of a titratable site, originally suggested for AAMD simulations [30], is also possible for CGMD simulations [26, 31]. pH alters the protonation states of aspartate, glutamate, and histidine in proteins [32]; thus, it changes a protein's activities.

Another feature that makes CGMD simulations very powerful is "reverse graining", to reconstruct all atom geometry from coarse-grained molecular geometry. Reverse graining is achieved by a simulated annealing method, which efficiently determines the minimum potential energy [33, 34]. Reverse graining enables a strategy to first observe any large conformational changes in a molecular system by CGMD and then analyze the detailed molecular interactions by AAMD.

Advantages and disadvantages

If a molecular system is appropriately constructed, then high-speed CGMD simulations can be used with nearly the same accuracy as those achieved with AAMD simulations. An example of a CGMD simulation of a human immunodeficiency virus (HIV) membrane protein is shown

(Figure 2). CGMD simulation predictions of binding free energies between proteins are more than 500 times faster than those with AAMD simulations, and with similar accuracy [13]. A CGMD simulation has the advantage of being fast; however, there are some disadvantages due to coarse graining.

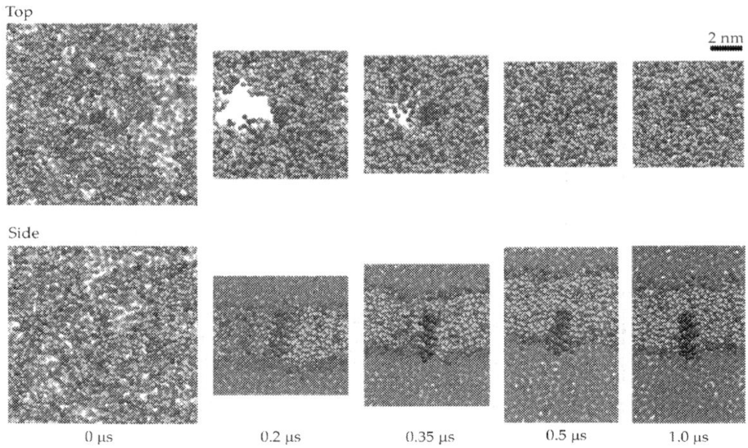

Figure 2: Example of a CGMD simulation. This molecular system consisted of one HIV-1 viral protein U (Vpu, PDB id: 2JPX), 30 POPEs, 30 POPCs, 30 DPPEs, 30 DPPCs, 30 galactosylceramides, 30 cholesterols, and 3000 CG water atoms. The total number of CG atoms was 5178. Lipids and water particles were initially randomly placed within a 12 × 12 × 12 nm box, after which the self-assembly of lipid molecules was observed. The position of the Vpu protein was constrained. The size of each picture corresponds to the size of a periodic boundary box. In the top views from 0.2 to 1.0 μs, water atoms were omitted for clarity. Simulation for 1 μs was performed using a personal laptop computer (2.1 GHz single core) with a computation time of about 9 h.

A disadvantage with the Martini CGMD model is that alterations in the secondary structures of proteins are beyond the scope of simulations. If alpha helix parameters are assigned to the amino acid residues of a protein, then these parameters do not change throughout the simulations. This can potentially affect the accuracy of CGMD simulations.

Stark et al. reported that the estimated protein-protein interaction between lysozyme and chymotrypsinogen with a CGMD simulation was greater than the experimentally derived value. They attributed this discrepancy to the intrinsic problems of CGMD parameters and proposed that down-scaling of van der Waals parameters could resolve this problem [11]. The key to a successful CGMD simulation is to set up the molecular system of interest based on sufficient knowledge regarding these issues.

APPLICATION OF CGMD

Although vast computational power enables more accurate AAMD to simulate larger molecular systems for a longer time [35, 36], CGMD simulations are still good at handling large molecular systems, such as a membrane tether consisting of 4 million particles [37], a fully solvated protein in a liposome [38], and for effectively sampling the configuration space of target molecules because of rapid calculation speed. An example of a conformational change in a mechanosensitive ion channel in response to membrane tension is shown (Figure 3), where a CGMD simulation can reproduce the dynamic actions of a protein and aid in explaining the observations derived from experiments [21]. Analysis of these simulation results leads to a better explanation of the experimentally observed phenomena based on the theories of physics and chemistry.

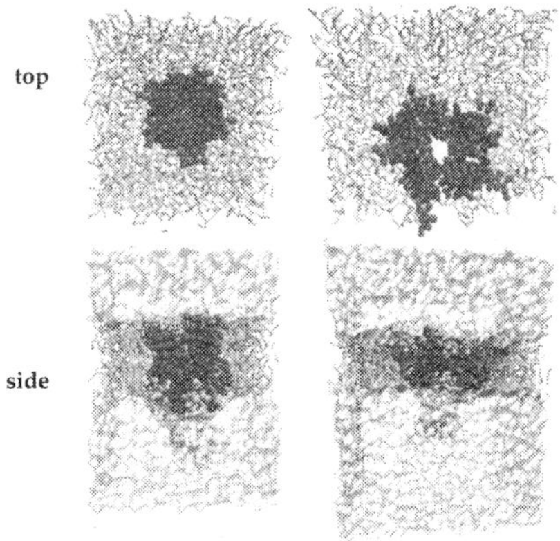

Figure 3: Opening of a bacterial mechanosensitive channel of large conductance (MscL, PDB accession number: 2OAR) in response to lipid bilayer stretch obtained by CGMD simulations. An MscL channel is embedded in a lipid bilayer consisting of POPC lipids, which are shown in stick representation. The simulation box is filled with CG water atoms. The channel's pore enlarged and the bilayer thickness decreased after applying stretch. Left: before applying tension, right: 4 μs after applying tension.

Simulations of large molecular systems

Protein folding is a problem that is scientifically interesting [39, 40] as well as medically important, given that protein misfolding is a cause of several diseases, such as Parkinson's disease [41]. Borgia et al. demonstrated that similarities in amino acid sequences between adjacent domains induced protein misfolding in multidomain proteins using CGMD simulations [42]. They used a Gō-like model in which each amino acid residue was represented as a single unified atom, and either an attractive or a repulsive interaction was defined between any two residues [43]. Water molecules and ions were not incorporated, drastically decreasing the degrees of freedom in their simulation system. This is a pr actical choice for effectively sampling protein structures among a vast conformational space in protein folding studies.

Protein aggregation, which is also a cause of certain diseases, including Alzheimer's disease, prion diseases, and Type II diabetes, is a phenomenon related to protein folding in that it can be regarded as an alternate folding pathway [44]. Aggregation begins from the nucleation of proteins. This nucleation process has been elucidated using coarse-grained simulations, including CGMD. Nguyen and Hall reported that an amorphous aggregate of peptides was formed first, after which an ordered nucleus was formed, resulting in the lateral addition of β-sheets [45].

In their CGMD model (designated PRIME), four pseudo atoms were mapped to each amino acid residue, hydrogen bonding was incorporated, a solvent was implicitly modeled, and discontinuous potential energy functions were employed [46]. Because the PRIME model allowed for alterations in the secondary structures of proteins, it was suitable for studying protein aggregation, as proteins undergo dynamic structural changes during this process. Discontinuous potential energy functions and implicit handling of solvents reduced the computational time and provided an advantage of sampling a vast conformational space.

A capsid is the protein shell of a virus that can be simulated using CGMD simulations [47, 48, 49, 50]. Some types of viruses, including HIV, pose a threat to the life of their hosts. Thus, unraveling the processes involved in virus proliferation, including capsid formation, is important. In a huge molecular system that includes a virus capsid typically comprising hundreds of proteins, one CGMD atom unites a large number of amino acid residues (e.g., a protein subunit), and solvents are not incorporated. Although the definition of CGMD atoms (i.e., identifying CG sites) can be deduced from the structural requirements of virus capsids [51], this can also be achieved by AAMD simulations using X-ray crystallographic structures to parameterize constants, such as Lennard-Jones potential parameters [48]. The effects of the lengths of polyions on the formation of a huge virus capsid can be explained using highly coarse-grained models [50]. In this case, the effects of electrostatic interactions are naturally incorporated. In addition, because the effects of pH and salt concentration on capsid formation are thought to be non-negligible, in future applications these should be incorporated in CGMD models for more detailed simulations.

CGMD simulations are also used for protein structure predictions. Although membrane proteins are important pharmacological targets, the three dimensional structures of most of these proteins remain unknown. Bucher et al. constructed a structural model for phospholipase A_2 (PLA_2) by combining homology modeling and CGMD simulations [18]. PLA_2 is a membrane protein that releases fatty acids when it hydrolyzes phospholipids. These play important roles in intracellular signal transduction and inflammatory processes that are associated with Alzheimer's disease, hypertensive heart failure, neurological diseases, and cancer.

The appropriate localization of a protein within a lipid bilayer is important for a protein's function. AAMD simulations require a large computational cost for the process of inserting a protein into a lipid bilayer, whereas CGMD simulations can handle this very efficiently. After the stabilization process of a protein within a lipid bilayer is simulated by CGMD, the coarse-grained geometry can be reverse grained to obtain all atom geometry. Then, extensive detailed analyses, including hydrogen bonding, can be achieved using AAMD simulations.

Cholesterol is a crucial component of mammalian cells, as it determines the structural, thermodynamic, and mechanical properties of lipid membranes [52]. Flip-flop, namely the exchange of cholesterol molecules inside lipid bilayers, is important for efficient cholesterol tracking. This relatively time-consuming event (estimated half-time of <1 s) can be simulated using the Martini CGMD forcefield [53].

Pharmacological/toxicological applications

Applications of CGMD to in silico drug design are in progress. Lewis et al. used CGMD simulations to search for the optimal structure for an antiatherogenic agent [16]. The interactions between oxidized low-density lipoproteins (LDL) and scavenger receptors (SR) on the cell surfaces of macrophages are important phenomena during arterial stiffening. Amphiphilic macromolecules (AM) are known to competitively inhibit the interactions between oxidized LDL and SR [16].

Because AMs can form highly variable structures, it is expensive to synthesize them chemically and to test their actions experimentally. Lewis et al. systematically generated different AM structures and estimated their actions using CGMD simulations. Using quantitative structure-activity relationships (QSAR) [54], a method for predicting the biological effects of a substance as per its chemical structure, they estimated the antiatherogenic effects of AMs based on their structures obtained from their CGMD simulations. AMs with high QSAR index values were experimentally confirmed to be effective because they exhibited strong interactions with SR and low oxidized LDL intake into macrophages.

CGMD is suitable for searching the feature spaces of several agents. Even when using CGMD that can track the dynamics of a molecular system on a relatively large time scale, direct observations of a chemical reaction (AM-SR interactions in the example above) are usually impossible. Therefore, using ingenious methods like QSAR is a key component for applying CGMD to in silico drug design.

If the target of a particular agent is located inside a cell, this agent should be able to pass through the lipid bilayer of a cell membrane. It has been shown that cationic nanoparticles (NPs) and cell-penetrating peptides (CPPs) are useful for delivering agents into cells. Determining the optimal size and the chemical composition of a carrier is crucial for developing a drug delivery system (DDS). Lin et al. suggested that the shapes of NPs and CPPs were important during their translocation through a lipid bilayer using CGMD simulations [4]. Furthermore, carbon nanotubes are used for DDS. Lee et al. used CGMD to analyze the processes involved when incorporating carbon nanotubes into a lipid bilayer [19].

For gene therapy, DNA must be transferred into a cell. Lipofection is a method used to transfer DNA by coating DNA with lipids. Khalid et al. analyzed the interactions between DNA and lipids using CGMD [23].

Protein-protein interactions

A major application of CGMD is analyzing protein-protein interactions. The association and dissociation dynamics of an antibody (immunoglobulin G, IgG) and its affinity ligand *Staphylococcus aureus* protein A (SpA) were analyzed using CGMD [21]. The binding status of IgG with SpA was evaluated using an index of conformational change, root mean square deviation (RMSD), and the potential energy between these proteins. In addition, the detailed contributions of each amino acid residue to the potential energy could be obtained. Liu et al. reported that the dissociation dynamics in response to changes in pH consisted of four phases based on an analysis of the hydrophobic interactions and electrostatic interactions between charged residues [21]. To properly describe the protein-protein interactions involved, the effects of pH and Coulomb's forces should be incorporated into a CGMD model, which in this case was the Martini forcefield.

Oligomerization and self-aggregation of proteins are perfect targets for CGMD simulations because these molecular systems are relatively large and chemical actions play important roles within these systems. G protein-coupled receptors (GPCR) are important pharmacological targets, including photon receptor rhodopsins and cardiac adrenergic receptors. The functions of these proteins are significantly modified by the formation of oligomers within a lipid bilayer. Mondal et al. found that hydrophobic mismatches between a protein and lipids were crucial for the oligomerization of β-adrenergic receptors [15]. Another example is the protein Ras, which is associated with cancerous transformation of cells, as Ras forms clusters in a structure-dependent manner [17].

EXPERIMENTAL VALIDATION OF CGMD SIMULATIONS

We will now discuss methods used for comparing the results obtained from simulations with those experimentally obtained by roughly categorizing these into structure-oriented properties and dynamics-oriented properties. Length is a simple structural metric of a molecule. López et al. compared the thickness of a glycolipid membrane with that experimentally obtained [22]. Khalid et al. compared the persistence length (index of polymer hardness) of coarse-grained duplex DNA obtained from simulations with that experimentally obtained [23].

The area per lipid is often used to analyze the properties of lipid membranes [22, 23, 26]. Lipid molecules can assume various phases, including liposomes, micelles, bilayer sheets, and hexagonal tubes, depending on environmental conditions like pH. Experimental observations of the hexagonal phase can be performed by freeze-fracture electron microscopy or X-ray diffraction. Dhalberg et al. compared the phase preferences and hexagonal spacing of lipid molecules obtained from simulations with those experimentally obtained [26]. Comparisons of the structural patterns of a molecular assembly discussed above are often used and applied to other cases, such as analyzing the pore diameters of ion channels and the distances between domains/molecules.

In addition to the structural properties, dynamic properties are often used to compare the results from simulations and experiments. Arkhipov et al. compared viral capsid hardness by atomic force microscopy [47]. Liu et al. analyzed the binding affinities between two proteins under different salt concentrations and pH values and compared these with experimental results [21].

With regard to the phase transitions of lipid molecules, the transition temperature is another property used for comparisons [26]. Khalid et al. confirmed that the concentration dependence of cationic lipids fit well with experimental results for DNA chain spacing [23]. Another sophisticated example for validating CGMD simulations is analyzing the solvent accessible surfaces of proteins [18]. This property can be measured by deuterium exchange mass spectrometry and compared with CGMD results.

CONCLUSIONS

There are certain biomolecular details that can only be revealed by CGMD simulations. CGMD is a powerful tool for simulating molecular systems with large spatial and temporal scales, where chemical interactions are crucial. Proteinprotein interactions, including self-assembly and oligomerization, proteinlipid interactions, and even analyzing the effects of artificial substances, such as carbon nanotubes on biological systems are targets for CGMD simulations. CGMD simulation results can be validated by comparing the structural and dynamic properties of a molecular system with those experimentally obtained. Uniform CGMD methods to handle lipids, proteins, and nucleic acids are under development.

As we have discussed in this article, applicable targets of CGMD simulations are rapidly expanding. However, CGMD alone is not sufficient to provide scientific and biological insights into the extensive molecular systems incorporated in a cell. A possible direction for the evolution of CGMD simulations is seamless cooperation with larger systems, such as the finite element method (FEM) based on continuum dynamics, and with smaller systems, such as AAMD. Approaches to associate simulations on different scales are currently in progress [55, 56, 57]. Combining CGMD and FEM is anticipated to be particularly useful for analyzing the mechanical properties of cellular components, such as the extracellular matrix, and their implications for cellular functions. Combinations of CGMD and AAMD will synergize each other's strengths; namely, the accuracy of AAMD and the speed of CGMD. In conclusion, CGMD simulations will contribute to resolving problems in more varied situations than ever before.

ACKNOWLEDGEMENTS

The authors are very grateful to Mr. Kazuya Saruwatari at Okayama University Medical School for his CGMD simulations of the MscL protein.

CONFLICTS OF INTEREST

The authors declare that there were no conflicts of interest associated with this study.

REFERENCES

1. Scuseria GE (1999) Linear Scaling Density Functional Calculations with Gaussian Orbitals. J Phys Chem A 103: 4782-4790.
2. Go N (1983) Theoretical Studies of Protein Folding. Annu Rev Biophys Bioeng 12: 183-210.
3. Doi K, Takeuchi H, Nii R, et al. (2013) Self-assembly of 50 bp poly(dA).poly(dT) DNA on highly oriented pyrolytic graphite via atomic force microscopy observation and molecular dynamics simulation. J Chem Phys 139: 085102.
4. Lin J, Alexander-Katz A (2013) Cell Membranes Open «Doors» for Cationic Nanoparticles/Biomolecules: Insights into Uptake Kinetics. ACS Nano 7: 10799-10808.
5. Tozzini V (2005) Coarse-grained models for proteins. Curr Opin Struct Biol 15: 144-150.
6. Cheon M, Chang I, Hall CK (2010) Extending the PRIME model for protein aggregation to all 20 amino acids. Proteins 78: 2950-2960.
7. Marrink SJ, Risselada HJ, Yefimov S, et al. (2007) The MARTINI force field: coarse grained model for biomolecular simulations. J Phys Chem B 111: 7812-7824.
8. Monticelli L, Kandasamy SK, Periole X, et al. (2008) The MARTINI coarse-grained force field: Extension to proteins. J Chem Theory Comput 4: 819-834.
9. Shih AY, Arkhipov A, Freddolino PL, et al. (2006) Coarse grained protein-lipid model with application to lipoprotein particles. J Phys Chem B 110: 3674-3684.
10. Takada S (2012) Coarse-grained molecular simulations of large biomolecules. Curr Opin Struct Biol 22: 130-137.
11. Stark AC, Andrews CT, Elcock AH (2013) Toward optimized potential functions for proteinprotein interactions in aqueous solutions: osmotic second virial coefficient calculations using the MARTINI coarse-grained force field. J Chem Theory Comput 9: 4176-4185.
12. Kopelevich DI (2013) One-dimensional potential of mean force underestimates activation barrier for transport across flexible lipid membranes. J Chem Phys 139: 134906.

13. May A, Pool R, van Dijk E, et al. (2013) Coarse-grained versus atomistic simulations: realistic interaction free energies for real proteins. Bioinformatics 30: 326-334.

14. Periole X, Knepp AM, Sakmar TP, et al. (2012) Structural determinants of the supramolecular organization of G protein-coupled receptors in bilayers. J Am Chem Soc 134: 10959-10965.

15. Mondal S, Johnston JM, Wang H, et al. (2013) Membrane Driven Spatial Organization of GPCRs. Sci Rep 3: 2909.

16. Lewis DR, Kholodovych V, Tomasini MD, et al. (2013) In silico design of anti-atherogenic biomaterials. Biomaterials 34: 7950-7959.

17. Li H, Gorfe AA (2013) Aggregation of lipid-anchored full-length H-Ras in lipid bilayers: simulations with the MARTINI force field. Plos One 8: e71018.

18. Bucher D, Hsu YH, Mouchlis VD, et al. (2013) Insertion of the Ca(2)(+)-independent phospholipase A(2) into a phospholipid bilayer via coarse-grained and atomistic molecular dynamics simulations. PLoS Comput Biol 9: e1003156.

19. Lee H (2013) Membrane penetration and curvature induced by single-walled carbon nanotubes: the effect of diameter, length, and concentration. Phys Chem Chem Phys 15:16334-16340.

20. Siuda I, Thogersen L (2013) Conformational flexibility of the leucine binding protein examined by protein domain coarse-grained molecular dynamics. J Mol Model 19: 4931-4945.

21. Liu FF, Huang B, Dong XY, et al. (2013) Molecular basis for the dissociation dynamics of protein a-immunoglobulin g1 complex. Plos One 8: e66935.

22. Lopez CA, Sovova Z, van Eerden FJ, et al. (2013) Martini Force Field Parameters for Glycolipids. J Chem Theory Comput 9: 1694-1708.

23. Khalid S, Bond PJ, Holyoake J, et al. (2008) DNA and lipid bilayers: self-assembly and insertion. J R Soc Interface 5 Suppl 3: S241-250.

24. Cruz VL, Ramos J, Melo MN, et al. (2013) Bacteriocin AS-48 binding to model membranes and pore formation as revealed by coarse-grained simulations. Biochim Biophys Acta 1828:2524-

2531.

25. Hess B, Kutzner C, van der Spoel D, et al. (2008) GROMACS 4: Algorithms for highly efficient, load-balanced, and scalable molecular simulation. J Chem Theory Comput 4: 435-447.
26. Dahlberg M (2007) Polymorphic phase behavior of cardiolipin derivatives studied by coarsegrained molecular dynamics. J Phys Chem B 111: 7194-7200.
27. López CA, Rzepiela AJ, de Vries AH, et al. (2009) Martini Coarse-Grained Force Field: Extension to Carbohydrates. J Chem Theory Comput 5: 3195-3210.
28. Yesylevskyy SO, Schafer LV, Sengupta D, et al. (2010) Polarizable Water Model for the Coarse-Grained MARTINI Force Field. Plos Comput Biol 6: e1000810.
29. Yoo J, Cui Q (2009) Curvature generation and pressure profile modulation in membrane by lysolipids: insights from coarse-grained simulations. Biophys J 97: 2267-2276.
30. Donnini S, Tegeler F, Groenhof G, et al. (2011) Constant pH Molecular Dynamics in Explicit Solvent with lambda-Dynamics. J Chem Theory Comput 7: 1962-1978.
31. Bennett WFD, Chen AW, Donnini S, et al. (2013) Constant pH simulations with the coarsegrained MARTINI model - Application to oleic acid aggregates. Can J Chem 91: 839-846.
32. Schonichen A, Webb BA, Jacobson MP, et al. (2013) Considering protonation as a posttranslational modification regulating protein structure and function. Annu Rev Biophys42: 289-314.
33. Brunger AT, Adams PD, Rice LM (1997) New applications of simulated annealing in X-ray crystallography and solution NMR. Structure 5: 325-336.
34. Kirkpatrick S, Gelatt CD, Jr., Vecchi MP (1983) Optimization by simulated annealing. Science 220: 671-680.
35. Nury H, Poitevin F, Van Renterghem C, et al. (2010) One-microsecond molecular dynamics simulation of channel gating in a nicotinic receptor homologue. Proc Natl Acad Sci U S A107: 6275-6280.

36. Jensen MO, Jogini V, Borhani DW, et al. (2012) Mechanism of voltage gating in potassium channels. Science 336: 229-233.
37. Baoukina S, Marrink SJ, Tieleman DP (2012) Molecular Structure of Membrane Tethers. Biophys J 102: 1866-1871.
38. Louhivuori M, Risselada HJ, van der Giessen E, et al. (2010) Release of content through mechano-sensitive gates in pressurized liposomes. Proc Natl Acad Sci USA 107: 19856-19860.
39. Dill KA, MacCallum JL (2012) The Protein-Folding Problem, 50 Years On. Science 338:1042-1046.
40. Clementi C (2008) Coarse-grained models of protein folding: toy models or predictive tools? Curr Opin Struct Biol 18: 10-15.
41. Gregersen N, Bross P, Vang S, et al. (2006) Protein Misfolding and Human Disease. Annu Rev Genomics Hum Genet 7: 103-124.
42. Borgia MB, Borgia A, Best RB, et al. (2011) Single-molecule fluorescence reveals sequencespecific misfolding in multidomain proteins. Nature 474: 662-665.
43. Yang S, Cho SS, Levy Y, et al. (2004) Domain swapping is a consequence of minimal frustration. Proc Natl Acad Sci USA 101: 13786-13791.
44. Wu C, Shea J-E (2011) Coarse-grained models for protein aggregation. Curr Opin Struct Biol21: 209-220.
45. Nguyen HD, Hall CK (2004) Molecular dynamics simulations of spontaneous fibril formation by random-coil peptides. Proc Natl Acad Sci USA 101: 16180-16185.
46. Voegler Smith A, Hall CK (2001) α-Helix formation: Discontinuous molecular dynamics on an intermediate-resolution protein model. Proteins 44: 344-360.
47. Arkhipov A, Roos WH, Wuite GJ, et al. (2009) Elucidating the mechanism behind irreversible deformation of viral capsids. Biophys J 97: 2061-2069.
48. Krishna V, Ayton GS, Voth GA (2010) Role of protein interactions in defining HIV-1 viral capsid shape and stability: a coarse-grained analysis. Biophys J 98: 18-26.
49. Grime JM, Voth GA (2012) Early stages of the HIV-1 capsid protein lattice formation. Biophys J 103: 1774-1783.

50. Zhang R, Linse P (2013) Icosahedral capsid formation by capsomers and short polyions. J Chem Phys 138: 154901.
51. Rapaport DC (2004) Self-assembly of polyhedral shells: A molecular dynamics study. Phys Rev E 70: 051905.
52. Khelashvili G, Harries D (2013) How sterol tilt regulates properties and organization of lipid membranes and membrane insertions. Chem Phys Lipids 169: 113-123.
53. Bennett WFD, MacCallum JL, Hinner MJ, et al. (2009) Molecular View of Cholesterol Flip- Flop and Chemical Potential in Different Membrane Environments. J Am Chem Soc 131:12714-12720.
54. Verma J, Khedkar VM, Coutinho EC (2010) 3D-QSAR in drug design--a review. Curr Top Med Chem 10: 95-115.
55. Roos WH, Gibbons MM, Arkhipov A, et al. (2010) Squeezing Protein Shells: How Continuum Elastic Models, Molecular Dynamics Simulations, and Experiments Coalesce at the Nanoscale. Biophys J 99: 1175-1181.
56. Chen X, Cui Q, Tang Y, et al. (2008) Gating Mechanisms of Mechanosensitive Channels of Large Conductance, I: A Continuum Mechanics-Based Hierarchical Framework. Biophys J95: 563-580.
57. Riniker S, van Gunsteren WF (2012) Mixing coarse-grained and fine-grained water in molecular dynamics simulations of a single system. J Chem Phys 137: 044120

PART II: COARSE-GRAINED MODELS FOR PROTEINS AND PEPTIDES

CHAPTER 2

A GENERIC FORCE FIELD FOR PROTEIN COARSE-GRAINED MOLECULAR DYNAMICS SIMULATION

Junfeng Gu [1], Fang Bai [2], Honglin Li [3] and Xicheng Wang [1]

[1]State Key Laboratory of Structural Analysis for Industrial Equipment, Department of Engineering Mechanics, Dalian University of Technology, Dalian 116023, China

[2]Faculty of Chemical, Environmental and Biological Science and Technology, Dalian University of Technology, Dalian 116023, China

[3]School of Pharmacy, East China University of Science and Technology, Shanghai 200237, China

ABSTRACT

Coarse-grained (CG) force fields have become promising tools for studies of protein behavior, but the balance of speed and accuracy is still a challenge in the research of protein coarse graining methodology. In this work, 20 CG beads have been designed based on the structures of amino acid residues, with which an amino acid can be represented by one or two beads, and a CG solvent model with five water molecules was adopted to ensure the consistence with the protein CG beads. The internal interactions in protein were classified according to the types of the interacting CG beads, and adequate potential functions were chosen and systematically parameterized to fit the energy distributions. The proposed CG force field has been tested on eight proteins, and each protein was simulated for 1000 ns. Even without any extra structure knowledge of the simulated proteins, the Cα root mean square deviations (RMSDs) with respect to their experimental structures are close to those of relatively short time all atom molecular dynamics simulations. However, our coarse grained force field will require further refinement to improve agreement with and persistence of native-like structures. In addition, the root mean square fluctuations (RMSFs) relative to the average structures derived from the simulations show that the conformational fluctuations of the proteins can be sampled.

Keywords: Coarse-grained; force field; molecular dynamics; protein

INTRODUCTION

Over the last 30 years, the Molecular Dynamics (MD) method has played an increasing important role in dynamic behavior simulation of biomolecule at the atomic level [1]. In numerous application areas such as structural biology, biophysics, biochemistry, enzymology, molecular biology and medicinal chemistry, *etc.*, MD has become a major routine research tool. By means of MD simulation, biomolecular structure, kinetics, and thermodynamics can be investigated, for example, macromolecular stability, conformational and allosteric properties, the role of dynamics in enzyme activity, molecular recognition and the properties of complexes, ion and small molecule transport, protein association, protein folding, and protein hydration [2]. However, All-Atom Molecular Dynamics (AA-MD) is restricted severely by available computational capabilities because of the need of large amount of computing resources. In the 1970s, a small protein (bovine pancreatic trypsin inhibitor, composed of about 500 atoms) was first simulated, and lasted only about 10 picoseconds with AA-MD, limited by the computing power at that time [3]. With the development of modern computer technology, high performance computing and molecular dynamics method, the application of AA-MD has made great progresses in both space scale and time scale. Nowadays,

AA-MD can simulate biomolecule system containing up to millions of atoms, with simulation time over microsecond level [4,5]. Despite this, AA-MD still cannot meet all the need of biomolecule research. Most dynamics and interactions within cells (e.g., protein-protein docking, rearrangement upon ligand binding, folding) occur on microsecond or even millisecond timescale, and usually involve large macromolecular aggregates. The simulation time of these processes is at least four to six orders of magnitude larger than the feasible time with AA-MD simulation, which has brought large barrier in the biomolecule simulation research.

In the past few years, Coarse-Grained Molecular Dynamics (CG-MD) methods for biomolecule have gained increasing attention [6–11]. The basic thought of CG-MD is to treat several or more atoms as a virtual particle (*i.e.*, so-called Coarse-Graining), so the huge quantity of degrees of freedom within complex biomolecule especially protein and the complexity of the corresponding force field will be decreased, therefore dramatically decreases the computational complexity of MD simulation. Various kinds of protein CG models and force fields have been introduced. Referenced to amino acid residue, protein CG models can be simply classified as multiple-point model [12–20], two-point model [21,22], one-point model [23–27] and much coarser multiple-residue model [28–30], and CG force fields varied from the simple harmonic potential to more realistic molecular force field. CG-MD has achieved plenty of research results, and has been applied in areas such as membrane [10,31,32], ion-channel [11,33], protein folding [34,35], and protein-protein interaction [36]. However, due to the limited speedup and reliability, the main available methods are difficult to be widely used in the simulation of large-scale biological systems to date, and the further development of CG-MD is still a challenge work for researchers. CG models need to be as simplified as possible in order to simulate more complicated biomolecules, while CG force field need to be as realistic as possible so that the kinetic behavior under AA-MD can be accurately reproduced. Current coarse graining methodologies are still not as predictive as AA-MD, because of the intrinsic difficulty in modeling the complex and diverse intra-molecular interactions with few parameters. Developing CG models and accurate force field for protein have become of great importance for studying large biological systems in both time and space scale.

As a representative CG force field, MARTINI has gained the most attention, and has been successfully applied to the simulations of protein and membrane systems. However, MARTINI still needs secondary structure restraints to maintain the stability of the native structure during the simulation, and the parameterization process of CG force field is too complicated and needs much experience, which usually needs quite considerable effort. Therefore, simpler and more efficient methods are continuously being researched. We report in our recent work on the improvement of CG-MD methodology. Novel CG models for protein simulation are designed, with which a residue is composed of only one or two beads, so the computational efficiency of MD can be improved significantly. A force filed based on the models is developed, based on known protein structures and AA-MD simulation results. Then the protein CG models and the force field are applied in MD simulations of eight small to medium size proteins. Finally, the simulation results are given and compared with those of AA-MD simulations and experimental values, indicating the effectiveness of the proposed CG models and force field.

RESULTS

Results of Bonded Potential Parameterization

In the bonded interactions, all the backbone beads are assumed to be the same, so the bond types can be classified as *B–B*and $B–S^i$. The statistical results of bond length distribution are shown in Figure 1. Figure 1A shows that the *B–B* bond length is distributed in a narrow area from 3.6 Å to 3.9 Å and centered on 3.8 Å, so 3.8 Å is adopted as the equilibrium stretching length L_{bond} in Equation 2 of *B–B*. Figure 1B shows the statistical results of distance distributions between 10 types of S_i beads and their backbone beads. Each distribution shows a similar character with *B–B*, but the equilibrium length of $B–S_i$ bond is S_i bead dependent. Table 1 summarizes the L_{bond} of each $B–S_i$ bond adopted in our force field. The stretching energy profile of bond is extracted from the distribution of bond length with Boltzmann conversion method, and fitting with Equation 2 to get the force constant. The *B–B* force constant adopts an approximate value 100,000 kJ nm^{-2}mol^{-1}, and the $B–S_i$ force constants adopt a mean value 5,000 kJ nm^{-2} mol^{-1} in our force field.

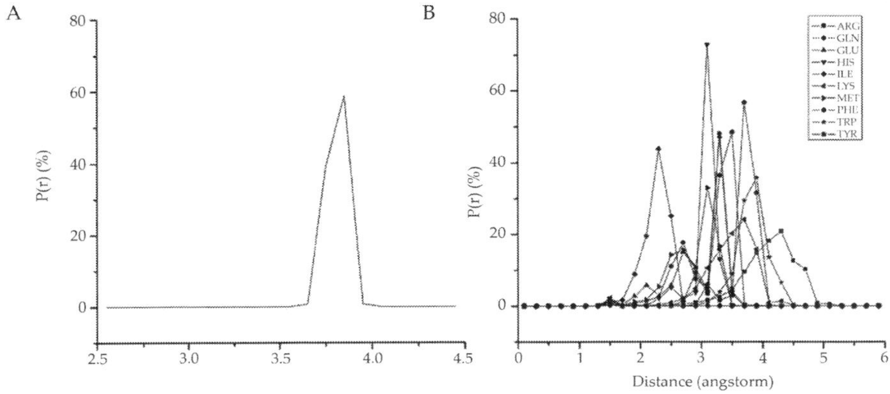

Figure 1: The bond length distribution of the *B–B* and *B–*S_i. *B* denotes the backbone bead, and S_i denotes the side-chain beads shown in distinct patterns.

Table 1: The equilibrium length between ten side-chain beads and their backbone beads

Bond	Length (nm)	Bond	Length (nm)
B–S_{ARG}	0.406	B–S_{LYS}	0.344
B–S_{GLN}	0.301	B–S_{MET}	0.287
B–S_{GLU}	0.295	B–S_{PHE}	0.333
B–S_{HIS}	0.307	B–S_{TRP}	0.381
B–S_{ILE}	0.226	B–S_{TYR}	0.371

The angles in CG protein system can be classified into three types: *B–B–B*, *B–B–*S_i and S_i*–B–B*, and angle bending energy profiles calculated from the probability distributions of these angels are shown in Figure 2, in which distinct colors and patterns are used to distinguish different S_i. Two minima at about 90 and 120 degrees can be found in energy profile of *B–B–B* angle, which correspond to the α-helix and β-sheet secondary structure. A similar pattern of energy profiles is observed in Figure 2B,C, and only one set of parameters is used for *B–B–*S_i (or S_i*–B–B*) bending potential function. Due to the coarse-graining, we have to neglect some specific characters in the structure or energy distribution, and focus on the common characters behind the details. For fitting with Equation 3, the mean value smooth technique is adopted to handle different profiles in *B–B–*S_i (or S_i*–B–B*), and the fitted potential function curves are also shown with solid curves in Figure 2. Gaussian parameters in Equation 3 obtained from the fitting process are given in the Supplementary Materials.

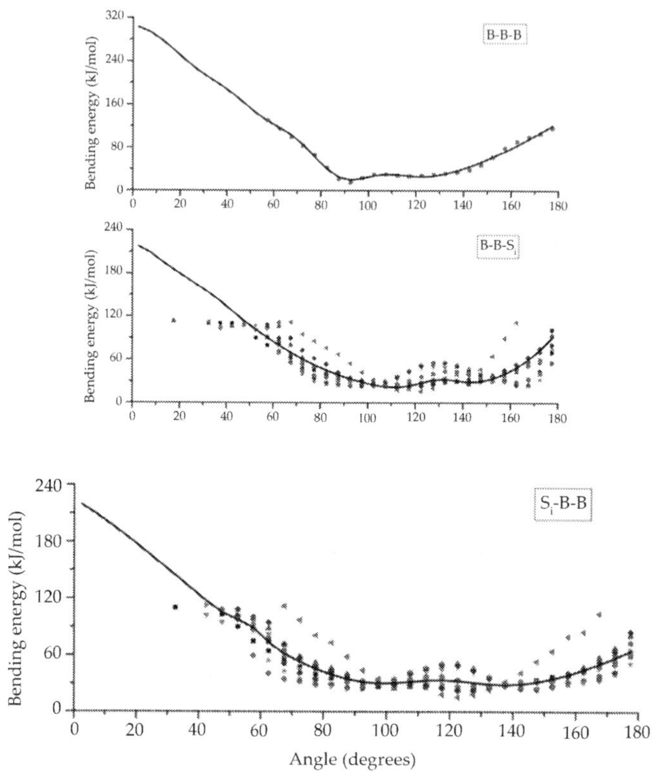

Figure 2: The angle bending energy profiles of B–B–B, B–B–S_i and S_i–B–B and the fitted potential function curves (black curves). B denotes the backbone bead, and S_i denotes the side-chain beads shown in distinct patterns and colors.

Similarly, the dihedral can be classified into four types: S_i–B–B–S_j, S_i–B–B–B, B–B–B–S_i and B–B–B–B. Figure 3 gives the pseudo-dihedral torsion energy profiles of each type, e.g., Figure 3A shows the 100 energy profiles of S_i–B–B–S_j. Each type is fitted with Equation 3, and the fitting results are also shown with solid curves in Figure 3. Gaussian parameters for torsion potential are also given in the Supplementary Materials.

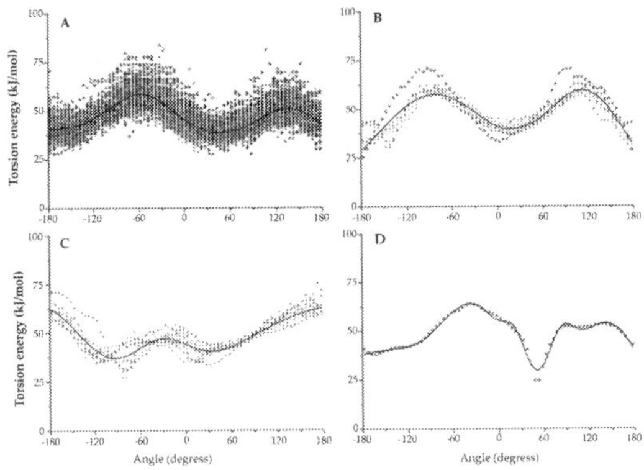

Figure 3: The dihedral torsion energy profiles of (**A**) S_i–B–B–S_j, (**B**) S_i–B–B–B, (**C**) B–B–B–S_i and (**D**) B–B–B–B and the fitted potential function curves (black curves). Bdenotes the backbone bead, and S_i/S_j denotes the side-chain beads shown in distinct patterns and colors.

Results of Non-Bonded Potential Parameterization

It is important to accurately describe the non-bonded interactions of 20 CG beads in order to study protein folding and protein-protein interactions. US as a sampling improving technique was used to get the PMF between two homologue CG beads, and PMF curve is fitted to Equation 5 for extracting the best van der Waals interaction potential parameters. Figure 4A gives the histograms of the configurations within the umbrella sampling windows, which indicates there is sufficient overlap between adjacent windows. Figure 4B gives the PMF against the distance of geometric center of two B_{ALA} beads, which have a minimum around 0.45 nm. However, when we made the statistical analyses of the distance distributions between two ALA amino acids on the above-mentioned protein structure database, the probability peak corresponding to the energy minimum was found around 0.55 nm. The reason for this inconsistency is that the CG bead is constrained by the surrounding beads while it is part of a protein, while is unrestricted in the US simulation. Most CG beads cannot be too close to each other in protein as in the US simulations, thus the short-range part of the PMF curve may not appropriate to model the non-bonded interactions in protein. However, the relatively long-distance interactions between CG beads are rarely affected by the environment in protein and can still be described by the PMF curves. Therefore, we made the statistical analyses of the distances for all 20 homologue CG bead pairs to determine the parameter c_{ij} in Equation 5 when the van der Waals potential is equal to zero as listed in Table 2. Equation 5 was fitted to the PMF curve for determining the van der Waals well depth parameter with determined parameter c_{ij}. Figure 5 gives the fitted results of CG beads B_{GLY}, B_{SER}, S_{GLU} and S_{ILE}. As in most cases, the position of the energy minimum determined by statistical c_{ij} is farther than that of the corresponding PMF curve, the fitting is only noticeable in the long-range part of the PMF, as shown in Figure 5 (B_{SER}, S_{GLU} and S_{ILE}).

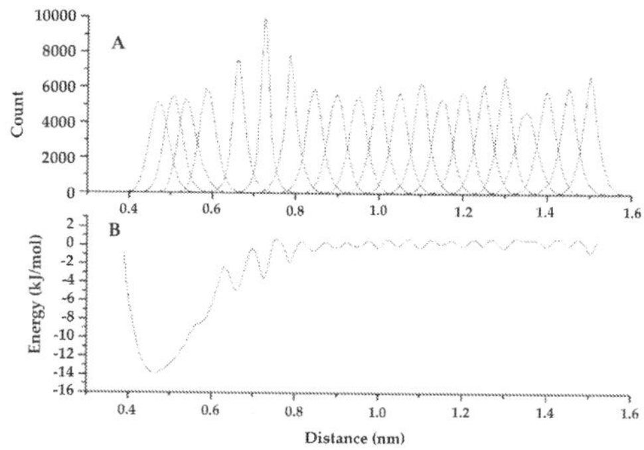

Figure 4: The histograms of the configurations within the umbrella sampling windows (**A**) and the potential of mean force against the distance of two ALA molecules (**B**).

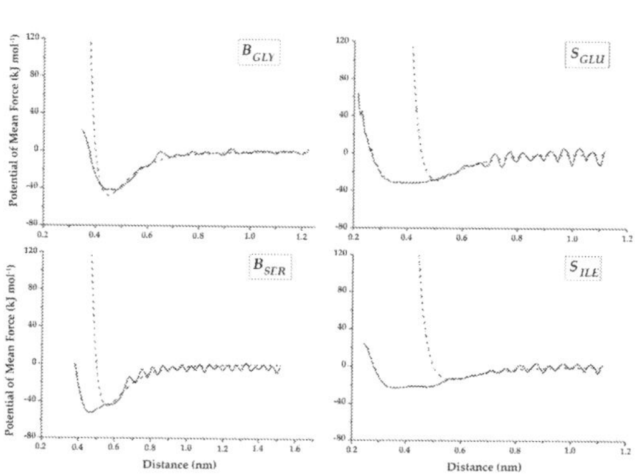

Figure 5: The potential of mean force between non-bonded homo pairs of coarse-grained (CG) beads (B_{GLY}, B_{SER}, S_{GLU} and S_{ILE}) against their distance, derived from umbrella sampling method with all-atom simulation (solid curves), and the van der Waals potential by fitting the potential of mean force with the Lennard-Jones function (dash curves).

Table 2: The finite distance c_{ij} when the van der Waals potential of two interacting beads is equal to zero

Interacting beads	Distance c_{ij} (nm)	Interacting beads	Distance c_{ij} (nm)
B_{ALA}–B_{ALA}	0.50	S_{ARG}–S_{ARG}	0.60
B_{ASN}–B_{ASN}	0.60	S_{GLN}–S_{GLN}	0.45
B_{ASP}–B_{ASP}	0.55	S_{GLU}–S_{GLU}	0.45
B_{CYS}–B_{CYS}	0.50	S_{HIS}–S_{HIS}	0.45
B_{GLY}–B_{GLY}	0.40	S_{ILE}–S_{ILE}	0.50
B_{LEU}–B_{LEU}	0.55	S_{LYS}–S_{LYS}	0.45
B_{PRO}–B_{PRO}	0.65	S_{MET}–S_{MET}	0.45
B_{SER}–B_{SER}	0.50	S_{PHE}–S_{PHE}	0.45
B_{THR}–B_{THR}	0.50	S_{TRP}–S_{TRP}	0.65
B_{VAL}–B_{VAL}	0.50	S_{TYR}–S_{TYR}	0.55

Verification of the Force Field

To verify our force field, several proteins solvated in water were coarse-grained and simulated for a relatively long time. During the simulations, the maintenance of experimental structures and other thermodynamic properties are deemed to be indications of the feasibility of force field for protein simulation.

The test protein group is composed of eight small to medium size proteins which are not included in the protein set used for bonded potential parameterization. These proteins have recently been used to examine the performance of a modified version of the CHARMM force fields [37], and part of them have been used to test the PACE CG force field [14], so they were chose to verify our force filed convenient for comparison. All the CG-MD simulations were based on the GROMACS 4.0.5 package [38]. First, the protein was coarse-grained based on the proposed CG model, and topology files were generated with our developed scripts. Then the CG protein was solvated in CG water molecules, and the system was energy minimized with the proposed CG force field. Worthy of mention is that the GROMACS does not provide a Gaussian function type interface in its topology files, so user supplied tabulated functions were used for calculating the energy of angle bending and dihedral torsion. After the energy minimization, the CG system was equilibrated for 200 ps and then submitted for a 1000 ns simulation, using the canonical NPT ensemble at 300 K and 1 bar pressure, and the detail information for the eight simulated protein systems are listed in Table 3.

Table 3: The simulation information of eight protein systems

System	PDB ID	Number of residues	Number of CG waters	Number of CG beads
Barstar	1BTA	89	939	1069
CheY	1CYE	129	1196	1375
Ubiquitin	1D3Z	76	1013	1124
FKBP12	1FKS	107	1264	1417
Barnase	1FW7	110	1157	1312
RNase H	1RCH	155	1982	2207
RNase A	2AAS	124	1126	1296
protein G	3GB1	56	887	963

Table 4 gives the simulation time and Cα RMSDs of eight proteins from their experimental structures derived from CG-MD simulations *versus* all-atom simulations. With the CG-MD methodology, eight proteins were all simulated for 1000 ns, and the average Cα RMSDs are varied from 0.316 to 0.415 nm, and the final RMSDs are between 0.323 and 0.431 nm. While with the all-atom simulation [37], eight proteins are simulated over 22–148 ns, and the average and final RMSDs is varied from 0.106 to 0.358 nm and 0.121 to 0.477 nm respectively. In general, the RMSDs with CG-MD are larger than those with AA-MD due to a longer simulation time and the roughness of our CG force field, but their values are comparable, and final RMSD of 1FKS is even lower. Thus, the experimental structures of proteins can be considered to be maintained with our CG force filed via long time MD simulations. Figure 6 gives the full trajectories of the Cα RMSDs of eight proteins. Most of the proteins reach their stable conformations within the first 100 ns and the Cα RMSDs are kept around 0.4 nm. In one case, the structure of 3GB1 is more stable and the Cα RMSDs are maintained around 0.32 nm, which mainly because few long loops are included in the native structure of the protein. It is noteworthy that the Cα RMSD trajectory of protein 2AAS has two distinct increases at around 460 and 800 ns. For analyzing the reason and investigating the conformation change during the simulation, the conformations of 2AAS at 0 ns, 250 ns, 450 ns, 480 ns, 750 ns and 1000 ns are sampled and plotted in Figure 7. From observation of the conformations at these 6 time points, skeleton structure of 2AAS is kept stable during the 1000 ns simulation, which also proves the native structure can be maintained with our CG force field. Comparing conformations at 450 ns and 480 ns, loop1 which is composed of residues 20–25 went through a large conformation change as indicated in Figure 7, which correspond to the distinct increase of Cα RMSD trajectory around 460 ns. The conformation change of residues 58–61 (labeled as loop2 in Figure 7) between 750 and 1000 ns corresponds to the RMSD value change around 800 ns. Both loop1 and loop2 are flexible loop regions located at the solvent-exposed surface of the protein, so they are less stable than the secondary structure and the hydrophobic core of the protein during the simulation.

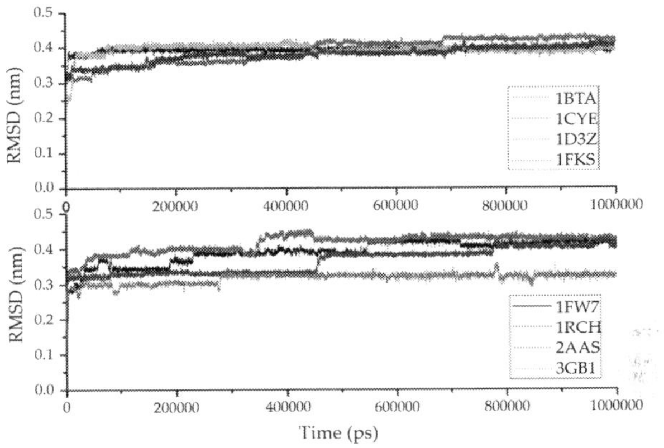

Figure 6: Resulting profiles of root mean square deviation of Cα carbons for eight proteins.

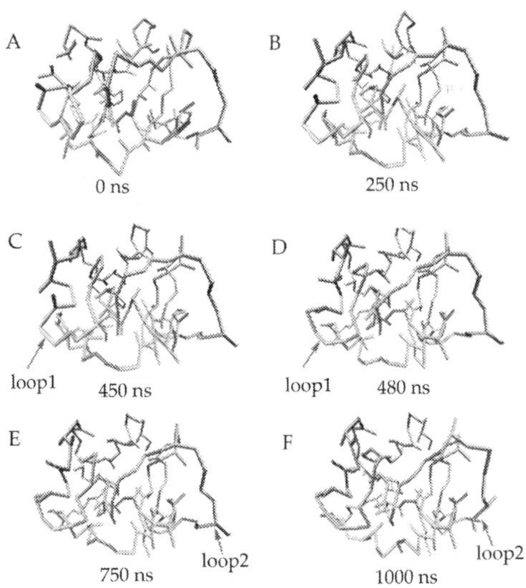

Figure 7: Snapshots of the 1000 ns coarse-grained molecular dynamics simulation for protein 2AAS at 0 ns (**A**), 250 ns (**B**), 450 ns (**C**), 480 ns (**D**), 750 ns (**E**) and 1000 ns (**F**).

Table 4: Resulting root mean square deviations from experimental structures of eight proteins during coarse-grained simulations compared with all-atom simulations (standard deviations are given in parentheses)

PDB	CG-MD			AA-MD *		
	Simulation length (ns)	Avg. Ca RMSD (nm)	Final Ca RMSD (nm)	Simulation length (ns)	Avg. Ca RMSD (nm)	Final Ca RMSD (nm)
1bta	1000	0.393(0.010)	0.396	142.9	0.134(0.016)	0.121
1cye	1000	0.389(0.036)	0.422	124.7	0.143(0.020)	0.170
1d3z	1000	0.394(0.020)	0.395	22.0	0.141(0.021)	0.128
1fks	1000	0.379(0.021)	0.415	143.5	0.358(0.074)	0.477
1fw7	1000	0.391(0.033)	0.408	148.0	0.171(0.015)	0.167
1rch	1000	0.415(0.025)	0.431	121.5	0.278(0.017)	0.289
2aas	1000	0.364(0.034)	0.400	148.3	0.249(0.043)	0.321
3gb1	1000	0.316(0.015)	0.323	50.0	0.106(0.020)	0.143

* The values of AA-MD are from reference 37.

Another question that interested us is whether the conformational fluctuations of a protein can be reasonably simulated with our CG model and force field. In the PDB file of an NMR model, the B-factor column for each atom contains a measure how much that atom position varies throughout the models in the ensemble, which provides an experimentally detectable measure of equilibrium dynamics. Figure 8 gives the Root Mean Square Fluctuations (RMSFs) relative to the averaged structures for protein 1BTA, 1D3Z, 1FKS and 3GB1, which provide B-factors in their NMR structures. The RMSFs simulated with CG-MD are compared with B-factors via conversion equation $RMSF^2 = 3 \times B/8/pi^2$, where B is the B-factor, which indicates the conformational stability degree. As shown in Figure 8, the RMSFs of protein 1D3Z and 3GB1 are consistent with the experimental values from a global perspective. However, at some locations of 1BTA and 1FKS, there are obvious inconsistencies between the simulated RMSFs and the experimental values: at residues 7–13, 15–20, 25–26 and 35–36 of protein 1BTA, the RMSFs are higher than the experimental values, while at residues 33–34, 40–44 and 84–91 of protein 1FKS, the situation is reversed. Through the analysis of protein structure and simulation trajectory, the above mentioned locations of 1BTA are either loops with lower curvature or ends of alpha helixes, while the locations of 1FKS are loops with higher curvature. The main reason of these conflicts is that the loop structure is mainly stabilized by the bonded interactions, while the bonded potentials adopted in our CG force field is a fitting of statistical average values due to the simplification. Therefore, loops with higher curvature are constrained by the bonded potential more strictly than they should be, while the situation of loops with lower curvature is opposite.

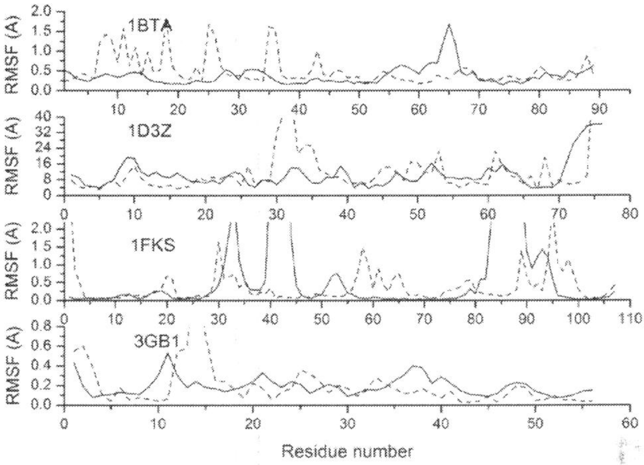

Figure 8: Resulting profiles of the residue root mean square fluctuations (dash curves) relative to averaged conformations compared with NMR experiments (solid curves) for proteins 1BTA, 1D3Z, 1FKS and 3GB1.

Efficiency of the Force Field

The main goal of coarse-graining is to improve the computational efficiency. For comparing the computational efficiency, the above mentioned eight testing proteins were simulated for another 10 ns with the proposed CG-MD methodology, AA-MD and MARTINI respectively. All the simulations were performed in serial on an Intel Xeon processor (2.4 GHz). The proteins were firstly centered in a box, the edge of which is 1 nm far from the molecules, and then solvated with water solvent. In all-atom simulations, GROMOS87 force filed and SPC water model were adopted, and a 2 fs time-step was used. In our and MARTINI CG-MD simulations, the corresponding CG water models and a 16 fs time-step were used. The energy of all the systems were minimized first, then equilibrated for 200 ps and submitted for a 10 ns simulation, using the canonical NPT ensemble at 300 K and 1 bar pressure. The simulation time is listed in Table 5. With each simulation method, the simulation time is proportional to the protein size (as listed in Table 3). The simulation time with our coarse-graining methodology is slightly less than that with MARTINI, which is mainly due to a coarser protein and water model.

Compared with AA-MD simulation, MARTINI and our CG-MD method can achieve about 75~100 speedup. When more complicated solvent model is adopted in AA-MD, such as TIP3P, the speedup will be more obvious. It seems that larger time-step adopted in the CG-MD is a direct factor relating to the speedup, but the profound reason is that the appropriate coarse-graining model can maintain the structure stability of the protein in a CG-MD simulation with a larger time-step.

Table 5: The efficiency of 10 ns simulations of eight proteins with three different simulation methodologies

PDB	The proposed CG-MD			MARTINI		AA-MD	
	Simulation time (s)	Avg. Ca RMSD (nm)	Avg. Ca RMSD in vacuum (nm)	Simulation time (s)	Avg. Ca RMSD (nm)	Simulation time (s)	Avg. Ca RMSD (nm)
1bta	3501	0.210	0.637	4002	0.341	313062	0.148
1cye	4309	0.292	0.440	4972	0.503	398432	0.148
1d3z	4032	0.283	0.416	4261	0.426	334203	0.185
1fks	4484	0.324	0.505	5242	0.378	436792	0.220
1fw7	4391	0.247	0.574	4902	0.400	388712	0.171
1rch	7330	0.337	0.681	7845	0.357	650507	0.234
2aas	4424	0.284	0.689	4801	0.421	387623	0.259
3gb1	3432	0.275	0.501	3721	0.339	274854	0.128

The average Cα RMSDs of all the simulations are also given in Table 5. The values of AA-MD here slightly differ with those provided in Table 4, which are got with a modified CHARMM force field and with a longer simulation period. With our simulations, the average RMSDs of AA-MD range from 0.128 to 0.259, which indicates AA-MD is most stable among these three simulation methodologies. The RMSDs with the proposed CG-MD methodology is lower than MARTINI for all eight proteins. Despite the fact that 10 ns is a relatively short simulation period, the results show that the proposed CG-MD methodology a comparable even better ability of native structure maintenance compared with the popular MARTINI CG force field. It should be noticed, the information of secondary structure is required in the simulations with MARTINI, while not required with the proposed method. It indicates that interactions in the CG protein structure can be balanced even without any extra structure restraints, and this makes the proposed model more suitable for simulating random or extended structures.

In addition, for evaluating the role of the CG solvent in the simulations, the testing proteins were also simulated in vacuum, and the average Cα RMSDs of the simulations are shown in Table 5. The RMSDs in vacuum range from 0.416 to 0.689, and are significantly higher than those of the simulations with CG solvent. The reason for this obvious difference is that the initial structures of the simulations are the native structures of proteins which are maintained in a solvent environment. The maintenance of the native structure is determined largely by the balance of the interactions among different amino acid residues with each other and with the aqueous solution surrounding the protein, and the solvent influences the conformation by competing with intramolecular interactions. The bonded potential in this work is derived from a statistical analysis of a representative protein set, so the solvent effect is incorporated in an implicit way. However, the van der Waals interactions are determined via simulations in vacuum, so the solvent effect is not incorporated in the potential. Therefore, the RMSD values to the initial structures in the simulations will be larger when the CG solvents are absent.

MATERIALS AND METHODS

The Coarse-Grained Protein Models

With our CG protein models, each amino acid is modeled by one or two beads according to their sizes. In total, 20 types of CG beads were designed for 20 amino acids as shown in Figure 9, which can be classified into two broad categories: backbone bead and side-chain bead. The backbone and side-chain beads can be denoted as B_i (i = ALA, ASN, ASP, CYS, GLY, LEU, PRO, SER, THR, VAL) and S_i (i = ARG, GLN, GLU, HIS, ILE, LYS, MET, PHE, TRP, TYR), respectively. Some amino acids are modeled only by one backbone bead due to their small side-chains, while others are modeled by one uniform backbone bead (Glycine bead B_{GLY}) and one distinct side-chain bead. All the CG beads are idealized as a sphere, and center of the backbone bead is located at the alpha-carbon atom, while the center of the side-chain bead is located at the geometric center of all its heavy atoms.

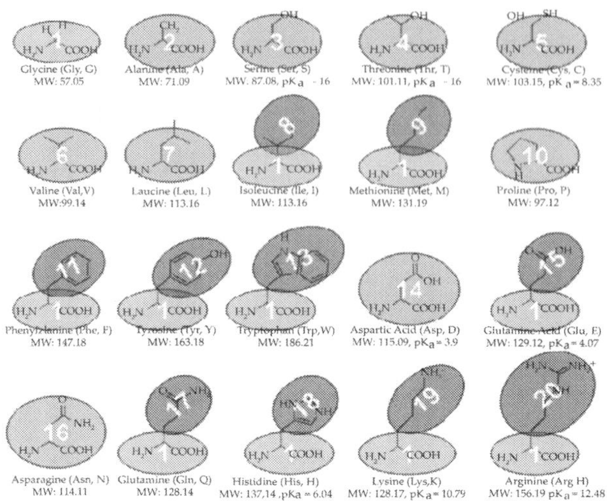

Figure 9: The coarse-grained models of 20 protein amino acids.

The Coarse-Grained Force Field

With the above mentioned protein CG models, the structure and internal interactions of a protein can be simplified as shown in Figure 10. The CG force field can be formulated as Equation 1:

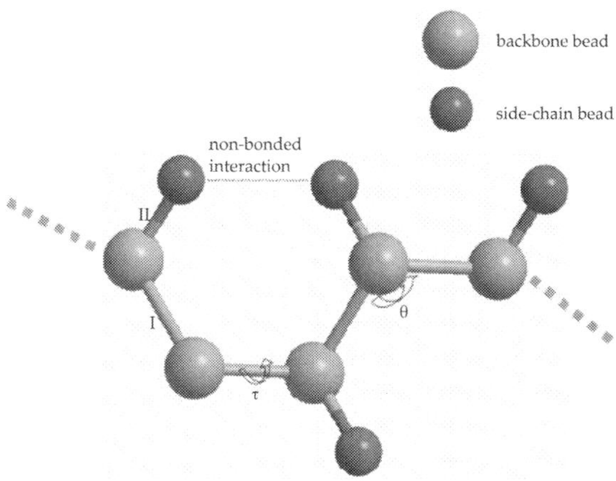

Figure 10: The coarse-grained protein model: I and II denote the backbone-backbone bead and backbone-side-chain bead bond stretching interaction respectively, θ denotes the virtual angle, and τ is the virtual dihedral angel.

$$U = U_{bond} + U_{angle} + U_{torsion} + U_{vdw} + U_{elec} \tag{1}$$

where U_{bond}, U_{angle} and $U_{torsion}$ are the stretching potential energy of a virtual bond, the potential energy of a virtual angle bending and the potential function of a dihedral angle about a rotating bond, respectively, which describe the bonded interactions between CG beads. U_{vdw} and U_{elec} describe the non-bonded interactions, which are the energy of van der Waals interactions and electrostatic interactions respectively.

The Bonded Potential and Parameterization

The virtual stretching interaction between two bonded CG beads can be described as a harmonic potential:

$$U_{bond} = \sum \frac{1}{2} K_{bond} (l - L_{bond})^2 \tag{2}$$

where K_{bond} and L_{bond} are the force constant and the equilibrium stretching length of a bond, respectively, which will be determined by fitting the energy distribution of the virtual bond. Due to the coarse-graining, U_{angle} and $U_{torsion}$ curves become more complex and irregular when compared with those of AA force field, and they are described with Gaussian distribution function:

$$U_{angle/torsion} = \sum_{i=1}^{N} a_i \exp[-(\frac{x - b_i}{c_i})^2] \tag{3}$$

where N, a_i, b_i and c_i are Gaussian parameters need to be determined in the parameterization process.

For correctly parameterize the bonded potential, we adopted a reduced and non-redundant set of protein structures used for fold recognition [39]. This set includes about 3600 structures chosen from the Protein Database Bank (PDB) [40], and the Root Mean Square Deviation (RMSD) of each structure is at least 6 Å to the rest of the structures in the set to avoid structure redundancy. Statistical analyses were performed against this protein set, and the resulting probability distributions were used to calculate Potential of Mean Force (PMF) via Boltzmann conversion method [13,41,42]:

$$U_i = -k_B T \ln(P_i) \tag{4}$$

where k_B is the Boltzmann constant, T is the temperature, and $P_i = n_i / n_{ref}$ is the probability of a property at value i, in which the reference number n_{ref} is the total number of the investigated internal coordinate obtained from the statistics of the above mentioned protein set.

The Non-Bonded Potential and Parameterization

Modeling the non-bonded potential is a key problem of constructing CG-MD force filed. As in a classical AA force field, we assume that the non-bonded interaction can be subdivided into two categories, *i.e.*, van der Waals interaction and electrostatic interaction. They can be formulated as sums of pairwise potential energy:

$$U_{vdw} = \sum_{i<j} 4\varepsilon_{ij} \left(\frac{c_{ij}^{12}}{r_{ij}^{12}} - \frac{c_{ij}^{6}}{r_{ij}^{6}} \right) \tag{5}$$

$$U_{elec} = \sum_{i<j} \frac{Q_i Q_j}{4\pi\varepsilon_0 \varepsilon_r r_{ij}} \tag{6}$$

where c_{ij} is the van der Waals interaction parameter, r_{ij} is the distance between CG beads i and j, and Q_i and Q_j are the charges of i and j. The strength of the van der Waals interaction is determined by the value of well depth ε_{ij} which depends on the types of the interacting CG beads and can be determined in the force field parameterization process for all the 20 types of CG beads. In the proposed force field, the electrostatic interaction is taken into account through distributed point charges, and four CG beads are treated as charged: backbone bead B_{ASP} and side-chain bead S_{GLU} are one unit negatively charged, and side-chain beads S_{ARG} and S_{LYS} are one unit positively charged. The electrostatic interaction between charged beads is calculated via Equation 6 with the relative dielectric constant $\varepsilon_r = 1$.

In the coarse-graining, a group of atoms are treated as a single bead, and the relative positions of these atoms are fixed, but in reality, their relative positions vary in all the time. PMF is defined as the potential that gives an average force over all the configurations of a given system, and is used here to characterize the non-bonded interactions between CG beads. Umbrella Sampling (US) method [43] based on AA-MD was applied on the AA molecules of 20 CG beads to get the van der Waals well depth parameter when one molecule interacts with itself, and Lorentz-Berthelot mixing rules were applied for getting the interaction parameters between different CG beads:

$$c_{ij}^{6} = (c_{ii}^{6} c_{jj}^{6})^{1/2}$$

$$c_{ij}^{12} = (c_{ii}^{12} c_{jj}^{12})^{1/2} \qquad (7)$$

To perform the US simulations, backbone beads are simulated with corresponding AA amino acids, and side-chain beads are replaced by analogous compounds, as listed in Table 6. The simulations were performed with GROMACS 4.0.5 package, using fully flexible molecules in vacuum and the canonical NVT ensemble, and with vanishing charge in order to capture the purely non-electrostatic interaction. The GROMOS87 force filed was applied to the molecules, and the temperature is kept at 300 K by coupling to a Berendsen thermostat. Two identical AA molecules of a CG bead were placed together and equilibrated for 2000 ps, then two molecules were pulled from their equilibrium position to 15 angstroms along a reaction coordinate via umbrella pulling with a constant pulling rate 0.001 nm ps^{-1}. The snapshots were saved every 1 ps, and the pulling distance was divided into subspaces every 0.5 angstrom. At last, US simulation was applied in every subspace for 10 ns, and the Weighted Histogram Analysis Method (WHAM) [44] was applied to accurately integrate the PMF of the non-bonded interaction between two homologue AA molecules.

Table 6: The corresponding analogous compounds of side-chain beads

Side-chain bead	Analogous compound	Side-chain bead	Analogous compound
S_{ARG}	*n*-propylguanidine	S_{LYS}	*n*-butylamine
S_{GLN}	propionamide	S_{MET}	methyl propyl sulfide
S_{GLU}	propionic acid	S_{PHE}	toluene
S_{HIS}	4-methylimidazole	S_{TRP}	3-methylindole
S_{ILE}	*n*-butane	S_{TYR}	p-cresol

Coarse-Grained Water Model and Parameterization

In protein simulation, quite often most of the computational cost is spent on calculating water-water intermolecular interactions rather than solute-water or solute-solute interactions, so the water coarse-graining will remarkably improve the efficiency of MD simulation. An appropriate water model should be constructed in the CG protein-solvent system, and the CG water model should be consistent with the protein CG models both in volume and mass, so the interactions between protein and solvent can be accurately reproduced. For determining the appropriate coarse-graining methodology of water molecules, we took a simulation of pure water with GROMACS 4.0.5 package with TIP3P water model, using GROMOS87 force field and the canonical NPT ensemble. The system was coupled to a temperature bath at 300 K and a barostat at 1 bar pressure, and was simulated for 1 ns. Statistical analysis with the simulation results showed that the average volume of five water molecules is about 140 $Å^3$ and the mass is 90 amu, which is consistent with the average volume 120 $Å^3$ and average mass 95 amu of the proposed CG protein beads. Therefore, the CG solvent model composed of five water molecules is adopted. The CG water bead is treated as neutral according to its total charge, so the interactions between CG water beads and other CG beads are mainly through van der Waals force. In order to determine the parameters of the van der Waals function for the CG water bead, every five nearest water molecules were clustered into a group with K-means algorithm, and the nearest distances between a group and the adjacent groups were calculated. According to the distribution probability, 0.51 nm is adopted for the parameter c_{ij} in Equation 5. Using identical settings with the previous AA-MD water simulation, CG water system was simulated with different ε_{ij}. For determining the best well depth parameter, the bulk density of the CG water system as a function of time was calculated and compared with the density variation of AA-MD. According to the comparison, $\varepsilon_{ij} = 6$ kJ mol^{-1} is the best value for reproducing the bulk phase density of water, and is adopted in our CG force field.

CONCLUSIONS

In this work, 20 CG beads for protein were constructed according to the characters of 20 amino acids, and a residue is composed of only one or two CG beads. Correspondingly, with the K-means method, a five-water coarse-grained solvent model was adopted to suit the CG protein model. A force field was developed for the CG protein and solvent model. For all the bonded interactions in protein, CG beads are divided into two types. All the combinations with these two types of the bonded interactions were analyzed on a known protein structure subset, and the resulting energy distributions were fitted to various potential functions to formulate the bonded interactions in CG-MD. The umbrella sampling method was used on the AA molecules of the CG beads to get the PMF of non-bonded interactions between CG beads, and the PMF was fitted to a Lennard-Jones function potential to describe the non-bonded interactions in CG-MD.

The CG model and force field were tested on eight small to medium size proteins. With the results analysis of the simulations, without any extra information of the simulated protein structure, the skeleton structure of the protein can be maintained during a long time equilibrium dynamics simulation with the proposed coarse-graining methodology. Comparison of the efficiency shows that the proposed CG-MD can make a 75~100 computing speedup relative to AA-MD, which is also higher than the popular MARTINI model. Meanwhile, the native structure of the proteins can be well preserved during the simulations. In addition, RMSFs of Cα atoms during the simulation show our CG-MD method can reasonably sampling the conformational fluctuations within a protein from a global perspective. However, the simulation results also indicate that the fluctuation of loop structures with a low curvature in protein may be overestimated in some proteins compared with experimental values, and the situation is converse when the loop structures have a high curvature because the simplification in the CG-MD. Further work is needed to investigate and carefully treat the loop structures in protein coarse-graining methodology.

ACKNOWLEDGMENTS

The authors gratefully acknowledge financial support for this work from the National Program on Key Basic Research Project (No. 2009CB918501), the Fundamental Research Funds for the Central Universities, and the National Natural Science Funds of China (No. 11202049 and 11072048).

REFERENCES

1. Gunsteren, W.F.; Dolenc, J. Biomolecular simulation: Historical picture and future perspectives. *Biochem. Soc. Trans*2008, *36*, 11–15.
2. Adcock, S.A.; McCammon, J.A. Molecular dynamics: Survey of methods simulating the activity of proteins. *Chem. Rev*2006, *106*, 1589–1615.
3. McCammon, J.A.; Gelin, B.R.; Kaplus, M. Dynamics of folded proteins. *Nature* 1977, *267*, 585–590.
4. Klepeis, J.L.; Lindorff-Larsen, K.; Dror, R.O.; Shaw, D.E. Long-timescale molecular dynamics simulations of protein structure and function. *Curr. Opin. Struct. Biol* 2009, *19*, 120–127.
5. Sanbonmatsun, K.Y.; Tung, C.-S. High performance computing in biology: Multimillion atom simulations of nanoscale systems. *J. Struct. Biol* 2007, *157*, 470–480.
6. Muller-Plathe, F. Coarse-graining in polymer simulation: From the atomistic to the mesoscopic scale and back. *ChemPhysChem* 2002, *3*, 754–769.
7. Kolinski, A.; Skolnick, J. Reduced models of proteins and their applications. *Polymer* 2004, *45*, 511–524.
8. Tozzini, V. Coarse-grained models for proteins. *Curr. Opin. Struct. Biol* 2005, *15*, 144–150.
9. Clementi, C. Coarse-grained models of protein folding: Toy models or predictive tools. *Curr. Opin. Struct. Biol* 2008, *18*, 10–15.
10. Lindahl, E.; Sansom, M.S. Membrane proteins: Molecular dynamics simulation. *Curr. Opin. Struct. Biol* 2008, *18*, 425–431.
11. Khalili-Araghi, F.; Gumbart, J.; Wen, P.; Sotomayor, M.; Tajkhorshid, E.; Schulten, K. Molecular dynamics simulations of membrane channels and transporters. *Curr. Opin. Struct. Biol* 2009, *19*, 128–137.
12. Monticelli, L.; Kandasamy, S.K.; Periole, X.; Larson, R.G.; Tieleman, D.P.; Marrink, S. The MARTINI coarse-grained force field: Extension to proteins. *J. Chem. Theory Comput* 2008, *4*, 819–834.
13. Ha-Duong, T. Protein backbone dynamics simulations using coarse-

grained bonded potentials and simplified hydrogen bonds. *J. Chem. Theory Comput* 2010, *6*, 761–773.

14. Han, W.; Wan, C.K.; Jiang, F.; Wu, Y.D. PACE force field for protein simulations. 1. Full parameterization of version1 and verification. *J. Chem. Theory Comput* 2010, *6*, 3373–3389.
15. Han, W.; Wan, C.K.; Jiang, F.; Wu, Y.D. PACE force field for protein simulations. 2. Folding simulations of peptides. *J. Chem. Theory Comput* 2010, *6*, 3390–3402.
16. Basdevant, N.; Borgis, D.; Ha-Duong, T. A coarse-grained protein-protein potential derived from an all-atom force field. *J. Phys. Chem. B* 2007, *111*, 9390–9399.
17. Han, W.; Wu, Y.D. Coarse-grained protein model coupled with a coarse-grained water model: Molecular dynamics study of polyalanine-based peptides. *J. Chem. Theory Comput* 2007, *3*, 2146–2161.
18. Bereau, T.; Deserno, M. Generic coarse-grained model for protein folding and aggregation. *J. Chem. Phys* 2009, *130*, 235106.
19. Han, W.; Wan, C.K.; Wu, Y.D. Toward a coarse-grained protein model coupled with a coarse-grained solvent model: Solvation free energies of amino acid side chains. *J. Chem. Theory Comput* 2008, *4*, 1891–1901.
20. DeVane, R.; Shinoda, W.; Moore, P.B.; Klein, M.L. Transferable coarse grain nonbonded interaction model for amino acids. *J. Chem. Theory Comput* 2009, *5*, 2115–2124.
21. Shih, A.Y.; Arkhipov, A.; Freddolino, P.L.; Schulten, K. A coarse grained protein-lipid model with application to lipprotein particles. *J. Phys. Chem. B* 2006, *110*, 3674–3684.
22. Zhou, J.; Thorpe, L.F.; Izvekov, S.; Voth, G.A. Coarse-grained peptide modeling using a systematic multiscale approach. *Biophys. J* 2007, *92*, 4289–4303.
23. Korkut, A.; Hendrickson, W.A. A force field for virtual atom molecular mechanics of proteins. *Proc. Natl. Acad. Sci. USA* 2009, *106*, 15667–15672.
24. Tozzini, V.; Rocchia, W.; McCammon, J.A. Mapping all-atom models onto one-bead coarse grained models: General properties

and applications to a minimal polypeptide model. *J. Chem. Theory Comput* 2006, *2*, 667–673.

25. Chang, C.A.; Trylska, J.; Tozzini, V.; McCammon, J.A. Binding pathways of ligands to HIV-1 protease: Coarse-grained and atomistic simulations. *Chem. Biol. Drug Des* 2007, *69*, 5–13.
26. Korkut, A.; Hendrickson, W.A. Computation of conformational transitions in proteins by virtual atom molecular mechanics as validated in application to adenylate kinase. *Proc. Natl. Acad. Sci. USA* 2009, *106*, 15673–15678.
27. Alemani, D.; Collu, F.; Cascella, M.; Peraro, M.D. A nonradial coarse-grained potential for proteins produces naturally stable secondary structure elements. *J. Chem. Theory Comput* 2010, *6*, 315–324.
28. Arkhipov, A.; Freddolino, P.L.; Schulten, K. Stability and dynamics of virus capsids described by coarse-grained modeling. *Structure* 2006, *14*, 1767–1777.
29. Arkhipov, A.; Yin, Y.; Schulten, K. Four-scale description of membrane sculpting by BAR domains. *Biophys. J* 2008, *95*, 2806–2861.
30. Arkhipov, A.; Freddolino, P.L.; Imada, K.; Namba, K.; Schulten, K. Coarse-grained molecular dynamics simulations of a rotating bacterial Flagellum. *Biophys. J* 2006, *91*, 4589–4597.
31. West, B.; Brown, F.L.H.; Schmid, B. Membrane-protein interactions in a generic coarse-grained model for lipid bilayers. *Biophys. J* 2009, *96*, 101–115.
32. Spijker, P.; Hoof, B.V.; Debertrand, M.; Markvoort, A.J.; Vaidehi, N.; Hilbers, P.A.J. Coarse grained molecular dynamics simulations of transmembrane protein-lipid systems. *Int. J. Mol. Sci* 2010, *11*, 2393–2420.
33. Treptow, W.; Marrink, S.; Tarek, M. Gating motions in voltage-gated potassium channels revealed by coarse-grained molecular dynamics simulations. *J. Phys. Chem. B* 2008, *112*, 3277–3282.
34. Guardiani, C.; Livi, R.; Cecconi, F. Coarse grained modeling and approaches to protein folding. *Curr. Bioinforma* 2010, *5*, 217–240.
35. Liwo, A.; Khalili, M.; Scheraga, H.A. Ab initio simulations of

protein-folding pathways by molecular dynamics with the united-residue model of polypeptide chains. *Proc. Natl. Acad. Sci. USA* 2005, *102*, 2362–2367.

36. Hall, B.; Sansom, M.S.P. Coarse-grained MD simulations and protein-protein interactions: The cohesion-dockerin system. *J. Chem. Theory Comput* 2009, *5*, 2465–2471.
37. Feig, M. Is alanine dipeptide a good model for representing the torsional preferences of protein backbone? *J. Chem. Theory Comput* 2008, *4*, 1555–1564.
38. Hess, B.; Kutzner, C.; Spoel, D.; Lindahl, E. GROMACS 4: Algorithms for highly efficient, load-balanced, and scalable molecular simulation. *J. Chem. Theory Comput* 2008, *4*, 435–447.
39. Meyerguz, L.; Grasso, C.; Kleinberg, J.; Elber, R. Computational analysis of sequence selection mechanisms. *Structure*2004, *12*, 547–557.
40. Berman, H.M.; Westbrook, J.; Feng, Z.; Gilliland, G.; Bhat, T.N.; Weissig, H.; Shindyalov, I.N.; Bourne, P.E. The protien data bank. *Nucl. Acids Res* 2000, *28*, 235–242.
41. Reith, D.; Putz, M.; Muller-plathe, F. Deriving effective mesoscale potentials from atomistic simulations. *J. Comput. Chem* 1997, *29*, 292–308.
42. Tozzini, V.; McCammon, J.A. A coarse grained model for the dynamics of flap opening in HIV-1 protease. *Chem. Phys. Lett* 2005, *413*, 123–128.
43. Torrie, G.M.; Valleau, J.P. Nonphysical sampling distributions in Monte Carlo free-energy estimation: Umbrella sampling. *J. Comput. Phys* 1977, *23*, 187–199.
44. Kumar, S.; Bouzida, D.; Swendsen, R.H.; Kollman, P.A.; Rosenberg, J.M. The weighted histogram analysis method for free-energy calculations on biomolecular. I. The method. *J. Comput. Chem* 1992, *13*, 1011–1021.

CHAPTER 3

INSIGHTS FROM COARSE-GRAINED GŌ MODELS FOR PROTEIN FOLDING AND DYNAMICS

Ronald D. Hills Jr. [1] **and Charles L. Brooks III** [1,2]

[1]Department of Molecular Biology and Kellogg School of Science and Technology, The Scripps Research Institute, 10550 N. Torrey Pines Rd. TPC6 La Jolla, CA 92037, USA

[2]Department of Chemistry and Biophysics Program, University of Michigan, 930 N. University Ave, Ann Arbor, MI 48109, USA

ABSTRACT

Exploring the landscape of large scale conformational changes such as protein folding at atomistic detail poses a considerable computational challenge. Coarse-grained representations of the peptide chain have therefore been developed and over the last decade have proved extremely valuable. These include topology-based Gō models, which constitute a smooth and funnel-like approximation to the folding landscape. We review the many variations of the Gō model that have been employed to yield insight into folding mechanisms. Their success has been interpreted as a consequence of the dominant role of the native topology in folding.

The role of local contact density in determining protein dynamics is also discussed and is used to explain the ability of Gō-like models to capture sequence effects in folding and elucidate conformational transitions.

Keywords: Protein folding; Gō models; coarse-graining; energy landscape; conformational transitions

INTRODUCTION

Advancing our understanding of biology in the 21st century will involve the structural and dynamical characterization of the countless molecular interactions in the cell. Much can be learned of the structural and dynamical properties of proteins, the workhorses of the cell, from their folding energy landscape. Molecular simulation has emerged as a valuable tool for the detailed exploration of the thermodynamic landscape responsible for guiding the unfolded ensemble through a collection of high energy intermediate structures in order to achieve the proper native structure. We examine recent insights obtained from a simple class of coarse-grained simulation methods for landscape characterization, known as Gō models.

Protein folding is a complex process involving large numbers of degrees of freedom. Therefore, performing traditional molecular dynamics simulations at atomistic detail poses a challenge if one wishes to exhaustively explore the folding landscape of a moderately sized protein with finite computational resources. For this reason coarse-grained representations of the polypeptide chain have been employed in folding simulations. Rather than representing each atom in the protein explicitly groups of atoms can be treated as a single coarse-grained site. In the most common variation, each residue corresponds to a single coarse-grained site placed at the alpha carbon [1–9], although models with multiple sites per residue have been employed [10–14]. Single-site per residue C_α models then make up beads on a string, which are held together with harmonic energy terms in the Hamiltonian for the resulting virtual bond and angle vibrational degrees of freedom. Virtual dihedral angles comprising four successive residues can be subjected to potential functions of varying forms, but all share a similar intent of mimicking backbone chirality and Ramachandran conformational preferences [15].

The underpinning of a Gō model, however, is its treatment of nonbonded interactions.

Early coarse-grained models of folding employed the HP recipe, in which hydrophobic residues interact with a constant favorable contact energy (denoted ε) and polar residues either don't interact or are repulsive. HP models enjoyed limited success as they predicted degenerate global energy minima and failed to identify the unique native state [16]. Current coarse-graining developments endeavor to augment such two-flavor models to contain up to 20 flavors, using distinct physicochemical types to represent each amino acid sidechain or multiple coarse-grained sites per residue for larger sidechains [17–20]. Parameterization of such models to ensure consistency with the all-atom thermodynamic ensemble is a challenging problem. Recent empirical implementations rely on dihedral or hydrogen bond restraints to maintain native secondary structure and are thus of limited use for studying protein conformational changes and folding [21–24]. Two ongoing approaches by the Voth [25] and Scheraga [26] groups offer a promising solution to this problem by employing direct parameterization with all-atom simulation data in order to obtain physically accurate peptide equilibria.

In lieu of a coarse-grained model with physically accurate nonbonded interactions, simplistic Gō models have been applied to the study of folding mechanisms with considerable success. An attractive term of the Lennard-Jones variety [12] is employed for pairwise contacts present in a protein's native structure, which consists of a shallow energy basin of depth ε that decays to zero for large distances. Residue-residue contacts are typically assigned for nonhydrogen sidechain atoms separated by less than a cutoff distance [12,27], and contacts can also be assigned for backbone hydrogen bonds [2]. All other (non-native) interactions experience a volume-exclusion repulsion. The topology-based approach of favorably distinguishing between native and non-native interactions is named after the early lattice simulation work of Nobuhiro Gō [28–36], and any potential that is biased toward a target structure can be referred to as Gō-like. In Section 2 we discuss the success of Gō models in describing folding. Section 3 follows with recent insights concerning the role of sequence in folding, and, finally, an outlook is presented for the coarse-graining of conformational transitions.

ROLE OF NATIVE CONTACTS IN FOLDING

Energy landscape theory posits that proteins have evolved a smooth and funnel-like thermodynamic landscape biased toward the native state, allowing them to fold to a unique stable structure on biological timescales [37–39]. A polypeptide does not have sufficient time to explore all possible conformations in search of the native state; therefore native interactions, whether local or nonlocal [40,41], must play a role in guiding the conformational search in the direction of native structure. As topology-centric Gō models let native interactions drive the folding, their large success has been interpreted as a validation of energy landscape theory.

The reduced degrees of freedom in a coarse-grained representation enable one to exhaustively explore the thermodynamic landscape for folding using standard molecular simulation techniques. In particular, high energy regions of conformational space can be characterized such as the transition state or folding intermediate(s) using either brute-force searching at the folding transition temperature or umbrella sampling with carefully chosen biasing restraints [42]. Gō model simulations can thus be readily compared with experimental phi-value analysis of the thermodynamic effects of point mutations. A phi-value is a qualitative measure of the amount of native-like structure a residue contains in the folding transition state [43]. Phi-values can be predicted for each residue in the protein by averaging over all conformations recorded in the simulation trajectory that correspond to the transition state ensemble. The distribution of native-like secondary and tertiary structure present in the transition state can in part describe the relative order of formation of structural elements, the location of the folding nucleus and hence the overall folding mechanism.

Experimental phi-values have been compared in the literature with predictions from simple C_α Gō models for a rapidly growing number of proteins [1–4,8,41,42,44–54]. Qualitative consensus is generally obtained between predictions and experiment for individual residue phi-values as well as the major features of the folding mechanism and folding rate [55]. Non-native interactions have been demonstrated to make some important contributions to the folding transition state and mechanism, and variants of the Gō algorithm have been modified to incorporate nonspecific interactions [4,10,13,56–62]. Notwithstanding, the widespread success of native-only Gō models in reproducing folding behavior as observed in experiment and atomistic simulation [63] suggests that the perturbations of non-native interactions on the structures of folding transition states and intermediates are minor in comparison to the dominating influence of native interactions on the funneled energy landscape [64].

SEQUENCE DEPENDENCE OF FOLDING

Over the last decade the folding community has emphasized the native topology as the main determinant of folding behavior [39,65]. The topology-centric paradigm has resulted from the rise of energy landscape theory, the prevalence and success of Gō modeling studies, and the notion of contact order. Defined as the average sequence separation between contacting residues in the native state, contact order is inversely correlated with experimental folding rates, allowing kinetic predictions to be made based on the local or nonlocal nature of complex backbone topologies. Small helical proteins are known to fold faster than large β-sheet proteins due to the more local nature of their contacts [66,67]. Nonetheless, it is a protein's sequence that determines the particular topology it chooses to adopt. Single residue mutations can significantly alter the kinetics of folding and the stability of the native state. Amyloid disease-causing mutations lead to partially unfolded forms that self-associate to form pathogenic oligomers and aggregates [68,69]. The precise role of sequence in folding remains to be understood.

Methods have been employed for extending simple Gō models to examine sequence effects in folding. Rather than weighting native contacts equally by assigning the same interaction energy ε to all contacts, energetic heterogeneity can be introduced by scaling residue-residue interactions according to physically motivated criteria. Native interactions have been optimized in Src-SH3, CI2, protein G and S6 using thermodynamic perturbation data from mutation experiments [64,70]. Clementi and coworkers optimized native and non-native contact energies in Src-SH3 using an iterative procedure to maximize the energy gap between the folded and unfolded states [57]. Dokholyan and coworkers scaled contact energies based on atomistic simulations with the CHARMM force field that identified a small set of critical residues in SOD1 folding [71]. Karanicolas and Brooks [2] employed a general set of residue-residue interactions using the tabulated potentials of Miyazawa and Jernigan (MJ) derived from statistical analysis of contacts in the Protein Data Bank [72]. Native contact energies have also been weighted by crystallographic B-factors [73] and hydrogen exchange protection factors [74] with success. One last method that can be considered to constitute a 'flavored' Gō model is the all-atom Gō approach [75–85]. The protein is represented at atomic resolution, and nonbonded atom-atom pairs within a cutoff distance are assigned a favorable contact potential. As larger sidechains are more likely to be assigned a higher number of atom-atom contacts, they are apt to interact more favorably than smaller sidechains.

Energetic heterogeneity adds ruggedness to the folding landscape and, more importantly, allows for the investigation of sequence effects in, for example, the folding of proteins with identical topology. The MJ-flavored model of Karanicolas and Brooks was used to investigate symmetry breaking in protein L and protein G, two proteins that share a common topology consisting of N- and C-terminal β-hairpins and a central α-helix. The balance between enthalpic interactions and chain entropy was observed to lead to N-terminally nucleated folding in protein L and C-terminal nucleation in protein G, consistent with experimental findings [2]. The same model was recently employed to compare the folding mechanisms for three members of the common flavodoxin fold [42,86,87]. CheY, NtrC and Spo0F each consist of five βα-repeats arranged into a central β-sheet and surrounding α-helices. Consistent with experimental findings for CheY [88,89], all three proteins were observed to obey an N-terminal nucleation mechanism in simulations due to abundant contacts in the N-terminus. A larger number of contacts in the C-terminal subdomain of Spo0F were observed to lead to some differences in its folding mechanism, however.

While Clementi *et al* had previously observed N-terminal nucleation in simulations of CheY with a simple Gō model [1], a more detailed analysis was undertaken by Hills *et al.* to compare the role of sequence in CheY homologs [42,86,87]. The free energy landscape was characterized as a function of the formation of the various secondary and tertiary structural elements within the 120-residue proteins in order to map out the relative order of structure formation. Though more rigorous computational methods exist for determining the folding transition state ensemble [90,91], using structural coordinates to monitor folding progress in simulations has proven valuable in the elucidation of complex folding mechanisms in large proteins when multiple reaction coordinates are needed to explain folding behavior [49,92,93]. Transition path sampling methods were recently employed to analyze the reaction coordinate for the 20-residue Trp-cage mini-protein [94]. Two structural coordinates were needed to best describe its parallel folding pathways of helix and tertiary contact formation. The fraction of native contacts formed, denoted Q, is a progress variable of considerable utility [92,95–97]. Folding in CheY and homologs was shown to be well described by native contact formation in the N- and C-terminal subdomains [42,86,87].

While CheY and NtrC contain an alanine-rich C-terminal cavity that is dynamic in the unphosphorylated state and important for function, in Spo0F this region is filled in with bulkier residues such as isoleucine and exhibits rigidity, having lost the C-terminal conformational allostery of its evolutionary homologs [98–101]. The enhanced stability and van der Waals contacts in this region were shown to lead to competitive frustration in the folding landscape (Figure 1). Whereas in CheY and NtrC the N-terminal subdomain formed easily and rapidly nucleated C-terminal folding, in Spo0F C-terminal contacts were seen to form ahead of and temporarily preclude N-terminal folding.

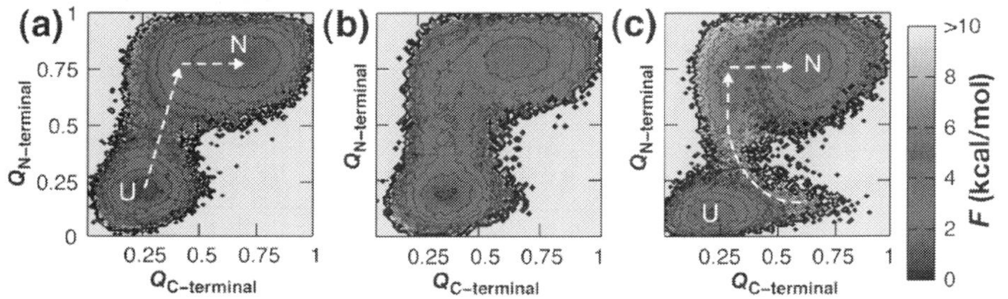

Figure 1: N-terminally nucleated folding landscapes for structural homologs CheY (a), NtrC (b) and Spo0F (c). The equilibrium free energy was computed as a function of the fraction of native contacts formed within the N- ($Q_{\text{N-terminal}}$) and C-terminal ($Q_{\text{C-terminal}}$) subdomains and is shown at the transition temperature at which the folded and unfolded states are equally populated [42,87].

Simulation results are in accord with experimental analysis of the three homologs; the kinetics of folding were found to be slowest in Spo0F, suggesting an interesting relationship between function and folding mechanism in the CheY family [87,88]. Comparison of the folding mechanisms for the CheY family, T4 lysozyme and interleukin-1β suggests that subdomain competition is a general property of multimodule proteins [42,47,102]. It is interesting to note that the contact energies assigned by the flavored Gō model were uniformly distributed throughout the three CheY-like proteins. The local contact density, defined as the number of native contacts per residue, could alone explain folding behavior, implying that an unflavored Gō model could also have possibly captured the main sequence effects in the three homologs. This notion is supported by a recent comparison study of results obtained from flavored and unflavored Gō models [103]. For β-sheet proteins in which tertiary contacts are well distributed throughout the protein, introducing energetic heterogeneity was observed to negligibly impact the folding mechanism; for α-helical bundles held together by only a handful of tertiary interactions, interaction heterogeneity was observed to alter folding.

CONCLUSIONS AND OUTLOOK

The role of contact density in dictating protein folding and dynamics is increasingly being recognized. Contact density has been used to explain B-factors [104] and downhill folding [54]. A comprehensive survey of experimental and simulation data for homodimers revealed that high interface hydrophobicity and a large ratio of interfacial to monomeric contacts are indicative of an obligate two-state association mechanism, in which dimerization precedes monomer folding [50,51]. Two-state dimers commonly have an intertwined or pseudo domain-swapped interface. Cooperativity in hydrophobic packing was shown to be sufficient for nucleating amyloid aggregation [105]. Residue fluctuations and the location of folding nuclei have been explained by treating proteins as small-world networks consisting of a regular lattice with a fraction of the edges between vertices replaced with new random connections [106–112]. Clusters of the branched aliphatic sidechains isoleucine, leucine and valine have been identified as particularly important in stability and folding [87,88,113,114]. In order to predict hydrogen exchange protection factors, a combinatorial algorithm was used to generate an ensemble of unfolded conformational states and the thermodynamic probability of each state was estimated using considerations of chain entropy and solvent exposed surface area [115]. A survey of large two-subdomain proteins revealed that native contact density can distinguish stable, nucleating subdomains from dynamic subdomains involved in function [86]. Correlations between contact density and folding/dynamics help to explain the success of coarse-grained models in biomolecular simulation. Nowhere is this success more surprising than in the case of analytical Ising models of folding that consider chain entropy and native contact enthalpy while neglecting the three-dimensional spatial representation of the peptide chain [116–120].

The role of native contacts in protein dynamics is best appreciated in the context of elastic network normal mode analysis (EN-NMA). To calculate vibrational modes in large systems proteins can be represented as an elastic network of C_α-beads in which residues within a cutoff distance experience harmonic interactions. The lowest frequency normal modes of vibration are then easily computed for the elastic network model. Often, a single or small subset of low frequency vibrations has been shown to reproduce the large amplitude structural rearrangements associated with biological function [19,121–123]. EN-NMA has successfully reproduced the functional dynamics of systems as diverse as adenylate kinase, myosin, RNA polymerase, the ribosome, GroEL-GroES and viral capsids [124,125]. The success of EN-NMA in capturing functional motions has been interpreted as a consequence of nature's exploitation of shape, topology and packing in the evolution of protein conformational dynamics [124,126–128].

A disadvantage of EN-NMA is that 'closed' to 'open' conformational transitions cannot be observed as they would involve the breaking of harmonic bonds in the network. Alternative methods exist for computing the minimum free energy path between two conformational states based on interpolation between the two reference elastic networks for each state. Optimization algorithms can be employed to find between the two global minima either a single minimum energy pathway [129] or an ensemble of low energy paths reflecting the ruggedness of protein landscapes [130]. The double-well approach has recently been employed successfully with Gō models, in which native interaction potentials were included for residue contacts present in either of two reference conformations [131–133]. The so-called dual, or switching, Gō model was used to study conformational transitions in myosin, Arc repressor and F1-ATPase.

The pervasive role of contact density can explain the robustness of the many varieties of Gō models that have been used to study folding. The results obtained with a Gō model are sensitive to the model parameters and various energy functional forms used [5,6,12,134–136]. A common means to safeguard against this is to compare folding results for multiple proteins obtained using the same model in light of available experimental data. Simulation results can also be influenced by the experimental structure from which the Gō model is derived, particularly the method used in its solution [137,138]. Note that NtrC exhibits a lower folding barrier than CheY and Spo0F in Figure 1, a result of having fewer native contacts overall and therefore a smoother landscape. The sparser native contacts in NtrC can be attributed to its being based on an NMR solution structure as opposed to the crystal structures for CheY and Spo0F [87]. It is also not uncommon for Gō models to be employed with native dihedral and/or angle restraints in addition to the typical nonbonded native interactions in order to make folding more cooperative [1,139]. Introducing a small desolvation barrier in the nonbonded native interaction potential has been employed as an alternate means of ensuring folding cooperativity [2,12,61,140].

Sulkowska and Cieplak recently compared a total of 62 different variations of the Gō potential in reproducing the experimental mechanical unfolding data of 28 proteins [12]. Several of the common functional forms employed for the native contact potential along with both uniform and heterogeneous energy scales were able to accurately reproduce the maximum unfolding force in stretching simulations ($R^2 \approx 0.84$). Most of the other models investigated came in a close second tier with $R^2 > 0.70$ for the correlation between predicted and experimental unfolding forces. Whitford *et al.* have compared results obtained from C_α and all-atom Gō models of three proteins [85]. While the C_α models neglected sidechain packing considerations, the correct global folding mechanisms were still obtained. Such studies suggest that the many variations of Gō models that have been employed to date are more alike than they are dissimilar.

Current non-Gō coarse-grained models depend on some form of secondary structure restraints to ensure native state stability [21–24]. Improving the physical accuracy of these non-Gō interactions will be a necessary step in removing biasing potentials from coarse-grained models so that large scale conformational transitions can be more readily studied. Native structure restraints aside, empirical coarse-graining techniques can often be considered Gō-like for their very nature of neglecting detailed atomistic interactions whilst favoring the essential interactions of the system [141–144]. In lieu of a physically accurate and transferable non-Gō coarse-grained model for proteins, Gō-like models will continue to find application in the study of protein folding and dynamics. The mechanical unfolding of titin, ubiquitin, bacteriorhodopsin and knotted transcarbamylase has recently been simulated using Gō potentials [76,145–149]. A Gō modeling survey of 7,510 structures in the Protein Data Bank was used to further delineate the topological determinants of mechanical stability [150]. A 297-residue outer membrane protease has been studied using a hybrid atomistic/coarse-grained approach in which the active site was represented with an atomistic force field and the protein scaffold was described using a Gō model [151–153]. Symmetrized Gō potentials have been employed for domain swapping and amyloid aggregation in which intermonomer residue contacts are treated identical to intramonomer native contacts [13,14,154]. Gō model simulations have been performed to explore the influence of macromolecular crowding agents [155,156] and confinement [62] on folding. Electrostatic interactions were incorporated between a Gō model of a transcription factor and coarse-grained DNA, which caused the protein to visit partially unfolded forms and suggested a fly casting mechanism of target recognition [157]. Lastly, in light of the connections with contact density, the link between function and folding mechanism is becoming a topic of considerable investigation [86,158–164].

ACKNOWLEDGMENTS

R.D.H. thanks the Burroughs Wellcome Fund La Jolla Interfaces in Science Training Program for financial support, the Center for Theoretical Biological Physics (ctbp.ucsd.edu) for providing a stimulating intellectual environment and D.A. Case for providing laboratory space. C.L.B. acknowledges support from NIH grant GM48807. Present address for Dr. Hills: Center for Biophysical Modeling and Simulation, University of Utah, 315 S 1400 E HEB2020, Salt Lake City, Utah 84112.

REFERENCES AND NOTES

1. Clementi, C; Nymeyer, H; Onuchic, JN. Topological and energetic factors: What determines the structural details of the transition state ensemble and "en-route" intermediates for protein folding? An investigation for small globular proteins. *J. Mol. Biol* 2000, *298*, 937–953.
2. Karanicolas, J; Brooks, CL, III. The origins of asymmetry in the folding transition states of protein L and protein G. *Protein Sci* 2002, *11*, 2351–2361.
3. Koga, N; Takada, S. Roles of native topology and chain-length scaling in protein folding: A simulation study with a Go-like model. *J. Mol. Biol* 2001, *313*, 171–180.
4. Paci, E; Vendruscolo, M; Karplus, M. Validity of Go models: Comparison with a solvent-shielded empirical energy decomposition. *Biophys. J* 2002, *83*, 3032–3038.
5. Prieto, L; de Sancho, D; Rey, A. Thermodynamics of Go-type models for protein folding. *J. Chem. Phys* 2005, *123*, 154903.
6. Prieto, L; Rey, A. Influence of the chain stiffness on the thermodynamics of a Go-type model for protein folding. *J. Chem. Phys* 2007, *126*, 166103.
7. Rhee, YM; Pande, VS. On the role of chemical detail in simulating protein folding kinetics. *Chem. Phys* 2006, *323*, 66–77.
8. Shea, JE; Onuchic, JN; Brooks, CL, III. Exploring the origins of topological frustration: Design of a minimally frustrated model of fragment B of protein A. *Proc. Natl. Acad. Sci. USA* 1999, *96*, 12512–12517.
9. Dokholyan, NV; Buldyrev, SV; Stanley, HE; Shakhnovich, EI. Discrete molecular dynamics studies of the folding of a protein-like model. *Fold. Des* 1998, *3*, 577–587.
10. Cheung, MS; Finke, JM; Callahan, B; Onuchic, JN. Exploring the interplay between topology and secondary structural formation in the protein folding problem. *J. Phys. Chem. B* 2003, *107*, 11193–11200.
11. Lam, AR; Borreguero, JM; Ding, F; Dokholyan, NV; Buldyrev, SV; Stanley, HE; Shakhnovich, E. Parallel foldng pathways in the SH3

domain protein. *J. Mol. Biol* 2007, *373*, 1348–1360.

12. Sulkowska, JI; Cieplak, M. Selection of optimal variants of Go-like models of proteins through studies of stretching. *Biophys. J* 2008, *95*, 3174–3191.
13. Ding, F; Dokholyan, NV; Buldyrev, SV; Stanley, HE; Shakhnovich, EI. Molecular dynamics simulation of the SH3 domain aggregation suggests a generic amyloidogenesis mechanism. *J. Mol. Biol* 2002, *324*, 851–857.
14. Barton, S; Jacak, R; Khare, SD; Ding, F; Dokholyan, NV. The length dependence of the PolyQ-mediated protein aggregation. *J. Biol. Chem* 2007, *282*, 25487–25492.
15. Kwiecinska, JI; Cieplak, M. Chirality and protein folding. *J. Phys.-Condes. Matter* 2005, *17*, S1565–S1580.
16. Yue, K; Fiebig, KM; Thomas, PD; Chan, HS; Shakhnovich, EI; Dill, KA. A test of lattice protein folding algorithms. *Proc. Natl. Acad. Sci. USA* 1995, *92*, 325–329.
17. Han, W; Wan, CK; Wu, YD. Toward a coarse-grained protein model coupled with a coarse-grained solvent model: Solvation free energies of amino acid side chains. *J. Chem. Theory Comput* 2008, *4*, 1891–1901.
18. Han, W; Wu, YD. Coarse-grained protein model coupled with a coarse-grained water model: Molecular dynamics study of polyalanine-based peptides. *J. Chem. Theory Comput* 2007, *3*, 2146–2161.
19. Sherwood, P; Brooks, BR; Sansom, MSP. Multiscale methods for macromolecular simulations. *Curr. Opin. Struct. Biol*2008, *18*, 630–640.
20. Tozzini, V. Coarse-grained models for proteins. *Curr. Opin. Struct. Biol* 2005, *15*, 144–150.
21. Bond, PJ; Sansom, MSP. Insertion and assembly of membrane proteins via simulation. *J. Am. Chem. Soc* 2006, *128*, 2697–2704.
22. Monticelli, L; Kandasamy, SK; Periole, X; Larson, RG; Tieleman, DP; Marrink, SJ. The MARTINI coarse-grained force field: Extension to proteins. *J. Chem. Theory Comput* 2008, *4*, 819–834.
23. Shih, AY; Arkhipov, A; Freddolino, PL; Schulten, K. Coarse grained

protein-lipid model with application to lipoprotein particles. *J. Phys. Chem. B* 2006, *110*, 3674–3684.

24. Yap, EH; Fawzi, NL; Head-Gordon, T. A coarse-grained alpha-carbon protein model with anisotropic hydrogen-bonding. *Proteins* 2008, *70*, 626–638.
25. Thorpe, IF; Zhou, J; Voth, GA. Peptide folding using multiscale coarse-grained models. *J. Phys. Chem. B* 2008, *112*, 13079–13090.
26. Makowski, M; Sobolewski, E; Czaplewski, C; Oldziej, S; Liwo, A; Scheraga, HA. Simple physics-based analytical formulas for the potentials of mean force for the interaction of amino acid side chains in water. IV. Pairs of different hydrophobic side chains. *J. Phys. Chem. B* 2008, *112*, 11385–11395.
27. da Silveira, CH; Pires, DEV; Minardi, RC; Ribeiro, C; Veloso, CJM; Lopes, JCD; Meira, W; Neshich, G; Ramos, CHI; Habesch, R; Santoro, MM. Protein cutoff scanning: A comparative analysis of cutoff dependent and cutoff free methods for prospecting contacts in proteins. *Proteins* 2009, *74*, 727–743.
28. Taketomi, H; Ueda, Y; Go, N. Studies on protein folding, unfolding and fluctuations by computer simulation. 1. Effect of specific amino acid sequence represented by specific inter-unit interactions. *Int. J. Pept. Protein Res* 1975, *7*, 445–459.
29. Go, N; Scheraga, HA. On the use of classical statistical mechanics in the treatment of polymer chain conformations. *Macromolecules* 1976, *9*, 535–542.
30. Ueda, Y; Go, N. Theory of large-amplitude conformational fluctuations in native globular proteins: Independent fluctuating site model. *Int. J. Pept. Protein Res* 1976, *8*, 551–558.
31. Go, N; Taketomi, H. Respective roles of short-range and long-range interactions in protein folding. *Proc. Natl. Acad. Sci. USA* 1978, *75*, 559–563.
32. Ueda, Y; Taketomi, H; Go, N. Studies on protein folding, unfolding and fluctuations by computer simulation. 2. A three-dimensional lattice model of lysozyme. *Biopolymers* 1978, *17*, 1531–1548.
33. Go, N; Taketomi, H. Studies on protein folding, unfolding and

fluctuations by computer simulation. 3. Effect of short-range interactions. *Int. J. Pept. Protein Res* 1979, *13*, 235–252.

34. Go, N; Taketomi, H. Studies on protein folding, unfolding and fluctuations by computer simulation. 4. Hydrophobic interactions. *Int. J. Pept. Protein Res* 1979, *13*, 447–461.
35. Go, N. Theoretical studies of protein folding. *Annu. Rev. Biophys. Bioeng* 1983, *12*, 183–210.
36. Taketomi, H; Kano, F; Go, N. The effect of amino acid substitution on protein folding and protein unfolding transition studied by computer simulation. *Biopolymers* 1988, *27*, 527–559.
37. Brooks, CL, III; Onuchic, JN; Wales, DJ. Statistical thermodynamics: Taking a walk on a landscape. *Science* 2001, *293*, 612–613.
38. Dill, KA; Ozkan, SB; Shell, MS; Weikl, TR. The protein folding problem. *Ann. Rev. Biophys* 2008, *37*, 289–316.
39. Onuchic, JN; Wolynes, PG. Theory of protein folding. *Curr. Opin. Struct. Biol* 2004, *14*, 70–75.
40. Ozkan, SB; Wu, GA; Chodera, JD; Dill, KA. Protein folding by zipping and assembly. *Proc. Natl. Acad. Sci. USA* 2007, *104*, 11987–11992.
41. Prieto, L; Rey, A. Influence of the native topology on the folding barrier for small proteins. *J. Chem. Phys* 2007, *127*, 175101.
42. Hills, RD, Jr; Brooks, CL, III. Subdomain competition, cooperativity, and topological frustration in the folding of CheY. *J. Mol. Biol* 2008, *382*, 485–495.
43. Fersht, AR; Sato, S. Phi-Value analysis and the nature of protein-folding transition states. *Proc. Natl. Acad. Sci. USA*2004, *101*, 7976–7981.
44. Chavez, LL; Gosavi, S; Jennings, PA; Onuchic, JN. Multiple routes lead to the native state in the energy landscape of the beta-trefoil family. *Proc. Natl. Acad. Sci. USA* 2006, *103*, 10254–10258.
45. Dokholyan, NV; Buldyrev, SV; Stanley, HE; Shakhnovich, EI. Identifying the protein folding nucleus using molecular dynamics. *J. Mol. Biol* 2000, *296*, 1183–1188.
46. Ferreiro, DU; Cho, SS; Komives, EA; Wolynes, PG. The energy landscape of modular repeat proteins: Topology determines folding

mechanism in the ankyrin family. *J. Mol. Biol* 2005, *354*, 679–692.
47. Gosavi, S; Chavez, LL; Jennings, PA; Onuchic, JN. Topological frustration and the folding of interleukin-1 beta. *J. Mol. Biol* 2006, *357*, 986–996.
48. Hubner, IA; Oliveberg, M; Shakhnovich, EI. Simulation, experiment, and evolution: Understanding nucleation in protein S6 folding. *Proc. Natl. Acad. Sci. USA* 2004, *101*, 8354–8359.
49. Karanicolas, J; Brooks, CL, III. Improved Go-like models demonstrate the robustness of protein folding mechanisms towards non-native interactions. *J. Mol. Biol* 2003, *334*, 309–325.
50. Levy, Y; Cho, SS; Onuchic, JN; Wolynes, PG. A survey of flexible protein binding mechanisms and their transition states using native topology based energy landscapes. *J. Mol. Biol* 2005, *346*, 1121–1145.
51. Levy, Y; Wolynes, PG; Onuchic, JN. Protein topology determines binding mechanism. *Proc. Natl. Acad. Sci. USA* 2004, *101*, 511–516.
52. Roy, M; Chavez, LL; Finke, JM; Heidary, DK; Onuchic, JN; Jennings, PA. The native energy landscape for interleukin-1 beta. Modulation of the population ensemble through native-state topology. *J. Mol. Biol* 2005, *348*, 335–347.
53. Settanni, G; Hoang, TX; Micheletti, C; Maritan, A. Folding pathways of prion and doppel. *Biophys. J* 2002, *83*, 3533–3541.
54. Zuo, GH; Wang, J; Wang, W. Folding with downhill behavior and low cooperativity of proteins. *Proteins* 2006, *63*, 165–173.
55. Kouza, M; Li, MS; O'Brien, EP; Hu, CK; Thirumalai, D. Effect of finite size on cooperativity and rates of protein folding. *J. Phys. Chem. A* 2006, *110*, 671–676.
56. Clementi, C; Plotkin, SS. The effects of nonnative interactions on protein folding rates: Theory and simulation. *Protein Sci* 2004, *13*, 1750–1766.
57. Das, P; Matysiak, S; Clementi, C. Balancing energy and entropy: A minimalist model for the characterization of protein folding landscapes. *Proc. Natl. Acad. Sci. USA* 2005, *102*, 10141–10146.
58. Gu, Z; Rao, MK; Forsyth, WR; Finke, JM; Matthews, CR. Structural

analysis of kinetic folding intermediates for a TIM barrel protein, indole-3-glycerol phosphate synthase, by hydrogen exchange mass spectrometry and Go model simulation. *J. Mol. Biol* 2007, *374*, 528–546.

59. Paci, E; Vendruscolo, M; Karplus, M. Native and non-native interactions along protein folding and unfolding pathways. *Proteins* 2002, *47*, 379–392.
60. Zarrine-Afsart, A; Wallin, S; Neculai, AM; Neudecker, P; Howell, PL; Davidson, AR; Chan, HS. Theoretical and experimental demonstration of the importance of specific nonnative interactions in protein folding. *Proc. Natl. Acad. Sci. USA* 2008, *105*, 9999–10004.
61. Zhang, Z; Chan, HS. Native topology of the designed protein Top7 is not conducive to cooperative folding. *Biophys. J*2009, *96*, L25–L27.
62. Griffin, MA; Friedel, M; Shea, JE. Effects of frustration, confinement, and surface interactions on the dimerization of an off-lattice beta-barrel protein. *J. Chem. Phys* 2005, *123*, 174707.
63. Zhou, YQ; Karplus, M. Folding thermodynamics of a model three-helix-bundle protein. *Proc. Natl. Acad. Sci. USA*1997, *94*, 14429–14432.
64. Matysiak, S; Clementi, C. Optimal combination of theory and experiment for the characterization of the protein folding landscape of S6: How far can a minimalist model go? *J. Mol. Biol* 2004, *343*, 235–248.
65. Du, R; Pande, VS; Grosberg, AY; Tanaka, T; Shakhnovich, E. On the role of conformational geometry in protein folding. *J. Chem. Phys* 1999, *111*, 10375–10380.
66. Ivankov, DN; Garbuzynskiy, SO; Alm, E; Plaxco, KW; Baker, D; Finkelstein, AV. Contact order revisited: Influence of protein size on the folding rate. *Protein Sci* 2003, *12*, 2057–2062.
67. Plaxco, KW; Simons, KT; Baker, D. Contact order, transition state placement and the refolding rates of single domain proteins. *J. Mol. Biol* 1998, *277*, 985–994.
68. Dyson, HJ; Wright, PE. Intrinsically unstructured proteins and their functions. *Nat. Rev. Mol. Cell Biol* 2005, *6*, 197–208.

69. Thirumalai, D; Klimov, DK; Dima, RI. Emerging ideas on the molecular basis of protein and peptide aggregation. *Curr. Opin. Struct. Biol* 2003, *13*, 146–159.
70. Sutto, L; Tiana, G; Broglia, RA. Sequence of events in folding mechanism: Beyond the Go model. *Protein Sci* 2006, *15*, 1638–1652.
71. Khare, SD; Ding, F; Dokholyan, NV. Folding of Cu, Zn superoxide dismutase and familial amyotrophic lateral sclerosis. *J. Mol. Biol* 2003, *334*, 515–525.
72. Miyazawa, S; Jernigan, RL. Residue-residue potentials with a favorable contact pair term and an unfavorable high packing density term, for simulation and threading. *J. Mol. Biol* 1996, *256*, 623–644.
73. Rao, MK; Chapman, TR; Finke, JM. Crystallographic B-factors highlight energetic frustration in aldolase folding. *J. Phys. Chem. B* 2008, *112*, 10417–10431.
74. Dixon, RDS; Chen, YW; Feng, D; Khare, SD; Prutzman, KC; Schaller, MD; Campbell, SL; Dokholyan, NV. New insights into FAK signaling and localization based on detection of a FAT domain folding intermediate. *Structure* 2004, *12*, 2161–2171.
75. Clementi, C; Garcia, AE; Onuchic, JN. Interplay among tertiary contacts, secondary structure formation and side-chain packing in the protein folding mechanism: All-atom representation study of protein L. *J. Mol. Biol* 2003, *326*, 933–954.
76. Kleiner, A; Shakhnovich, E. The mechanical unfolding of ubiquitin through all-atom Monte Carlo simulation with a Go-type potential. *Biophys. J* 2007, *92*, 2054–2061.
77. Linhananta, A; Boer, J; MacKay, I. The equilibrium properties and folding kinetics of an all-atom Go model of the Trp-cage. *J. Chem. Phys* 2005, *122*, 114901.
78. Linhananta, A; Zhou, YQ. The role of sidechain packing and native contact interactions in folding: Discontinuous molecular dynamics folding simulations of an all-atom Go model of fragment B of Staphylococcal protein A. *J. Chem. Phys* 2002, *117*, 8983–8995.
79. Meinke, JH; Hansmann, UHE. Protein simulations combining an all-atom force field with a Go term. *J. Phys.-Condes. Matter* 2007, *19*, 285215.

80. Shimada, J; Kussell, EL; Shakhnovich, EI. The folding thermodynamics and kinetics of crambin using an all-atom Monte Carlo simulation. *J. Mol. Biol* 2001, *308*, 79–95.
81. Shimada, J; Shakhnovich, EI. The ensemble folding kinetics of protein G from an all-atom Monte Carlo simulation. *Proc. Natl. Acad. Sci. USA* 2002, *99*, 11175–11180.
82. Zhou, YQ; Linhananta, A. Thermodynamics of an all-atom off-lattice model of the fragment B of Staphylococcal protein A: Implication for the origin of the cooperativity of protein folding. *J. Phys. Chem. B* 2002, *106*, 1481–1485.
83. Luo, ZL; Ding, JD; Zhou, YQ. Temperature-dependent folding pathways of pin1 WW domain: An all-atom molecular dynamics simulation of a Go model. *Biophys. J* 2007, *93*, 2152–2161.
84. Luo, ZL; Ding, JD; Zhou, YQ. Folding mechanisms of individual beta-hairpins in a Go model of Pin1 WW domain by all-atom molecular dynamics simulations. *J. Chem. Phys* 2008, *128*, 225103.
85. Whitford, PC; Noel, JK; Gosavi, S; Schug, A; Sanbonmatsu, KY; Onuchic, JN. An all-atom structure-based potential for proteins: Bridging minimal models with all-atom empirical forcefields. *Proteins* In press. 2009.
86. Hills, RD, Jr; Brooks, CL, III. Coevolution of function and the folding landscape: Correlation with density of native contacts. *Biophys. J* 2008, *95*, L57–L59.
87. Hills, RD, Jr; Kathuria, SV; Wallace, LA; Matthews, CR; Brooks, CL, III. Functional diversity and shifting cores of stability modulate the folding mechanisms of flavodoxin fold proteins. In *J Mol Biol*; 2009; Pending publication.
88. Kathuria, SV; Day, IJ; Wallace, LA; Matthews, CR. Kinetic traps in the folding of beta alpha-repeat proteins: CheY initially misfolds before accessing the native conformation. *J. Mol. Biol* 2008, *382*, 467–484.
89. LopezHernandez, E; Serrano, L. Structure of the transition state for folding of the 129 aa protein CheY resembles that of a smaller protein, CI-2. *Fold. Des* 1996, *1*, 43–55.
90. Du, R; Pande, VS; Grosberg, AY; Tanaka, T; Shakhnovich, ES. On the transition coordinate for protein folding. *J. Chem. Phys* 1998, *108*,

334–350.

91. Snow, CD; Rhee, YM; Pande, VS. Kinetic definition of protein folding transition state ensembles and reaction coordinates. *Biophys. J* 2006, *91*, 14–24.
92. Cho, SS; Levy, Y; Wolynes, PG. P versus Q: Structural reaction coordinates capture protein folding on smooth landscapes. *Proc. Natl. Acad. Sci. USA* 2006, *103*, 586–591.
93. Piana, S; Laio, A. Advillin folding takes place on a hypersurface of small dimensionality. *Phys. Rev. Lett* 2008, *101*, 208101.
94. Juraszek, J; Bolhuis, PG. Rate constant and reaction coordinate of Trp-cage folding in explicit water. *Biophys. J* 2008, *95*, 4246–4257.
95. Shakhnovich, E. Protein folding thermodynamics and dynamics: Where physics, chemistry, and biology meet. *Chem. Rev* 2006, *106*, 1559–1588.
96. Shakhnovich, E; Farztdinov, G; Gutin, AM; Karplus, M. Protein folding bottlenecks: A lattice Monte-Carlo simulation. *Phys. Rev. Lett* 1991, *67*, 1665–1668.
97. Faccioli, P. Characterization of protein folding by dominant reaction pathways. *J. Phys. Chem. B* 2008, *112*, 13756–13764.
98. Sola, M; Lopez-Hernandez, E; Cronet, P; Lacroix, E; Serrano, L; Coll, M; Parraga, A. Towards understanding a molecular switch mechanism: Thermodynamic and crystallographic studies of the signal transduction protein CheY. *J. Mol. Biol* 2000, *303*, 213–225.
99. Stock, AM; Guhaniyogi, J. A new perspective on response regulator activation. *J. Bacteriol* 2006, *188*, 7328–7330.
100. Varughese, KI; Tsigelny, I; Zhao, HY. The crystal structure of beryllofluoride Spo0F in complex with the phosphotransferase Spo0B represents a phosphotransfer pretransition state. *J. Bacteriol* 2006, *188*, 4970–4977.
101. Volkman, BF; Lipson, D; Wemmer, DE; Kern, D. Two-state allosteric behavior in a single-domain signaling protein. *Science* 2001, *291*, 2429–2433.
102. Cellitti, J; Bernstein, R; Marqusee, S. Exploring subdomain cooperativity in T4 lysozyme II: Uncovering the C-terminal subdomain as a hidden intermediate in the kinetic folding

pathway. *Protein Sci* 2007, *16*, 852–862.

103. Cho, SS; Levy, Y; Wolynes, PG. Quantitative criteria for native energetic heterogeneity influences in the prediction of protein folding kinetics. *Proc. Natl. Acad. Sci. USA* 2009, *106*, 434–439.
104. Halle, B. Flexibility and packing in proteins. *Proc. Natl. Acad. Sci. USA* 2002, *99*, 1274–1279.
105. Hills, RD, Jr; Brooks, CL, III. Hydrophobic cooperativity as a mechanism for amyloid nucleation. *J. Mol. Biol* 2007, *368*, 894–901.
106. Aftabuddin, M; Kundu, S. Hydrophobic, hydrophilic, and charged amino acid networks within protein. *Biophys. J*2007, *93*, 225–231.
107. Atilgan, AR; Akan, P; Baysal, C. Small-world communication of residues and significance for protein dynamics. *Biophys. J* 2004, *86*, 85–91.
108. Bode, C; Kovacs, IA; Szalay, MS; Palotai, R; Korcsmaros, T; Csermely, P. Network analysis of protein dynamics. *FEBS Lett* 2007, *581*, 2776–2782.
109. Dokholyan, NV; Li, L; Ding, F; Shakhnovich, EI. Topological determinants of protein folding. *Proc. Natl. Acad. Sci. USA* 2002, *99*, 8637–8641.
110. Greene, LH; Higman, VA. Uncovering network systems within protein structures. *J. Mol. Biol* 2003, *334*, 781–791.
111. Higman, VA; Greene, LH. Elucidation of conserved long-range interaction networks in proteins and their significance in determining protein topology. *Physica A* 2006, *368*, 595–606.
112. Vendruscolo, M; Dokholyan, NV; Paci, E; Karplus, M. Small-world view of the amino acids that play a key role in protein folding. *Phys Rev E* 2002, *65*. 061910-1-4.
113. Kathuria, SV; Matthews, CR. Clusters of isoleucine, leucine and valine side chains define cores of stability in globular proteins: Sequence determinants of structure, stability and folding. 2009; Pending publication.
114. Wu, Y; Vadrevu, R; Kathuria, S; Yang, XY; Matthews, CR. A tightly packed hydrophobic cluster directs the formation of an off-pathway sub-millisecond folding intermediate in the alpha subunit of

tryptophan synthase, a TIM barrel protein. *J. Mol. Biol* 2007, *366*, 1624–1638.

115. Hilser, VJ; Freire, E. Structure-based calculation of the equilibrium folding pathway of proteins. Correlation with hydrogen exchange protection factors. *J. Mol. Biol* 1996, *262*, 756–772.
116. Alm, E; Baker, D. Prediction of protein-folding mechanisms from free-energy landscapes derived from native structures. *Proc. Natl. Acad. Sci. USA* 1999, *96*, 11305–11310.
117. Galzitskaya, OV; Finkelstein, AV. A theoretical search for folding/unfolding nuclei in three-dimensional protein structures. *Proc. Natl. Acad. Sci. USA* 1999, *96*, 11299–11304.
118. Karanicolas, J; Brooks, CL, III. The importance of explicit chain representation in protein folding models: An examination of Ising-like models. *Proteins* 2003, *53*, 740–747.
119. Munoz, V; Eaton, WA. A simple model for calculating the kinetics of protein folding from three-dimensional structures. *Proc. Natl. Acad. Sci. USA* 1999, *96*, 11311–11316.
120. Takada, S. Go-ing for the prediction of protein folding mechanisms. *Proc. Natl. Acad. Sci. USA* 1999, *96*, 11698–11700.
121. Doruker, P; Jernigan, RL; Bahar, I. Dynamics of large proteins through hierarchical levels of coarse-grained structures. *J. Comput. Chem* 2002, *23*, 119–127.
122. Tama, F; Sanejouand, YH. Conformational change of proteins arising from normal mode calculations. *Protein Eng*2001, *14*, 1–6.
123. Petrone, P; Pande, VS. Can conformational change be described by only a few normal modes? *Biophys. J* 2006, *90*, 1583–1593.
124. Tama, F; Brooks, CL, III. Symmetry, form, and shape: Guiding principles for robustness in macromolecular machines. *Annu. Rev. Biophys. Biomolec. Struct* 2006, *35*, 115–133.
125. Yang, L; Song, G; Jernigan, RL. How well can we understand large-scale protein motions using normal modes of elastic network models? *Biophys. J* 2007, *93*, 920–929.
126. Bagci, Z; Jernigan, RL; Bahar, I. Residue packing in proteins: Uniform distribution on a coarse-grained scale. *J. Chem. Phys* 2002, *116*, 2269–2276.

127. Keskin, O; Jernigan, RL; Bahar, I. Proteins with similar architecture exhibit similar large-scale dynamic behavior. *Biophys. J* 2000, *78*, 2093–2106.

128. Lu, MY; Ma, JP. The role of shape in determining molecular motions. *Biophys. J* 2005, *89*, 2395–2401.

129. Maragakis, P; Karplus, M. Large amplitude conformational change in proteins explored with a plastic network model: Adenylate kinase. *J. Mol. Biol* 2005, *352*, 807–822.

130. Chu, JW; Voth, GA. Coarse-grained free energy functions for studying protein conformational changes: A double-well network model. *Biophys. J* 2007, *93*, 3860–3871.

131. Best, RB; Chen, YG; Hummer, G. Slow protein conformational dynamics from multiple experimental structures: The helix/sheet transition of arc repressor. *Structure* 2005, *13*, 1755–1763.

132. Koga, N; Takada, S. Folding-based molecular simulations reveal mechanisms of the rotary motor F-1-ATPase. *Proc. Natl. Acad. Sci. USA* 2006, *103*, 5367–5372.

133. Takagi, F; Kikuchi, M. Structural change and nucleotide dissociation of myosin motor domain: Dual Go model simulation. *Biophys. J* 2007, *93*, 3820–3827.

134. Cieplak, M; Hoang, TX. Universality classes in folding times of proteins. *Biophys. J* 2003, *84*, 475–488.

135. Kaya, H; Chan, HS. Solvation effects and driving forces for protein thermodynamic and kinetic cooperativity: How adequate is native-centric topological modeling? *J. Mol. Biol* 2003, *326*, 911–931.

136. Kaya, H; Liu, ZR; Chan, HS. Chevron Behavior and isostable enthalpic barriers in protein folding: Successes and limitations of simple Go-like modeling. *Biophys. J* 2005, *89*, 520–535.

137. Prieto, L; Rey, A. Simulations of the protein folding process using topology-based models depend on the experimental structure. *J. Chem. Phys* 2008, *129*, 115101.

138. Rey-Stolle, MF; Enciso, M; Rey, A. Topology-based models and NMR structures in protein folding simulations. *J Comp Chem* In press. 2009.

139. Settanni, G; Cattaneo, A; Maritan, A. Role of native-state topology in the stabilization of intracellular antibodies. *Biophys. J* 2001, *81*, 2935–2945.

140. Cheung, MS; Garcia, AE; Onuchic, JN. Protein folding mediated by solvation: Water expulsion and formation of the hydrophobic core occur after the structural collapse. *Proc. Natl. Acad. Sci. USA* 2002, *99*, 685–690.

141. Bellesia, G; Shea, JE. Self-assembly of beta-sheet forming peptides into chiral fibrillar aggregates. *J. Chem. Phys* 2007, *126*, 245104.

142. Fawzi, NL; Kohlstedt, KL; Okabe, Y; Head-Gordon, T. Protofibril assemblies of the arctic, dutch, and flemish mutants of the Alzheimer's A beta(1–40) peptide. *Biophys. J* 2008, *94*, 2007–2016.

143. Nguyen, HD; Hall, CK. Spontaneous fibril formation by polyalanines; Discontinuous molecular dynamics simulations. *J. Am. Chem. Soc* 2006, *128*, 1890–1901.

144. Nguyen, HD; Reddy, VS; Brooks, CL, III. Deciphering the kinetic mechanism of spontaneous self-assembly of icosahedral capsids. *Nano Lett* 2007, *7*, 338–344.

145. Cieplak, M; Filipek, S; Janovjak, H; Krzysko, KA. Pulling single bacteriorhodopsin out of a membrane: Comparison of simulation and experiment. *Biochim. Biophys. Acta-Biomembr* 2006, *1758*, 537–544.

146. Cieplak, M; Hoang, TX; Robbins, MO. Folding and stretching in a Go-like model of titin. *Proteins* 2002, *49*, 114–124.

147. Sulkowska, JI; Sulkowski, P; Szymczak, P; Cieplak, M. Stabilizing effect of knots on proteins. *Proc. Natl. Acad. Sci. USA* 2008, *105*, 19714–19719.

148. Sulkowska, JI; Kloczkowski, A; Sen, TZ; Cieplak, M; Jernigan, RL. Predicting the order in which contacts are broken during single molecule protein stretching experiments. *Proteins* 2008, *71*, 45–60.

149. Li, MS; Kouza, M; Hu, CK. Refolding upon force quench and pathways of mechanical and thermal unfolding of ubiquitin. *Biophys. J* 2007, *92*, 547–561.

150. Sulkowska, JI; Cieplak, M. Mechanical stretching of proteins:

A theoretical survey of the Protein Data Bank. *J. Phys.-Condes. Matter* 2007, *19*, 283201.

151. Neri, M; Anselmi, C; Carnevale, V; Vargiu, AV; Carloni, P. Molecular dynamics simulations of outer-membrane protease T from E-coli based on a hybrid coarse-grained/atomistic potential. *J. Phys.-Condes. Matter* 2006, *18*, S347–S355.
152. Neri, M; Anselmi, C; Cascella, M; Maritan, A; Carloni, P. Coarse-grained model of proteins incorporating atomistic detail of the active site. *Phys. Rev. Lett* 2005, *95*, 218102.
153. Neri, M; Baaden, M; Carnevale, V; Anselmi, C; Maritan, A; Carloni, P. Microseconds dynamics simulations of the outer-membrane protease T. *Biophys. J* 2008, *94*, 71–78.
154. Yang, SC; Levine, H; Onuchic, JN. Protein oligomerization through domain swapping: Role of inter-molecular interactions and protein concentration. *J. Mol. Biol* 2005, *352*, 202–211.
155. Homouz, D; Perham, M; Samiotakis, A; Cheung, MS; Wittung-Stafshede, P. Crowded, cell-like environment induces shape changes in aspherical protein. *Proc. Natl. Acad. Sci. USA* 2008, *105*, 11754–11759.
156. Pincus, DL; Thirumalai, D. Crowding effects on the mechanical stability and unfolding pathways of ubiquitin. *J. Phys. Chem. B* 2009, *113*, 359–368.
157. Levy, Y; Onuchic, JN; Wolynes, PG. Fly-casting in protein-DNA binding: Frustration between protein folding and electrostatics facilitates target recognition. *J. Am. Chem. Soc* 2007, *129*, 738–739.
158. Ferreiro, DU; Hegler, JA; Komives, EA; Wolynes, PG. Localizing frustration in native proteins and protein assemblies. *Proc. Natl. Acad. Sci. USA* 2007, *104*, 19819–19824.
159. Gosavi, S; Whitford, PC; Jennings, PA; Onuchic, JN. Extracting function from a beta-trefoil folding motif. *Proc. Natl. Acad. Sci. USA* 2008, *105*, 10384–10389.
160. Jager, M; Zhang, Y; Bieschke, J; Nguyen, H; Dendle, M; Bowman, ME; Noel, JP; Gruebele, M; Kelly, JW. Structure-function-folding relationship in a WW domain. *Proc. Natl. Acad. Sci. USA* 2006, *103*, 10648–10653.

161. Miyashita, O; Onuchic, JN; Wolynes, PG. Nonlinear elasticity, proteinquakes, and the energy landscapes of functional transitions in proteins. *Proc. Natl. Acad. Sci. USA* 2003, *100*, 12570–12575.

162. Miyashita, O; Wolynes, PG; Onuchic, JN. Simple energy landscape model for the kinetics of functional transitions in proteins. *J. Phys. Chem. B* 2005, *109*, 1959–1969.

163. Whitford, PC; Miyashita, O; Levy, Y; Onuchic, JN. Conformational transitions of adenylate kinase: Switching by cracking. *J. Mol. Biol* 2007, *366*, 1661–1671.

164. Karanicolas, J; Brooks, CL, III. Integrating folding kinetics and protein function: Biphasic kinetics and dual binding specificity in a WW domain. *Proc. Natl. Acad. Sci. USA* 2004, *101*, 3432–3437.

CHAPTER 4

COARSE-GRAINED MODELING OF PEPTIDE DOCKING ASSOCIATED WITH LARGE CONFORMATION TRANSITIONS OF THE BINDING PROTEIN: TROPONIN I FRAGMENT–TROPONIN C SYSTEM

Jacek Wabik, Mateusz Kurcinski, Andrzej Kolinski and Rino Ragno

Laboratory of Theory of Biopolymers, Faculty of Chemistry, University of Warsaw, Pasteura 1, 02-093 Warsaw, Poland

ABSTRACT

Most of the current docking procedures are focused on fine conformational adjustments of assembled complexes and fail to reproduce large-scale protein motion. In this paper, we test a new modeling approach developed to address this problem. CABS-dock is a versatile and efficient tool for modeling the structure, dynamics and interactions of protein complexes. The docking protocol employs a coarse-grained representation of proteins, a simplified model of interactions and advanced protocols for conformational sampling. CABS-dock is one of the very few tools that allow unrestrained docking with large conformational freedom of the receptor. In an example application we modeled the process of complex assembly between two proteins: Troponin C (TnC) and the N-terminal helix of Troponin I (TnI N-helix), which occurs *in vivo* during muscle contraction. Docking simulations illustrated how the TnC molecule undergoes significant conformational transition on complex formation, a phenomenon that can be modeled only when protein flexibility is properly accounted for. This way our procedure opens up a new possibility for studying mechanisms of protein complex assembly, which may be a supporting tool for rational drug design.

Keywords: Protein docking; flexible docking; coarse-grained modeling; molecular mechanisms

INTRODUCTION

The development of new drugs is one of the most challenging tasks in science today. Combined efforts of the pharmaceutical industry, academic researchers and biotech companies have not only improved the process of drug design, but also contributed to advances in science itself. Computational modeling and simulations have become integral procedures in introducing new drugs to the market and it is safe to assume that the role of theoretical computations in this field will increase in the future [1].

Protein activity in organisms involves interactions with other biomolecules, which makes them perfect targets for rationally designed drugs. After the Human Genome Project [2] was completed, a large number of new drug targets were expected to be found [3]. It soon became evident that a protein's crystallographic structure alone is, in general, insufficient to understand its function [4]. What in fact is needed, is a complete description of the structure and dynamics of the protein along with its interacting partners. Such a demand from the pharmaceutical industry has recently forced transition in the development of theoretical methods for protein examination: from protein structure prediction to the modeling of whole protein complexes. Moreover, the emphasis in protein modeling is nowadays being put more on obtaining a dynamic picture of a protein system than a simple static structure.

Molecular docking is a technique for predicting the structure of a molecular complex, given the structures of its individual components. There is a great number of software packages available for protein-ligand docking, including DOCK [5], AutoDock [6], HADDOCK (High Ambiguity Driven protein-protein DOCKing) [7], Flex [8], Glide [9], GOLD [10], ZDOCK [11], and others. Docking methods play an important role in drug design, although several important challenges of the docking approach still remain unsolved [12]. Only for a few examples of protein-protein complexes is it possible to obtain high quality models. In most cases docking must be combined with restraints, evolutionary information, or binding site predictions.

It is also problematic to incorporate adequate flexibility to a model of protein docking. In general, when a pair of proteins form a complex, there is a degree of structural rearrangement between unbound and bound conformations. In some cases these changes are marginal and docking can be done using methods based on rigid-body shape complementarity, such as FTDOCK (Fourier Transform Dock) [13] and ZDOCK [11]. These methods exhaustively scan a six-dimensional conformational space of translations and rotations of one molecule *vs.* the other. The rigid-body assumption is naturally a shortcoming for proteins that exhibit large-scale conformational changes on binding. In some cases even small backbone deformations could be essential [14].

The ultimate goal in the development of docking methods is to dock fully flexible molecules without any prior knowledge of the shape and location of the binding site, starting from random conformation of the ligand. This could be achieved in Molecular Dynamics (MD) simulations, but it is highly computationally demanding. MD docking would take an impractical amount of time, even using the most powerful supercomputers. So far, some degree of flexibility only has been handled in a small number of models [15]. For example, ATTRACT employs a coarse-grained model with Normal Mode Analysis [16], whereas the Rosetta's 'fold-and-dock' algorithm uses experimental data: nuclear magnetic resonance shifts and residual dipolar couplings [17].

CABS-dock is a new modeling tool for the efficient modeling of structure and dynamics of protein and peptide complexes. Recently, we demonstrated that the method enables very realistic modeling of fully unrestrained docking of peptide chains with unlimited conformational freedom to a very flexible receptor but restricted to near-native fluctuations [18]. CABS-dock provides a variable, pre-defined level of protein chain flexibility. In the present study, it is used for docking a fully flexible peptide ligand to the receptor with fully flexible linkers between semi-flexible domains. This way large-scale conformational transitions of the receptor can be modeled. Since CABS-dock uses a Monte Carlo dynamics sampling method, this approach can be used not only for predicting native conformations, but also to investigate the mechanism of folding and binding processes. Additionally, we used the CABS-flex [19] tool for investigating near-native fluctuations of a protein complex. It has been shown recently that the results obtained using this tool are consistent with Molecular Dynamics 10 ns simulations [20]. Applications of both modeling tools provide a possibility for the realistic comparison of short-range fluctuations of a receptor molecule alone and a receptor molecule in complex with a peptide.

CABS-dock and CABS-flex are based on the CABS model initially developed for globular protein structure prediction [21]. It has performed exceptionally well in the CASP6 (Critical Assessment of protein Structure Prediction) experiment, where the Kolinski-Bujnicki group was ranked as one of a couple of the best predictors from more than a hundred leading groups [22,23,24,25]. Since then, CABS has been successfully used in a number of other applications, such as protein fragment reconstruction [26], modeling of folding pathways [27,28], protein-peptide docking [29,30,31], modeling of protein flexibility [32], or folding and binding of intrinsically disordered proteins [33]. The main features of the CABS model are briefly described in the Methods Section.

As a proof of concept of fully flexible peptide docking with large conformational changes of the receptor, we study protein-peptide docking in a complex between a fragment of skeletal troponin I and troponin C (pdb code: 1A2X [34]). Troponin is a complex of three proteins: troponin C (TnC), troponin I (TnI) and troponin T (TnT). It plays a crucial role in the regulation of skeletal and cardiac muscle contraction [35]. Troponin binds at regular intervals along the actin polymer on the thin filament. TnT is responsible for binding the whole complex to a tropomyosin molecule. TnC forms the core of the complex: it binds to TnI with variable strength dependent on Ca^{2+} concentration.

The main sites of TnC/TnI interaction are located in the hydrophobic pocket of the N-terminal and C-terminal domains of TnC (TnC N-domain, TnC C-domain). TnC binds two calcium ions with high affinity. Another two calcium ions are bound at a low affinity rate, triggering exposure of the TnC N-domain hydrophobic surface. During the muscle contraction and relaxation cycle Ca^{2+} concentration changes, which in turn triggers conformational transition in the TnC regulatory N-domain. When Ca^{2+} concentration is low, the hydrophobic pocket on the N-domain remains closed, blocking this interaction site for TnI binding. In this conformational state, troponin prevents the approach of myosin heads toward actin [34]. At high Ca^{2+} concentration, the hydrophobic pocket of the N-domain is accessible for TnI binding [36], which may also be co-induced by the cardiac TnI molecule [37].

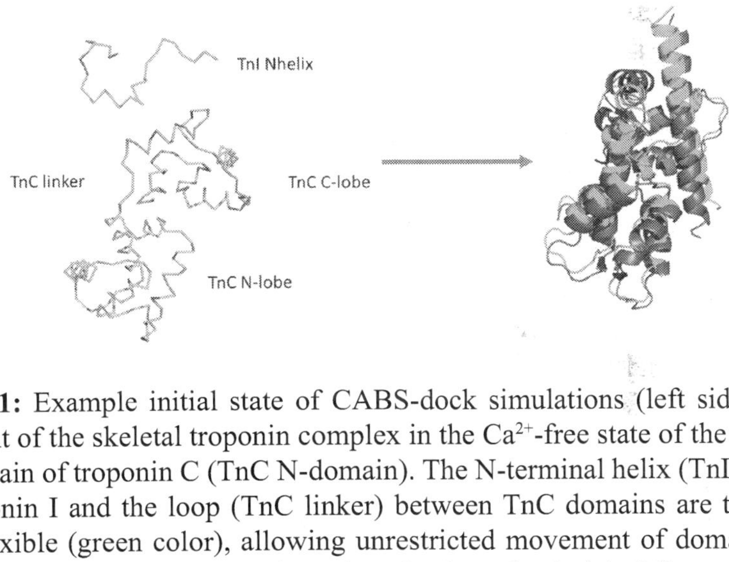

Figure 1: Example initial state of CABS-dock simulations (left side) of the fragment of the skeletal troponin complex in the Ca^{2+}-free state of the N-terminal domain of troponin C (TnC N-domain). The N-terminal helix (TnI N-helix) of troponin I and the loop (TnC linker) between TnC domains are treated as fully flexible (green color), allowing unrestricted movement of domains. The remaining part of the structure (two domains in teal color) is fully mobile with domain structures restricted to near-native fluctuations. On the right side the best conformation ($RMSD_{bound}$ = 3.2 Å and $RMSD_{domains}$ = 2.5 Å) from the lowest temperature replica (green color) is shown, aligned with the crystallographic structure (pdb code: 1A2X [34], red color).

The second site of interaction between TnC and TnI is located in the hydrophobic pocket of the structural C-domain. The TnI/TnC complex at this end is stable regardless of the calcium ion levels in the sarcoplasm. The interaction of TnC and TnI is important, because it is partly responsible for anchoring troponin to tropomyosin [38]. Mutations that weaken these interactions in cardiac troponin can cause disorders in the cardiac muscle contraction and relaxation cycle [38].

In the fragment of the skeletal troponin complex [34], the TnI N-terminal helix interacts not only with the TnC C-domain, but also partly with the TnC N-domain. The TnC domains (Figure 1) are connected by a flexible linker [34,39,40,41,42] and they change their relative positions during complex formation. Movement of such scale cannot be captured by rigid or semi-flexible docking methods.

RESULTS

During the Replica Exchange Monte Carlo (REMC) coarse-grained simulation we obtained a large cluster of near-native conformations in the lowest temperature replica (T = 1.5). From all conformations at this temperature, *ca.* 80% near-native conformations were observed (Figure 2). We assume that the distribution of conformational states in this replica roughly corresponds to distribution at physical room temperature. With increasing temperature the fraction of near-native conformations decreases continuously, reaching about zero with the highest temperature replica. In the T = 2.0 replica, *ca.*40% near-native conformations were observed, about half of the value seen at the lowest temperature. This is near the transition temperature, where the largest number of transitions between conformational states could be observed.

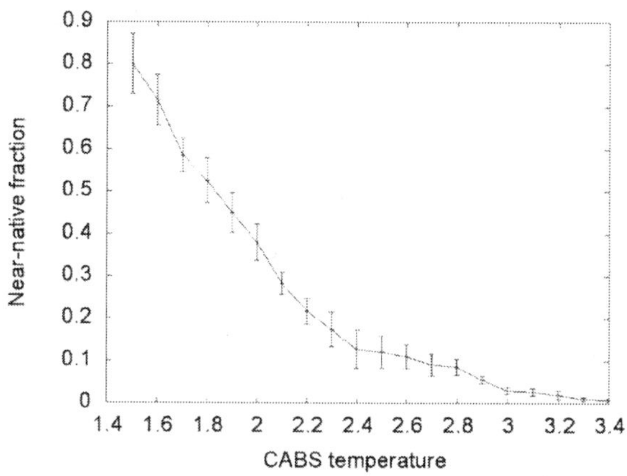

Figure 2: Transition curve of the TnI N-helix (N-terminal helix of troponin I) docking process to skeletal troponin C (pdb code: 1A2X). Conformations for which RMSD (root-mean-square deviation) was below 6 Å were considered

near-native. RMSD was calculated on Cα atoms of the troponin C complex with the TnI N-helix *vs.*crystallographic structures.

Because both TnC domains are connected with a flexible linker [34,39,40,41,42], fluctuations of the system are considerable (Figure 3a). In the lowest temperature replica, there are three main clusters of near-native conformations with the main cluster around $RMSD_{bound} \sim 3$ Å (root-mean-square deviation) and $RMSD_{domains} \sim 4.5$ Å.

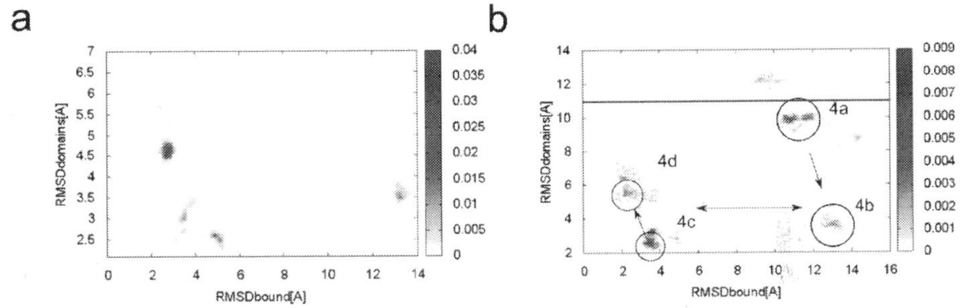

Figure 3: Conformational density map plotted as a function of $RMSD_{bound}$ (root-mean-square deviation) and $RMSD_{domains}$ to the crystallographic structure of skeletal TnC (troponin C) with the N-terminal helix of troponin I (PDB:1A2X [34]) at (**a**) T = 1.5; (**b**) T = 2.0. In the Figure 3b the main clusters are marked with references to Figure 4. The blue horizontal line indicates the $RMSD_{domains}$ value of the crystallographic structure of unbound TnC (PDB: 1TOP [43]).

To analyze transition states observed on protein folding and binding we also isolated clusters of conformations in a 2.0 CABS temperature replica. In Figure 3b we marked four main clusters, which dominate in this replica. In the first cluster representative conformation (Figure 4a) the TnI N-helix is bound to both TnC domains but it is not connected to the hydrophobic pocket of any domain. In Figure 5 it is shown that the TnI N-helix interacts with the amino acids of the TnC C-domain but the TnI N-helix does not have access to the hydrophobic residues of the TnC C-domain. In this unbound state, troponin C is in a collapsed conformation. This transition state of the receptor is not similar to its crystallographic structure in a complex with the TnI N-helix ($RMSD_{domains} = 9.4$ Å) and it is not present in the lowest temperature replica. In Figure 3b we also notice a bunch of other receptor structures without the ligand inside the hydrophobic pocket. For these conformations $RMSD_{domains}$ is larger than 6 Å. It shows that binding the peptide is essential for stabilizing the receptor conformation.

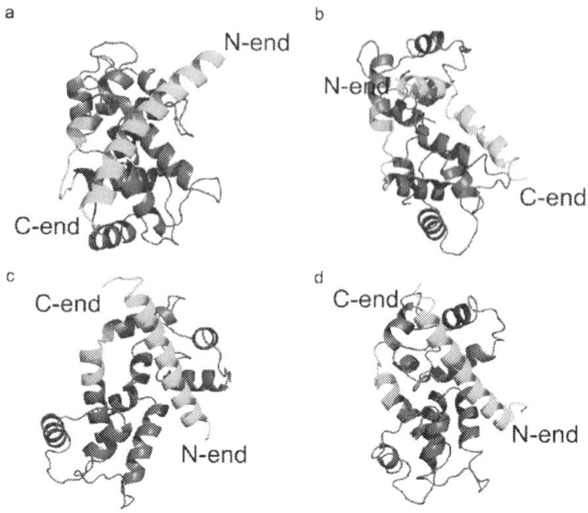

Figure 4: Representative conformations of the four main clusters in the transition temperature replica (T = 2.0) from the CABS-dock simulation of skeletal troponin. It shows transition states in the docking process of the TnI N-helix (N-terminal helix of troponin I) to troponin C. RMSD (root-mean-square deviation) values of the structures are: (**a**) $RMSD_{bound}$ = 11.2 Å and $RMSD_{domains}$ = 9.4 Å; (**b**) $RMSD_{bound}$ = 13.1 Å and $RMSD_{domains}$ = 4.8 Å; (**c**) $RMSD_{bound}$ = 3.6 Å and $RMSD_{domains}$ = 2.6 Å; (**d**) $RMSD_{bound}$ = 3.4 Å and $RMSD_{domains}$ = 4.5 Å.

In most of the unbound conformations, the peptide has a helical shape. It suggests that the ligand reaches its final structure before binding into the hydrophobic pocket.

In another state (Figure 4b), the TnI N-helix is partially bound to the hydrophobic pocket of the TnC C-domain, but with the helix bound in the opposite direction compared to the native state. In this case, TnI amino acids from the N-terminal part of the helix (residues 8–14) interact with the hydrophobic patch of the TnC C-domain (Figure 5). Such a state is unstable because there are only a few hydrophobic contacts with the TnC C-domain. What is more, the ligand interacts with both TnC domains. We assume that it is one of the alternative states during the folding-binding process. It is noted that interactions between the TnI N-helix with TnC in this conformation also stabilize the arrangement of TnC domains. When the TnI N-helix is inside the hydrophobic pocket of the TnC C-domain, it prevents the TnC domains from approaching each other.

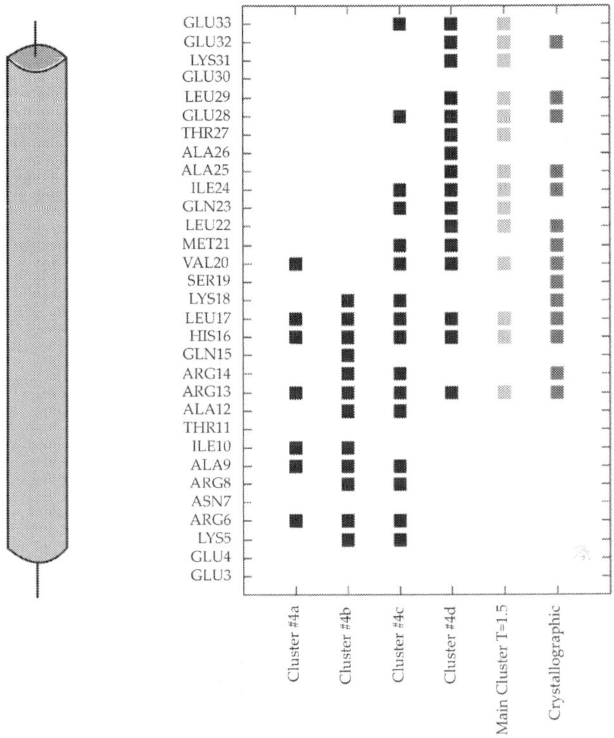

Figure 5: Contacts of each residue of the TnI N-helix (N-terminal helix of troponin I) with the C-terminal domain of troponin C for each cluster representative from Figure 4 (black). Contacts for the largest cluster representative at T = 1.5 are colored green and contacts for the crystallographic structure are colored red.

In conformations 4c and 4d, the TnI N-helix is positioned in the hydrophobic pocket of the TnC C-domain in the proper direction. In both conformations the secondary structure of the helix is stabilized. Nevertheless, in transition state 4c, the TnI N-helix is more shifted in the hydrophobic pocket compared to the crystallographic structure. In structure 4c, the TnI N-helix interacts with the TnC C-domain mainly through TnI residues 12–18. The number of similar conformations with a shifted helix inside the hydrophobic pocket largely decreases in the lowest temperature replica. Structures in the largest cluster in the lowest replica are similar to the structure in Figure 4d. This conformation has the TnI N-helix positioned similarly to the crystallographic structure. The shift of the helix inside the hydrophobic pocket implicates a change in the structure of troponin C and alters non-specific electrostatic interactions (Figure 6) of polar residues (e.g., Arg13, Arg14, Arg8, Asn7, Glu4) of the TnI N-helix with residues of the TnC N-domain. These interactions deviate because of the flexibility of the N-terminal residues of the TnI N-helix and flexibility of the TnC linker. They are not essential to bind the TnI N-helix to TnC but they determine the TnC structure.

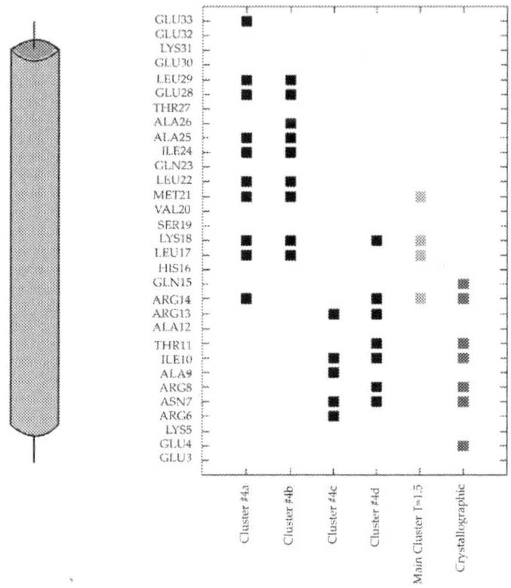

Figure 6: Contacts of each residue of the N-terminal helix of troponin I with the N-terminal domain of troponin C for each cluster representative from Figure 4 (black). Contacts for the largest cluster representative at T = 1.5 are colored green and contacts for the crystallographic structure are colored red.

To examine the effect of ligand binding on TnC domain arrangement more precisely, we also use the CABS-flex program. In the first CABS-flex simulation, the starting point is TnC alone in the state from the crystallographic structure of the complex [34]. It is shown that the initial shape of unbound TnC is destabilized (Figure 7, dotted line). The radius of gyration for TnC without the ligand is lower compared to TnC bound to the TnI N-helix (Figure 7, solid line) and fluctuates in the range of 16 Å to 16.8 Å. It is consistent with the experimental data for which the maximum radius of gyration distribution is located near 19 Å [41]. Additionally, the distances between residues M25-C98 for structures obtained during CABS-flex simulation of TnC without the ligand (Figure 8) are consistent with fluorescence energy transfer studies [44].

In the CABS-flex simulation of TnC without the ligand, we do not observe any TnC structures for which the radius of gyration is larger than 17 Å, and it is inconsistent with X-ray solution scattering data [41]. It results from the fact that the CABS-flex program is designed to show local fluctuations only, which occur in the nanosecond time-scale.

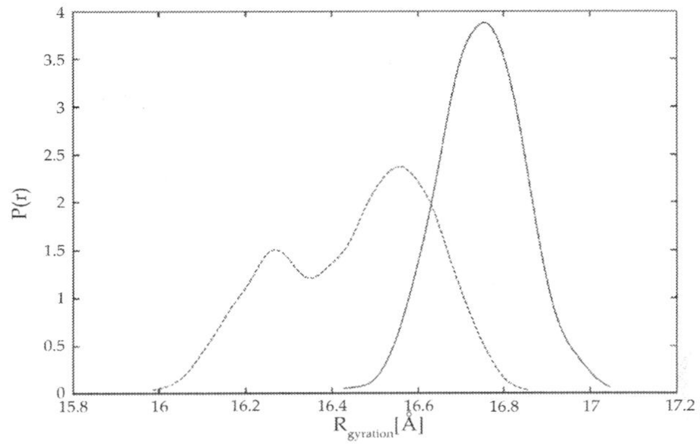

Figure 7: Probability density function of the radius of gyration estimated by the Kernel Density Estimator in the CABS-flex simulations of TnC (troponin C) with the N-terminal helix of troponin I (PDB:1a2x) (solid line) and TnC alone (dotted line).

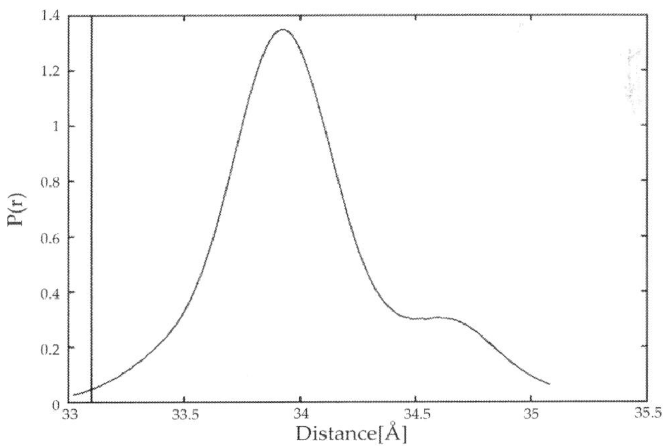

Figure 8: Probability density function of the distance between residues M25-C98 in CABS-flex simulation of troponin C estimated by the Kernel Density Estimator. The vertical line shows the value of the distance measured in fluorescence energy transfer studies [44].

DISCUSSION

The above simulations show that CABS-dock simulation of TnI N-helix flexible docking to TnC converges to the structures, which are near to the crystallographic structure of the complex [34]. The structure with the lowest RMSD ($RMSD_{bound}$ = 3.2 Å and $RMSD_{domains}$ = 2.5 Å) from the lowest temperature replica is shown in Figure 1 and it is aligned with the experimental structure (PDB:1A2X).

The TnC domains in the final complex are close to each other, which enables interactions of the TnI N-helix with both of them. This arrangement is different from the structures of unbound TnC [41,42,43,44].

The crystallographic data of Ca^{2+}-saturated troponin C [45] and the crystallographic structure with Ca^{2+} ions bound only to the TnC C-domain [43] indicate that TnC without TnI has extended structure. In this case linker residues between the TnC domains form a helix. Nevertheless, NMR (nuclear magnetic resonance) studies on the TnC structure show that in the Ca^{2+}-saturated state TnC forms a whole range of structures [42] and the linker between the domains retains its flexibility. When other kinds of experiments are compared, such as X-ray solution scattering [41] or fluorescence energy transfer [44], it is noted that the dominating state of unbound troponin C is collapsed, with the radius of gyration below 20 Å [41].

The results of CABS-dock simulations presented in this paper indicate that binding the TnI N-helix is essential to obtain TnC domain arrangement similar to the crystallographic structure of the complex [34]. In Figure 3a it is shown that CABS-dock simulation converges mainly to a cluster of structures with $RMSD_{domains}$ around 4.5 Å. For the structures where the TnI N-helix is not bound to the hydrophobic pocket, TnC has a completely different shape from the crystallographic structure (Figure 3b). In the highest temperature replicas, unbound states dominate. This causes increasing mutual rearrangements of the two receptor domains, but due to the model limitations, the unbound native structure is not clearly represented even at this temperature. Nevertheless, as shown in Figure 9, the structural mobility of the receptor is significant. Experimental structures of unbound skeletal troponin C [42,43,45,46] are also far from the structure of skeletal troponin C with bound TnI [34]. All these findings show that the binding of the TnI N-helix to TnC determines to some extent the arrangement of the TnC domains. This bias to a particular TnC shape is partly caused by non-specific electrostatic interactions of a fragment of the TnI N-helix with the TnC N-domain. The position of the domains is also quite similar to the crystallographic structure when the TnI N-helix is bound in reverse direction in the hydrophobic pocket. Nevertheless, this transition state is not stable because there are not enough hydrophobic contacts to interact with the TnC C-domain and the helical structure of the TnI N-helix is destabilized (Figure 4b and Figure 5).

Finally, it has also been shown that the CABS-dock program is a useful alternative to classical docking approaches, especially for peptide docking to significantly flexible receptors. In this paper we described docking simulations of just one selected system, however many other complexes in which ligand binding causes conformational transitions in receptor structures [47] are already known. Peptide binding to other calmodulin-like proteins is a very important step in calcium signal transduction [48]. Our approach could be used in these similar systems.

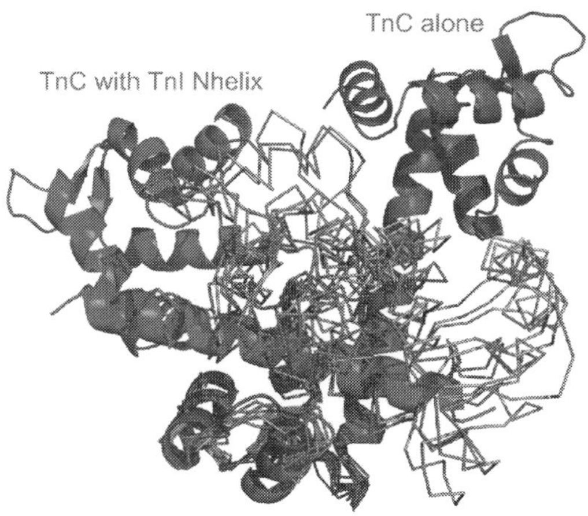

Figure 9: Superposed crystallographic structures of the TnC (troponin C) and troponin I (red color) and unbound TnC [43] (blue) compared with the unbound TnC structures observed in the CABS-dock simulation (black thin lines), where for clarity of conformational changes the TnC C-domain (C-terminal domain of troponin C) only is superposed onto the experimental structures. The unbound structures are cluster representative conformations from one of the temperature replicas (T = 2.0) for which the TnI N-helix (N-terminal helix of troponin I) does not interact with the hydrophobic pocket of the TnC C-domain.

METHODS

CABS Model

The CABS model employs a coarse-grained representation of protein structures and Monte Carlo sampling schemes enabling very fast simulations of the long-time dynamics of protein systems. Its name is an acronym of the first letters of pseudo atoms used to represent a single amino acid, that is: carbon α (CA), carbon β (B) and side chain (S). An additional pseudo atom, defined in the geometrical center of the virtual Cα-Cα bond, is used for defining the main chain hydrogen bonds. CABS uses a simple cubic lattice to store the coordinates of Cα atoms to speed up calculations. The remaining atoms are located off the lattice and "follow" movements of the main chain. The lattice constant is arbitrarily set to 0.61 Å, which translates to a roughly 0.35 Å average error made on casting a protein structure onto the lattice. At these settings there are 800 possible Cα-Cα pseudo bonds, which is a high enough number to prevent any lattice-associated anisotropy.

The knowledge-based force field of the CABS has been derived by statistical analysis of the local regularities observed in protein crystallographic structures, using the Boltzmann inversion. CABS force field consists of five potentials. The first one is the sequence-independent local (close in sequence) terms and biases responsible for stiffening the otherwise very flexible Cα trace. The second potential is sequence-dependent local force derived from the distribution of distances between ith and i + 2nd, i + 3rd and i + 4th Cα atoms. The third one is sequence-dependent pairwise potential for side chain-side chain contact interactions. Attractive part of this potential depends on the mutual orientations of the interacting side-chains and local geometry of the backbone.

Other terms are sequence-independent and mimic excluded volume of the united atoms representing the coarse-grained protein chains. The effects of the solvent and electrostatic interactions are implicitly encoded into the side chain-side chain potential. A detailed description of the force field of the CABS model (also used by CABS-fold [49]) could be found in the literature [21].

Conformational sampling is carried out using Monte Carlo dynamics, controlled by the asymmetric Metropolis scheme. For better sampling coverage of the conformational space, simulations are run parallel in multiple replicas at different temperatures, exchanged between replicas at regular intervals.

Due to its knowledge-based statistical force field, the temperature of the model is a dimensionless variable proportional to real temperature, although the proportionality parameter needs to be defined at a selected reference state, for instance at a transition observed in the real system in the simulations.

CABS-Dock Procedure

CABS-dock is a multiscale docking protocol built around the CABS model. It consists of the steps showed in Figure 10.

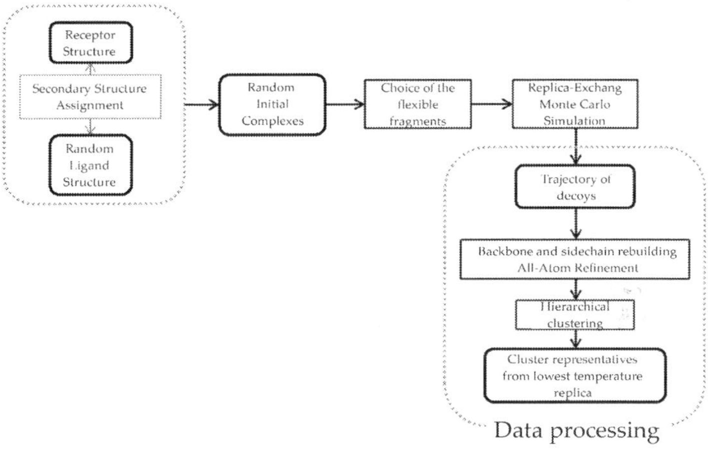

Figure 10: Multiscale docking protocol using the CABS-dock program.

Firstly, the 3D structure of the molecule marked as the receptor (TnC, Figure 1) is projected onto the lattice using least-square fitting. The initial state of the receptor is the same as in the complex with the TnI N-helix (PDB: 1A2X [34]). For the peptide ligand, a TnI N-helix-30 residues (Figure 1) random structure is generated and placed in a random location on the sphere around the geometrical center of the receptor at 20 Å from receptor surface. For each replica, the ligand was positioned at different starting position and conformation.

The next stage is selection of flexible segments present in the docking simulation. CABS-dock allows a variable level of flexibility for different fragments of the receptor. For example, linkers between the domains may be modeled with full flexibility; the active site in a semi-flexible manner (with weak restraints) and the rest of the protein may fluctuate around its native conformation or remain almost completely rigid. Choice of the flexible receptor fragments conditions the results of docking. Larger number of flexible fragments can be put to predict dynamics of the system. In specific cases, it is necessary for successful docking of a peptide to the receptor [18].

In our current work, we applied distance restraints on pairs of atoms within both troponin C domains. It keeps the structural fluctuations of both domains near their individual structures taken from the crystallographic structure of the entire protein (PDB: 1A2X [34]) at the low Ca^{2+} concentration. Structural fluctuations for the lowest temperature replica within both TnC domains are in the range of 1 Å. We did not use any cross-domain distance restraints. TnI (ligand) and the loops between the TnC domains were treated as fully flexible. Such simulation set-ups enabled mutual reorientation of the both TnC domains, indicated by large RMSD (from the crystallographic structure) changes of the whole receptor.

Replica exchange Monte Carlo simulation is run with 20 replicas in a CABS temperature range of 1.5–3.4. In summary, it gives 2,000,000 MC steps and it took ca. 1.5 day of calculation on the single CPU. Such relatively low computational cost is due to about 4 orders of magnitude faster sampling of protein conformational space by coarse-grained CABS in comparison to all-atom MD simulations.

For further analysis, we take the second half of the CABS trajectory. These models from temperature T = 1.5 and T = 2.0 replicas are clustered using the g_cluster tool from the Gromacs package [50] with the linkage method and 1.5 Å cutoff. Subsequently, the main-chain of each structure is rebuilt and minimized by the MODELLER tool [51].

CABS-Flex Simulations

CABS-flex is a CABS-based protocol applied to observe fluctuations around the near-native state [19]. In this model scaling force field components, distance restraints and time of simulations are adjusted to achieve very good correlation with Molecular Dynamics 10 ns simulations [20]. Here we used it for investigating TnC domain fluctuations. With the aim of performing it, we ran two simulations. In the first one, the starting point was TnC structure complexed with the TnI N-helix (PDB: 1A2X) and the second simulation started from unbound TnC in the same state as in the complex with the TnI N-helix (PDB: 1A2X [34]). For analysis we took the second half of the trajectory.

Data Analysis

Three measures were used to compare the docking results with the crystallographic structure (PDB: 1A2X [34]):

- $RMSD_{domains}$—RMSD calculated on the Cα atoms of both troponin C domains *vs*. crystallographic structure (PDB: 1A2X [34]).
- $RMSD_{bound}$—RMSD calculated on the Cα atoms of the TnC C-terminal domain and the TnI N-terminal helix: residues 3–33.
- RMSD—RMSD calculated on the Cα atoms of TnC and the TnI N-terminal helix: residues 3–33. Conformations for which RMSD was below 6 Å were considered near-native.

To provide receptor-ligand contact analysis, a contact between any two residues was defined as follows: The distance between any two heavy atoms from each residue was smaller than 4.5 Å. In Figure 5 each point shows the contact of a particular TnI residue with any residue from the TnC C-domain. This approach shows which fragment of the TnI N-helix interacts with the TnC C-domain.

In Figure 6 and Figure 7, we estimated the probability density function of radius of gyration and residue-residue distances. Therefore, we used the Kernel Density Estimator.

The Kernel estimator of an unknown probability density function $f(x)$ for a given vector of N observations x_i, $i \in [1, N]$ can be defined as follows:

$$\hat{f}_h(x) = \frac{1}{Nh}\sum_{i=0}^{N} K\left(\frac{(x - x_i)}{h}\right) \qquad (1)$$

where K is kernel function and h is the bandwidth (smoothing parameter).

In our calculation we used the standard Gaussian function with mean zero and variance 1. For more details see the Bioshell [52] manual.

CONCLUSIONS

The paper describes the application of the CABS-dock protocol to a difficult problem of protein-peptide complex assembly. Starting from near-native structures of the receptor and completely random structures of the peptide (located at random positions around the receptor structure), we performed REMC simulations of free docking. The docking simulations and the clustering of the manifold of generated conformations led to a well-defined structure of the complex, in good agreement with the available crystallographic structure [34]. Experimental studies also show that the linker between the N-terminal and the C-terminal domain is flexible [34,39,40,41,42]. In accordance with these observations, our simulations revealed significant oscillation of the receptor structure around its native state. It was also shown that the relative position of the TnC domains was dependent on the position of the TnI N-helix.

To investigate the influence of peptide binding on the structure of the whole complex, we also performed CABS-flex simulations, a tool for the computational study of protein domain dynamics. The simulations showed that the binding of the peptide fixed to a large extent the troponin C structure seen in the bound state, while the structures of the receptor without any ligand were closer to the unbound experimental structure of troponin C [41,44] and exhibited higher flexibility.

Finally, let us note that CABS-dock method enables free search (without predefined binding site location) of the peptide docking pathways with properly predefined ranges of the receptor structure fluctuations [18], which gives a possibility of significant domain reorientations for large receptors, as has been shown in this paper. CABS-dock can be particularly useful for studying systems where binding of a peptide induces opening of the binding site of the receptor and therefore is associated with large conformational changes of the receptor [47]. Such free and flexible docking is beyond the ranges of applicability of classical methods.

ACKNOWLEDGMENTS

The authors thank Dominik Gront for the help in calculations of probability density functions.

The authors acknowledge support from the Foundation for Polish Science TEAM project (TEAM/2011-7/6), co-financed by the European Regional Development Fund operated within the Innovative Economy Operational Program. Calculations were performed in the Interdisciplinary Center for Mathematical and Computational Modeling (ICM) of the University of Warsaw (Grant G43-9).

AUTHOR CONTRIBUTIONS

J.W. and M.K. performed simulations. J.W. analyzed the data. M.K. and A.K. designed CABS-dock protocol. J.W., M.K., and A.K. wrote the manuscript.

REFERENCES AND NOTES

1. Alonso, H.; Bliznyuk, A.A.; Gready, J.E. Combining docking and molecular dynamic simulations in drug design. *Med. Res. Rev.* 2006, *26*, 531–568.
2. Burley, S.K.; Almo, S.C.; Bonanno, J.B.; Capel, M.; Chance, M.R.; Gaasterland, T.; Lin, D.; Sali, A.; Studier, F.W.; Swaminathan, S. Structural genomics: Beyond the human genome project. *Nat. Genet.* 1999, *23*, 151–157.
3. Overington, J.P.; Al-Lazikani, B.; Hopkins, A.L. How many drug targets are there? *Nat. Rev. Drug Discov.* 2006, *5*, 993–996.
4. Alberts, B.; Johnson, A.; Lewis, J.; Raff, M.; Roberts, K.; Walter, P. Analyzing Protein Structure and Function. In Molecular Biology of the Cell; Garland Science: New York, NY, USA, 2002; (Eds).
5. Kuntz, I.D.; Blaney, J.M.; Oatley, S.J.; Langridge, R.; Ferrin, T.E. A geometric approach to macromolecule-ligand interactions. *J. Mol. Biol.* 1982, *161*, 269–288.
6. Morris, G.M.; Huey, R.; Lindstrom, W.; Sanner, M.F.; Belew, R.K.; Goodsell, D.S.; Olson, A.J. AutoDock4 and AutoDockTools4: Automated docking with selective receptor flexibility. *J. Comput.*

Chem. 2009, *30*, 2785–2791.

7. Dominguez, C.; Boelens, R.; Bonvin, A.M.J.J. HADDOCK: A protein-protein docking approach based on biochemical or biophysical information. *J. Am. Chem. Soc.* 2003, *125*, 1731–1737.
8. Claußen, H.; Buning, C.; Rarey, M.; Lengauer, T. FlexE: Efficient molecular docking considering protein structure variations. *J. Mol. Biol.* 2001, *308*, 377–395.
9. Friesner, R.A.; Banks, J.L.; Murphy, R.B.; Halgren, T.A.; Klicic, J.J.; Mainz, D.T.; Repasky, M.P.; Knoll, E.H.; Shelley, M.; Perry, J.K.; *et al.* Glide: A new approach for rapid, accurate docking and scoring. 1. Method and assessment of docking accuracy. *J. Med. Chem.* 2004, *47*, 1739–1749.
10. Verdonk,M.L.;Cole,J.C.;Hartshorn,M.J.;Murray,C.W.;Taylor,R.D. Improved protein-ligand docking using GOLD. *Proteins* 2003, *52*, 609–623.
11. Chen, R.; Weng, Z. Docking unbound proteins using shape complementarity, desolvation, and electrostatics. *Proteins*2002, *47*, 281–294.
12. Pons, C.; Grosdidier, S.; Solernou, A.; Pérez-Cano, L.; Fernández-Recio, J. Present and future challenges and limitations in protein-protein docking. *Proteins* 2010, *78*, 95–108.
13. Katchalski-Katzir, E.; Shariv, I.; Eisenstein, M.; Friesem, A.A.; Aflalo, C.; Vakser, I.A. Molecular surface recognition: determination of geometric fit between proteins and their ligands by correlation techniques. *Proc. Natl. Acad. Sci. USA* 1992, *89*, 2195–2199.
14. Ehrlich, L.P.; Nilges, M.; Wade, R.C. The impact of protein flexibility on protein-protein docking. *Proteins* 2005, *58*, 126–133.
15. B-Rao, C.; Subramanian, J.; Sharma, S.D. Managing protein flexibility in docking and its applications. *Drug Discov. Today* 2009, *14*, 394–400.
16. Zacharias, M. ATTRACT: Protein-protein docking in CAPRI using a reduced protein model. *Proteins* 2005, *60*, 252–256.
17. Das, R.; André, I.; Shen, Y.; Wu, Y.; Lemak, A.; Bansal, S.; Arrowsmith, C.H.; Szyperski, T.; Baker, D. Simultaneous prediction of protein folding and docking at high resolution. *Proc. Natl. Acad.*

Sci. USA 2009, *106*, 18978–18983.

18. Kurcinski, M.; Jamroz, M.; Blaszczyk, M.; Kolinski, A.; Kmiecik, S. CABS-dock web server for the flexible docking of peptides to proteins without prior knowledge of the binding site. *Nucleic Acids Res.* 2015.
19. Jamroz, M.; Kolinski, A.; Kmiecik, S. CABS-flex: Server for fast simulation of protein structure fluctuations. *Nucleic Acids Res.* 2013, *41*, W427–W431.
20. Jamroz, M.; Orozco, M.; Kolinski, A.; Kmiecik, S. Consistent view of protein fluctuations from all-atom molecular dynamics and coarse-grained dynamics with knowledge-based force-field. *J. Chem. Theory Comput.* 2013, *9*, 119–125.
21. Kolinski, A. Protein modeling and structure prediction with a reduced representation. *Acta Biochim. Pol.* 2004, *51*, 349–371.
22. Wang, G.; Jin, Y.; Dunbrack, R.L. Assessment of fold recognition predictions in CASP6. *Proteins* 2005, *61* (Suppl. 7), 46–66.
23. Tress, M.; Ezkurdia, I.; Graña, O.; López, G.; Valencia, A. Assessment of predictions submitted for the CASP6 comparative modeling category. *Proteins* 2005, *61* (Suppl. 7), 27–45.
24. Vincent, J.J.; Tai, C.H.; Sathyanarayana, B.K.; Lee, B. Assessment of CASP6 predictions for new and nearly new fold targets. *Proteins* 2005, *61* (Suppl. 7), 67–83.
25. Koliński, A.; Bujnicki, J.M. Generalized protein structure prediction based on combination of fold-recognition with de novo folding and evaluation of models. *Proteins* 2005, *61* (Suppl. 7), 84–90.
26. Boniecki, M.; Rotkiewicz, P.; Skolnick, J.; Kolinski, A. Protein fragment reconstruction using various modeling techniques. *J. Comput. Aided Mol. Des.* 2003, *17*, 725–738.
27. Kmiecik, S.; Kolinski, A. Characterization of protein-folding pathways by reduced-space modeling. *Proc. Natl. Acad. Sci. USA* 2007, *104*, 12330–12335.
28. Wabik, J.; Kmiecik, S.; Gront, D.; Kouza, M.; Koliński, A. Combining coarse-grained protein models with replica-exchange all-atom molecular dynamics. *Int. J. Mol. Sci.* 2013, *14*, 9893–9905.
29. Kurcinski, M.; Kolinski, A. Theoretical study of molecular

mechanism of binding TRAP220 coactivator to Retinoid X Receptor alpha, activated by 9-cis retinoic acid. *J. Steroid Biochem. Mol. Biol.* 2010, *121*, 124–129.

30. Kurcinski, M.; Kolinski, A. Steps towards flexible docking: modeling of three-dimensional structures of the nuclear receptors bound with peptide ligands mimicking co-activators' sequences. *J. Steroid Biochem. Mol. Biol.* 2007, *103*, 357–360.
31. Kurcinski, M.; Kolinski, A. Hierarchical modeling of protein interactions. *J. Mol. Model.* 2007, *13*, 691–698.
32. Jamroz, M.; Kolinski, A.; Kmiecik, S. CABS-flex predictions of protein flexibility compared with NMR ensembles. *Bioinformatics* 2014, *30*, 2150–2154.
33. Kurcinski, M.; Kolinski, A.; Kmiecik, S. Mechanism of folding and binding of an intrinsically disordered protein as revealed by ab initio simulations. *J. Chem. Theory Comput.* 2014, *10*, 2224–2231.
34. Vassylyev, D.G.; Takeda, S.; Wakatsuki, S.; Maeda, K.; Maéda, Y. Crystal structure of troponin C in complex with troponin I fragment at 2.3-A resolution. *Proc. Natl. Acad. Sci. USA* 1998, *95*, 4847–4852.
35. Metzger, J.M.; Westfall, M.V. Covalent and noncovalent modification of thin filament action: The essential role of troponin in cardiac muscle regulation. *Circ. Res.* 2004, *94*, 146–158.
36. Vinogradova, M.V.; Stone, D.B.; Malanina, G.G.; Karatzaferi, C.; Cooke, R.; Mendelson, R.A.; Fletterick, R.J. Ca(2+)-regulated structural changes in troponin. *Proc. Natl. Acad. Sci. USA* 2005, *102*, 5038–5043.
37. Sia, S.K.; Li, M.X.; Spyracopoulos, L.; Gagne, S.M.; Liu, W.; Putkey, J.A.; Sykes, B.D. Structure of Cardiac Muscle Troponin C Unexpectedly Reveals a Closed Regulatory Domain. *J. Biol. Chem.* 1997, *272*, 18216–18221.
38. Baryshnikova, O.K.; Robertson, I.M.; Mercier, P.; Sykes, B.D. The dilated cardiomyopathy G159D mutation in cardiac troponin C weakens the anchoring interaction with troponin I. *Biochemistry* 2008, *47*, 10950–10960.
39. Abbott, M.B.; Gaponenko, V.; Abusamhadneh, E.; Finley, N.; Li, G.; Dvoretsky, A.; Rance, M.; Solaro, R.J.; Rosevear, P.R.

Regulatory domain conformational exchange and linker region flexibility in cardiac troponin C bound to cardiac troponin I. *J. Biol. Chem.* 2000, *275*, 20610–20617.

40. Blumenschein, T.M.A.; Stone, D.B.; Fletterick, R.J.; Mendelson, R.A.; Sykes, B.D. Calcium-dependent changes in the flexibility of the regulatory domain of troponin C in the troponin complex. *J. Biol. Chem.* 2005, *280*, 21924–21932.
41. Heidorn, D.B.; Trewhella, J. Comparison of the crystal and solution structures of calmodulin and troponin C. *Biochemistry* 1988, *27*, 909–915.
42. Slupsky, C.M.; Sykes, B.D. NMR solution structure of calcium-saturated skeletal muscle troponin C. *Biochemistry*1995, *34*, 15953–15964.
43. Satyshur, K.A.; Pyzalska, D.; Greaser, M.; Rao, S.T.; Sundaralingam, M. Structure of chicken skeletal muscle troponin C at 1.78 A resolution. *Acta Crystallogr. D Biol. Crystallogr.* 1994, *50*, 40–49.
44. Cheung, H.C.; Wang, C.K.; Garland, F. Fluorescence energy transfer studies of skeletal troponin C proximity between methionine-25 and cysteine-98. *Biochemistry* 1982, *21*, 5135–5142.
45. Houdusse, A.; Love, M.L.; Dominguez, R.; Grabarek, Z.; Cohen, C. Structures of four Ca^{2+}-bound troponin C at 2.0 A resolution: Further insights into the Ca^{2+}-switch in the calmodulin superfamily. *Structure* 1997, *5*, 1695–1711.
46. Wang, X.; Mercier, P.; Letourneau, P.J.; Sykes, B.D. Effects of Phe-to-Trp mutation and fluorotryptophan incorporation on the solution structure of cardiac troponin C, and analysis of its suitability as a potential probe for *in situ* NMR studies. *Protein Sci.* 2005, *14*, 2447–2460.
47. Kenakin, T. Ligand-selective receptor conformations revisited: The promise and the problem. *Trends Pharmacol. Sci.*2003, *24*, 346–354.
48. Perochon, A.; Aldon, D.; Galaud, J.P.; Ranty, B. Calmodulin and calmodulin-like proteins in plant calcium signaling. *Biochimie* 2011, *93*, 2048–2053.

49. Blaszczyk, M.; Jamroz, M.; Kmiecik, S.; Kolinski, A. Cabs-fold: Server for the de novo and consensus-based prediction of protein structure. *Nucleic Acids Res.* 2013, *41*, W406–W411.

50. Pronk, S.; Páll, S.; Schulz, R.; Larsson, P.; Bjelkmar, P.; Apostolov, R.; Shirts, M.R.; Smith, J.C.; Kasson, P.M.; Van Der Spoel, D.; *et al.* GROMACS 4.5: A high-throughput and highly parallel open source molecular simulation toolkit. *Bioinformatics* 2013, *29*, 845–854.

51. Eswar, N.; Webb, B.; Marti-Renom, M.A.; Madhusudhan, M.S.; Eramian, D.; Shen, M.Y.; Pieper, U.; Sali, A. Comparative protein structure modeling using MODELLE. *Curr. Protoc. Protein Sci.* 2007, *2*. Chapter 2, Unit 2.9.

52. Gront, D.; Kolinski, A. Utility library for structural bioinformatics. *Bioinformatics* 2008, *24*, 584–585.

CHAPTER 5

COARSE GRAINED MOLECULAR DYNAMICS SIMULATIONS OF TRANSMEMBRANE PROTEIN-LIPID SYSTEMS

Peter Spijker [1], Bram van Hoof [1,2], Michel Debertrand [1,2], Albert J. Markvoort [1], Nagarajan Vaidehi [2] and Peter A.J. Hilbers [1]

[1]Department of Biomedical Engineering, Eindhoven University of Technology, 5600 MB, Eindhoven, The Netherlands

[2]Division of Immunology, Beckman Research Institute of the City of Hope, 1500 E. Duarte Road, Duarte, CA-91010, USA

ABSTRACT

Many biological cellular processes occur at the micro- or millisecond time scale. With traditional all-atom molecular modeling techniques it is difficult to investigate the dynamics of long time scales or large systems, such as protein aggregation or activation. Coarse graining (CG) can be used to reduce the number of degrees of freedom in such a system, and reduce the computational complexity. In this paper the first version of a coarse grained model for trans membrane proteins is presented. This model differs from other coarse grained protein models due to the introduction of a novel angle potential as well as a hydrogen bonding potential. These new potentials are used to stabilize the backbone. The model has been validated by investigating the adaptation of the hydrophobic mismatch induced by the insertion of WALP-peptides into a lipid membrane, showing that the first step in the adaptation is an increase in the membrane thickness, followed by a tilting of the peptide.

Keywords: Molecular dynamics; coarse grained; transmembrane proteins; α-helices; WALP-peptides; hydrophobic mismatch

INTRODUCTION

The cell membrane maintains the essential differences between the cytoplasmic and the extracellular environment and additionally has the task to transport signals or substances from one side to the other. By doing so, the membrane is a rather complex system on its own, consisting mainly of lipid molecules and membrane proteins. How a membrane protein is associated with the lipid membrane reflects the function of the protein. Transmembrane proteins can function on both sides of the membrane or transport molecules across it. Cell-surface receptors are transmembrane proteins that bind water-soluble signal molecules in the extracellular space and generate different intracellular signals.

The largest family of cell surface receptors is formed by the G-protein coupled receptors (GPCRs), which are responsible for most transmembrane signal transduction by hormones and neurotransmitters, as well as for the senses of vision, smell and taste [1–3]. All of the GPCRs belong to a large family of homologous, seven-pass transmembrane proteins. The interaction between the receptor and its target protein is mediated by a third protein, a trimeric GTP-binding protein (G-protein). When a ligand binds to the extracellular domain of the GPCR, the receptor undergoes a conformational change that enables the activation of the G-protein [4].

The historical lack of 3D crystal structures has lead to the development of different computational methods in order to predict the structure of a GCPR [5–7]. More recently, the activation pathway of different GPCR complexes have been investigated using molecular dynamics (MD), where the solvent and membrane environment is taken into account [8–11].

Using MD simulation techniques at the atomistic level allows for the exploration of ligand positioning and the initialization of protein activation. However, due to the computational demand, it is not feasible to model the entire cascade of protein activation with atomistic detail. To partially overcome these limitations it is possible to use a coarse grained (CG) model instead of the atomistic one.

With a CG model typically several atoms are grouped into one particle, while the interaction properties of the new particle are modeled as close as possible to those of the original atoms it represents. By grouping atoms together, coarse graining reduces the degrees of freedom and, therefore, the computational cost. Moreover, due to the removal of high-frequency motions, such as the vibrational movements of hydrogen atoms, larger steps can be made in MD simulations, allowing longer simulations. Simultaneously, the dynamics in the CG system is likely to be further increased as additional friction and noise from eliminated degrees of freedom are absent. Still, an important point with coarse graining a system is with determining the effective interactions between the CG particles.

One of the first applications of CG methods has been the MD simulation of the aggregation of amphiphiles [12]. Since then many different applications for coarse graining have been studied, ranging from polymer chains [13–19] to lipid systems [20–24]. We have used lipid CG methods to study many interesting properties of bilayer systems, such as the spontaneous formation of vesicles [23], the fusion and fission of vesicles [25–27], as well as shape changes of vesicles due to changing conditions in the environment [28,29].

More recently coarse graining methods have been applied to more complex systems such as small peptides or even larger proteins [30–33], as well as viral capsids [34,35]. However, all coarse graining methods differ in their approach with respect to the number of atoms per CG particle as well as in the way interactions are described. In this paper the first version of a new CG model for proteins is being developed, which differs from other models due to the introduction of a novel angle potential with two equilibrium angles as well as a hydrogen bonding potential. These new potentials are used to stabilize the backbone without the need to use different particle types for different parts of the protein backbone. In this approach we differ from other CG models of transmembrane proteins in which one on forehand has to decide for each backbone particle type which are, for instance, in a helical or in a loop conformation [32,33,36,37]. It must be emphasized that the model presented in this paper is still in development and inclusion of new features and reparametrization are likely to occur. Moreover, the main focus of the model is on a qualitative description of transmembrane peptides and proteins (consisting mainly of α-helices), but the CG model is also tested against globular proteins. Finally, to prove the possible applicability of the new CG model, it is used to investigate membrane inserting properties of a large set of WALP-peptides. All MD simulations in this paper are performed using our in-house developed code [23].

COMPUTATIONAL MODEL

As mentioned in the introduction, one of the steps involved in designing a CG model is to choose the mapping of the original atoms to the CG particles. Different levels of coarse graining are possible, ranging from only omitting the hydrogen atoms (the united atoms approach) to much higher level CG particles combining several original atoms. However, the most common approach is to group four non-hydrogen atoms into one CG particle [20,22,23]. Thus, for instance in the case of lipid membranes, dipalmitoylphosphatidylcholine (DPPC) is represented by four particles in the head group (representing the choline, phosphate and glycerol backbone) and two hydrocarbon tails, each consisting of four tail particles.

In order to keep the CG protein particles in the new protein model in the same range of mass and size with respect to the CG bilayer models, amino acids are mapped into two CG interaction sites, one for the backbone and one for the sidechain. The backbone particle is located at the position of C_α in the all-atom model, whereas the sidechain particle is located along the C_α-C_β bond vector, with its specific position depending on the specific residue type of the amino acid, which is discussed further on in the paper. The size of the CG particles resembles the size of the underlying original atoms. In Figure 1 the mapping of phenylalanine to the CG model is schematically depicted.

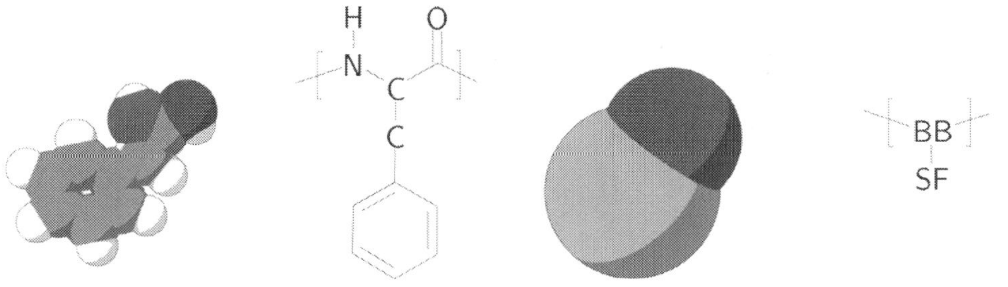

(a) All-atom representation (b) CG representation

Figure 1: Mapping of the all-atom representation of phenylalanine (figures to the left, both schematic and as chemical composition) to its CG representation (figures to the right). The center of the CG backbone particle is determined by the position of the C_α atom, whereas the CG sidechain particle is positioned along the C_α-C_β bond vector. The brackets represent the repetitive unit.

The mass of all particles is computed directly from the chemical composition and the size of the particles is computed from the occupied volumes of the residues [38]. Because of the chosen spherical nature of the CG particles, the Van derWaals radius of each particle is computed directly from its volume. The radii of the backbone particles are all assumed to be equal to that calculated from glycine. To obtain the volumes of the sidechain particles of the other residues, the volume of glycine is subtracted from the total residue volume. Since no occupied volume for arginine has been reported, its volume is estimated to be equal to that of phenylalanine which has an equal number of non-hydrogen atoms. This method to obtain the size of the CG particles is similar to that used by Levitt [39]. However, different to Levitt, it is chosen to use particles that are less soft, with respect to their interactions, in order to make them more compatible with existing water and lipid CG models. In Figure 2 a schematic overview of the sizes of all residue types in the CG protein model is depicted, including the lipid and water parameters from the lipid model by Markvoort et al. [23].

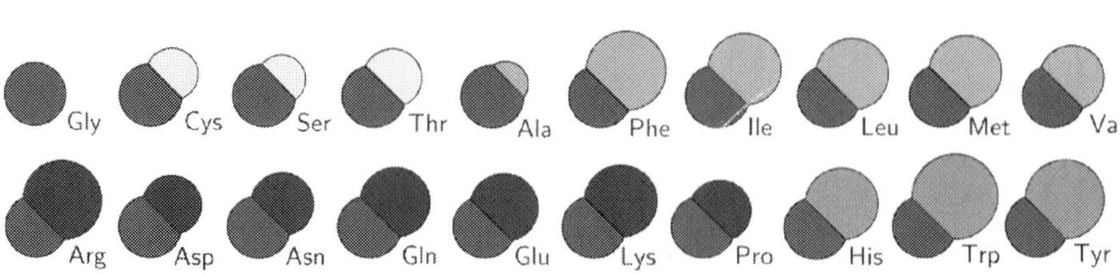

Figure 2: A schematic overview of all the residue types used in the CG protein model. The sizes of all particles are proportional with respect to each other. All residues (except for Gly) can be assigned to any of the four groups (as discussed in the text): the neutral group (Cys, Ser, Thr), the apolar group (Ala, Phe, Ile, Leu, Met, Val), the polar group (Arg, Asp, Asn, Gln, Glu, Lys, Pro), and the aromatic group (His, Trp, Tyr). Each of these groups is indicated by a different color.

All CG protein particle types are denoted either by their full or abbreviated name (e.g., isoleucine or Ile) or by two letters, where the backbone particle notation starts with a B, and the sidechain particle notation with a S. Depending on the type of amino acid the sidechain particle resembles, the second letter is added in accordance with the one letter abbreviation for the residue (thus, SF for phenylalanine). For the backbone particle three types are distinguished, only differing in mass: the default type (BB) and two types to account for the proline residue (BP) and the glycine residue (BG, which has no sidechain particle attached).

Force Field Development

Besides the topological mapping of the original atoms into CG particles, it is also necessary to derive interaction parameters (the force field) for all of the CG particles; both bonded and non-bonded interactions. This is a difficult and tedious process, since the problem is highly multi-dimensional.

The general force field equation that describes all contributions to the potential energy U of the coarse grained system is given by

$$V_{\text{total}} = \sum^{N_{\text{pairs}}} V_{\text{LJ}}(r_{ij}) + \sum^{N'_{\text{pairs}}} V_{\text{LJHB}}(r_{ij}) + \sum^{N_{\text{bonds}}} V_{\text{harm}}(r_{ij}) + \sum^{N_{\text{angles}}} V_{\text{harm}}(\theta_{ijk}) + \sum^{N_{\text{double angles}}} V_{\text{double}}(\theta_{ijk}) \tag{1}$$

where the first two sums over all pairs (or a subset, indicated by the prime, as explained further on in this paper) in the system account for the nonbonded interactions (a standard Lennard-Jones (LJ) potential or modified version (LJHB) as explained further on), the third sum all interactions due to atoms that are bonded (a standard harmonic potential), and the fourth and fifth sums for all possible bonded angle triplets (either a standard harmonic or double angle potential). Which angles are subject to which potentials (harmonic or double angle) is explained further on in this paper.

In the current CG model for the bonded interactions only bond and angle terms are considered, both in their harmonic form. Torsion interactions, either dihedral or improper, are not present in the model, in contrast to other CG protein models [33,40]. Although torsion interactions are normally used in order to stabilize the secondary structure of the protein, using these interactions also means that it has to be decided on forehand by using particle typing which part of the protein is going to be, for instance, helical, extended or in a loop. Therefore, the torsional interactions are discarded in the current model, but in order to account for the ability to maintain the secondary structure a force field extension is developed for non-bonded interactions, which is explained further on.

First the parameters of the CG force field with respect to the bonded interactions are discussed, followed by a discussion for the non-bonded interactions. Parameters for lipid and water molecules are taken from an existing bilayer model [23]. Because the new CG transmembrane protein model is based upon Markvoort's bilayer model, the time scale of diffusion of molecules in the system is of similar size, leading to a similar time scale for the simulations [25].

Bonded Interactions

Average bond lengths for CG backbone particles are determined from all-atom MD simulations of several nanoseconds of both rhodopsin [41] and the β_2AR [8]. For both trajectories the all-atom representation of the backbone has been mapped into the CG representation. Subsequently, the distances between neighboring CG backbone particles are determined. This gives a probability distribution for the bond length for CG backbone particles with an average of 0.384 nm. Since both simulations are of transmembrane proteins (consisting mainly of α-helices), the bond length distribution could be biased toward α-helices. Because the main goal is to develop a CG transmembrane protein model the bias is acceptable. Furthermore, the computed bond length is in good agreement with previous studies where the same CG bond length is based on crystal structures of many different proteins found in the protein data bank [39,42].

Unfortunately, the all-atom simulations do not hold sufficient data to accurately determine the bond lengths between CG backbone particles and sidechain particles. Therefore, these bond lengths are based on previously determined bond lengths by averaging across entries in the protein data bank [43]. The bond length associated with the disulfide bond has been obtained from the previously mentioned all-atom MD simulations.

Bond strengths are initially based on the constants reported by Wallqvist [42] and, for the disulfide bond, by Levitt [39]. However, because in the current model different masses are used, the constants have been adjusted in such a way, that the bond oscillation frequencies are the same.

In order to incorporate rigidity and allow for the anti-conformation of the hydrocarbon tail the CG lipid model only uses harmonic angle interactions between tail particles. Thus, the equilibrium angle for lipid particles is set to 180°, with a force constant of 5.408 kJ/mol/rad^2. The harmonic angle potential allows only one equilibrium angle to be set. Thus, the angle along the backbone of the protein would also only have one equilibrium angle using this potential. However, the backbone of a protein can be in very different conformations, for instance as α-helices or random coils, for which the angles along the backbone are very different (roughly around 90° and 120° respectively). Introducing different particle types, with different angle potentials between them, for each of these two possible conformations, would allow the different conformations, but it would also impose a-priori knowledge upon the model. A different approach is to allow two equilibrium angles in the potential energy function for the angle, and thus allow different conformations using only one particle type. To this end a novel angle potential with two equilibrium angles has been developed and is named the double angle potential.

The double angle potential is defined as a fourth power polynomial, which is determined by five parameters given in the force field parameter set. These five parameters are: the first equilibrium angle θ_1, the second equilibrium angle θ_2, the potential energy for the first equilibrium angle $V_A(\theta_1)$, the potential energy for the second equilibrium angle $V_A(\theta_2)$, and the maximal potential energy for the barrier between the two equilibrium angles $V_A(\xi)$, where ξ is the angle belonging to the maximum of the barrier. It must be noted that not ξ, but the height of the maximum is given in the force field parameter set. In Figure 3(a) the general shape of the double angle potential is shown, indicating all parameters discussed above. Others have also used fourth power polynomials to describe similar angles, but either with equal well depths [42] or with only one equilibrium angle [44].

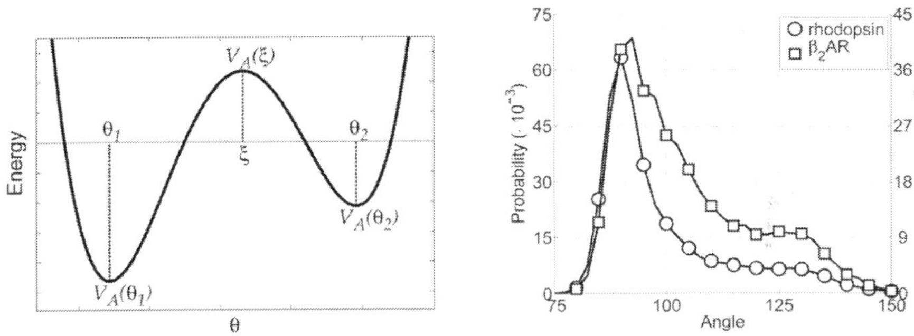

(a) Double angle potential

(b) Angle distributions

Figure 3: In (a) a schematic depiction of the double angle potential is shown and in (b) the distribution of angles along the backbone chain as found in all-atom MD simulations for rhodopsin and the β_2AR are depicted. The distribution for rhodopsin is on the left axis and for the β_2AR on the right axis.

The functional form of the double angle potential is given by

$$V_A(\theta_{ijk}) = A\,g(\theta_{ijk}) + D \tag{2}$$

where A and D are both constants given by

$$A = \frac{V_A(\xi) - V_A(\theta_1)}{g(\xi) - g(\theta_1)} \tag{3}$$

and

$$D = \frac{g(\theta_1)\,V_A(\xi) - g(\xi)\,V_A(\theta_1)}{g(\theta_1) - g(\xi)} \tag{4}$$

and $g(\theta_{ijk})$ the fourth power polynomial expressed as

$$g(\theta_{ijk}) = \tfrac{1}{4}\theta_{ijk}^4 - \tfrac{1}{3}(\theta_1 + \theta_2 + \xi)\,\theta_{ijk}^3 + \tfrac{1}{2}(\theta_1\theta_2 + \theta_1\xi + \theta_2\xi)\,\theta_{ijk}^2 - (\theta_1\theta_2\xi)\,\theta_{ijk} \tag{5}$$

Furthermore, the angle belonging to the maximum of the barrier obeys the requirement $\theta_1 < \xi < \theta_2$ and the energies for the minima satisfy $V_A(\xi) > V_A(\theta_1)$ and $V_A(\xi) > V_A(\theta_2)$. The value for ξ can be determined from all known parameters using the iterative Newton–Raphson method, which requires only a few steps under normal conditions. In the supplementary materials a complete derivation of the double angle potential is given.

Similar to determining the bond parameters for the CG particles, both all-atom MD simulations are used again to extract the probability distribution for the angles between CG backbone particles, see Figure 3(b). Furthermore, the Boltzmann inversion method [14,16,45] has been used to obtain estimates for values of the potential energy at any of the three angles. For the CG backbone particles this gives the equilibrium angles at 91.25° and 123.25°, with potential energies of 0 kJ/mol and 23.0 kJ/mol, respectively. By default the lowest energy in the double angle potential is set to zero, because the offset of the potential is arbitrary. The barrier energy is set at 23.7 kJ/mol. From these choices it can be seen that the model is developed for α-helices, mainly because the equilibrium angle around 90° is highly favored.

The angle between CG sidechain particles and two backbone particles (a B-B-S angle) is also governed by the double angle potential. This angle is introduced to restrain the movement of the sidechain particles. The two equilibrium angles are set such that the backbone-sidechain bond is the bisector outward of the underlying angle in the backbone. Thus, the equilibrium angles are chose to be 118° and 135°, both with a corresponding zero potential value. The barrier in between is very small (only 0.003 kJ/mol) allowing sidechain particles to adapt their position along the backbone relatively easy.

Non-Bonded Interactions

To account for interactions between CG particles belonging to different molecules and for long range interactions within a molecule not dealt with by the bonded interactions, an accurate description of non-bonded interactions is necessary. Between all CG particles, except those that are directly bonded or connected through an angle, both Van der Waals and Coulomb interactions are modeled by the Lennard–Jones potential [23]. When only repulsive interactions are to be taken into account an adaptation of the Lennard-Jones potential is used: the Weeks–Chandler–

Andersen potential [46], which is effectively a Lennard-Jones potential truncated at the point of minimal energy and shifted to remain continuous. For distances larger than the truncation distance the potential is zero.

Within the CG protein model three non-bonded interaction classes can be distinguished: protein-protein, environment-protein, and environment-environment. Since the model is developed for transmembrane proteins, the latter interaction class consists mainly of lipid and water interactions. For the interactions between lipid and water particles the force field given by Markvoort et al. [23] is used.

The second interaction class, environment-protein, deals with all interactions of the lipids and water with the CG protein particles. These interactions are especially important in the case of transmembrane proteins, since in that case the environment is heterogeneous. Ulmschneider et al. have derived implicit potentials for amino acids in a membrane [47]. In their work a potential of mean force for each amino acid is derived from statistical data on the occurrence of the amino acid in the membrane region of the transmembrane proteins. This splits all amino acids into four clearly distinguishable groups:

- **Polar:** a group favoring a water surroundings, containing arginine, aspartic acid, glutamic acid, lysine, asparagine, glutamine and proline, as well as the backbone particles including glycine.
- **Apolar:** a group favoring the center of the membrane, containing alanine, isoleucine, methionine, leucine, phenylalanine and valine.
- **Aromatic:** a group favoring the interface region between the hydrophobic core of the lipid membrane and the water, containing histidine, tryptophan and tyrosine.
- **Neutral:** a group with no significant preference, containing cysteine, threonine and serine.

From the shapes of the potential of mean force across the lipid membrane a qualitative estimate is made of the required interaction energy to recreate these shapes. Thus, the general behavior of the sidechain particles is more important than the inclusion of chemical detail. For the polar group the interaction with water is favorable, so all cross interaction energies with water are set to 1.97 kJ/mol. The interactions with the interface region of the lipid membrane (the lipid head particles) is not unfavorable, but not very favorable either and is, therefore, set to 0.50 kJ/mol, except for glycine, whose interaction energy is set to 1.00 kJ/mol. The interactions with the hydrophobic core (the lipid tail particles) are unfavorable and, accordingly, are set to be repulsive only.

On the other hand with the apolar group the diversity between the residues is larger, so some distinctions have to be made. For all residues in the group the interaction with water is unfavorable and is therefore set to be repulsive. The interactions with the lipid tail particles is set to be 1.00 kJ/mol for methionine and 1.97 kJ/mol elsewise. The interactions with the lipid head groups are neither strong nor weak for most residues in this group and is therefore set to 0.50 kJ/mol, except for phenylalanine. Phenylalanine has a much wider potential of mean force and, thus, seems to have a rather favorable interaction with lipid head particles, bringing its interaction strength to 1.00 kJ/mol.

The aromatic residues, histidine, tryptophan and tyrosine, only favor the interface region, so their interaction strengths with the lipid head particles is set to 1.97 kJ/mol. All other environment interactions for this group are set to be repulsive only. The last of the four groups, the neutral group, has no preference for any region, and, therefore, all interaction strengths are set to 0.50 kJ/mol.

The only interaction class not discussed so far is the protein-protein interaction class. These interactions are derived from the work by Liwo et al. [40]. However, since Liwo does not only use Lennard-Jones potentials, but also potential functions based both on distance and orientation, some adaptations to the parameters are necessary. These alterations have been made iteratively in the initial stage of development of the CG protein model. Finally, the interactions of backbone particles with sidechain particles are estimated from the assumption of the backbone particles being hydrophilic.

As mentioned before, the new CG protein model excludes torsional interactions, but to account for the most important structural element in α-helices, H-bonds, some alterations to the non-bonded interactions are required. In regular α-helices the backbone particles which are above each other have the opportunity to form H-bonds. In terms of the chain sequence this means that the backbone particles on position i and i +

4 can form an H-bond, leading to an α-helix. Therefore the distance distributions for these backbone pairs have been determined from the previously used all-atom simulations. In Figure 4(a) these distributions are depicted. From these distributions it can be seen that there exists an average H-bond length of 0.61 nm (measured between the C_α-atoms in the all-atom model). In order to maintain α-helices in the CG model, this distribution has to be reproduced. However, this is not possible using only the Lennard–Jones potential, because the distance for which the interaction energy is lowest is equal to 0.448 nm.

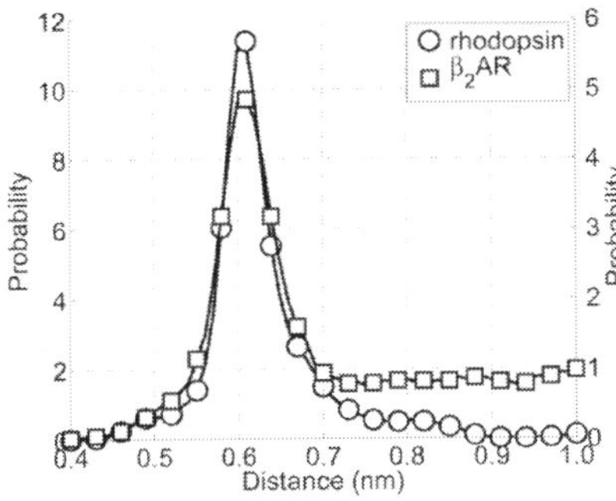

(a) H-bond distribution

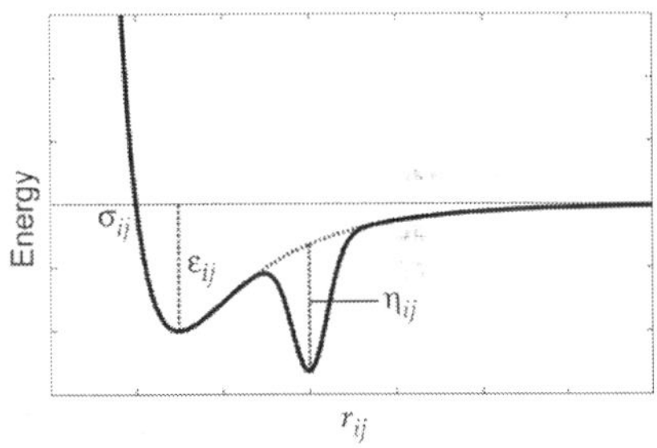

(b) LJ-H-bond potential

Figure 4: In (a) the distribution of potential H-bonding CG backbone particles along the backbone chain (i to i + 4) is depicted, whereas in (b) the newly developed Lennard-Jones H-bond potential is shown. The dashed line in (b) indicates the curve belonging to a regular Lennard-Jones potential. The important parameters are indicated in the figure.

To overcome this problem an alternative type of the Lennard-Jones potential has been developed, which is only active between backbone particles. The alternative description consists of a regular Lennard-Jones potential, $V_{LJ}(r_{ij})$, combined with an inverted Gaussian curve, $V_{HB}(r_{ij})$, giving $V_{LJHB}(r_{ij}) = V_{LJ}(r_{ij}) + V_{HB}(r_{ij})$, with $V_{LJHB}(r_{ij})$ the Lennard–Jones H-bond (LJHB) potential. The Gaussian curve accounts for the H-bond part of the potential, since it can be used to introduce a second minimum in the potential at the distance at which the backbone particles should be in case an H-bond is formed. The H-bond contribution is defined as

$$V_{\mathrm{HB}}(r_{ij}) = -\eta_{ij}\, \mathrm{e}^{-\frac{1}{2}(r_{ij}-\mu_{ij})^2/\kappa_{ij}^2} \tag{6}$$

where μ_{ij} is the location of the H-bond minimum, κ_{ij} determines the width of the H-bond well, and η_{ij} represents the well depth of the H-bond minimum. In Figure 4(b) a schematic representation of the LJHB potential is shown. Both parameters μ_{ij} and κ_{ij} can be determined from the distribution shown previously in Figure 4(a), and in the case of backbone-backbone H-bond interactions these are 0.61 nm and 0.015 nm, respectively. The depth of $V_{HB}(r_{ij})$ is determined from computing the Coulomb contributions of the atoms the backbone particles represent involved in the H-bond maintaining the α-helix. Using the same all-atom MD simulations, the Coulomb energies for all of these possible H-bond forming pairs have been computed at the level of the CG particles. The parameter η_{ij} is chosen in such a way that the well depth of the LJHB potential is in agreement with the computed values for the original Coulomb interactions. This gives for the well depth parameter of the backbone-backbone interaction an energy of 15.0 kJ/mol. It is important to realize that this LJHB potential is only active between CG backbone particles. Furthermore, the LJHB potential acts between any pair of backbone particles (except for those separated by one or two bonds). Thus, the LJHB potential has the ability to form α-helices, but only if the backbone particles are close to the hydrogen bonding distance μ_{ij}. Otherwise, it simply acts as a normal Lennard-Jones potential.

Comparison to Existing CG Protein Force Fields

In the previous section a complete description of the new CG protein model is given. However, this is not the first type of CG protein model that has been developed; over the past few years all kind of different force fields have been reported in literature.

The first CG protein models that have been developed mainly focused on the problem of protein folding [38–40,43,48]. The modeling of these folding processes is very complex and computationally time demanding, and, therefore, using CG models can be very beneficial. Recently, different CG models have been used to investigate protein folding again [30,31,49]. In most of these models the backbone of the protein is represented by all of its heavy atoms along the chain, which is a clear difference with the model introduced in this paper, which only has CG backbone particles at the C_α positions. On the other hand, the sidechain particles are modeled as one CG particle as well.

A model that uses one CG particle for the backbone and one for the sidechain is the model developed by Voth et al.[32,50,51]. From this point of view it is very comparable to the present model. However, Voth et al. use the force matching algorithm [32,52,53] to determine free-form potentials from all-atom simulations for all interactions between the CG particles. The derived free-form potentials are then used as tabulated input for their model. Although this method allows for rapid determination of an underlying force field, it has to be derived again from all-atom simulations for every new molecular structure, for every new mixture and for every new state point.

The CG protein model which currently receives most attention is the model developed by Marrink et al. [33]. Recently Bond et al. have used this model to investigate peptide insertion into a membrane, as well as the behavior of membrane channel proteins [37,54–56]. Shih et al. have used Marrink's model to take a closer look at discoidal lipoprotein particles [36]. A model similar to Marrink's has recently been proposed in a study by Han et al. where they investigate poly-alanine based peptides [57].

In Marrink's model the protein backbone is mapped into one CG particle as well, but the sidechains are not necessarily represented by one side-chain particle. For instance, the sidechain parts of arginine and lysine are modeled by two CG particles, whereas the ring based residues have either three or four CG particles. Moreover, besides glycine, also alanine is modeled as a backbone particle only. The difference between our and Marrink's model is not only in the number of CG particles, but also in their size. Whereas the CG particles in the new model all have different sizes, depending on their occupied volumes, in Marrink's model all particles have the same diameter (0.264 nm), except for the ring based particles (0.241 nm). Similarly, in Marrink's model the masses for all CG particles are the same (72 amu), with the ring based particles (45 amu) as an exception again. Furthermore, instead of having different interaction types for each residue, in the model of Marrink all coarse grained amino acids are made of the similar building blocks as used in their lipid model [24], and, thus, the same coarse grained particle types are used to describe both lipids and amino acids. This is contrary to the present model, where different particle types are used for lipids and amino acids.

Although in our model 20 different particle types are necessary to describe the residues, Marrink's model uses about 15 different building blocks to create the CG representation of all 20 residues. So the number of different particle types between both models is comparable leading to a similar complexity.

The major difference with respect to the modeling of the bonded interactions between Marrink's model and our model is that Marrink included torsion interactions combined with backbone particle typing to fix the protein's local secondary structure, while we discard torsion interactions explicitly. In a preliminary version of their model artificially bonds were added to constrain the conformation and orientation of α-helices [58]. In our model the secondary structure has to arise from the conformational freedom allowed by the double angle and the LJHB potentials.

Because Marrink's model generally uses smaller and more particles to represent the CG sidechains, their force constants for bonds are smaller than the bond force constants in our model, which has larger and softer particles, but, consequently, needs to reduce the flexibility of the bonds. Furthermore, to avoid numerical instabilities arising from fast fluctuations, some backbone-sidechain bonds in Marrink's model are constrained.

The different limitations between each of these CG protein models show that the applicability of a specific model might be limited to specific cases.

PERFORMANCE ANALYSIS OF THE COMPUTATIONAL MODEL

To investigate the performance of the CG model two different series of MD simulations are carried out, the first being an analysis of the protein-environment non-bonded interactions, and the second a set of five different large protein complex simulations, each simulated for tens of nanoseconds.

Evaluating whether the non-bonded interactions between the protein and the environment correspond to the potentials of mean force in the model by Ulmschneider [47] is important. Therefore, MD simulations for each protein particle type being pulled through a lipid membrane are performed. The lipids are modeled as H4T4T4-molecules (representing

DPPC), which means that each lipid molecule has two tails of 4 T-particles each and a head group of 4 H-particles. These so-called steered MD simulations differ from normal MD simulations with respect to the fact that particles (in this case the protein particles) are pulled slowly but steadily through the simulation box. However, the particles themselves are not directly pulled, but instead they are connected to a phantom particle with a harmonic spring.

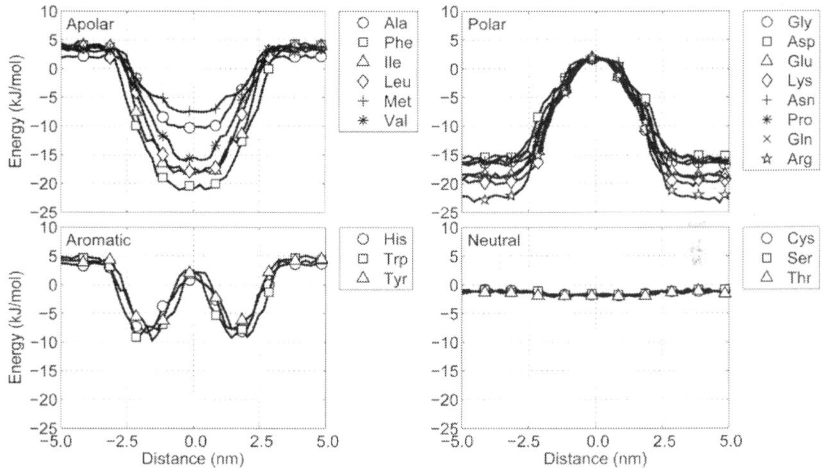

Figure 5: Non-bonded interaction energy between the CG protein particles and the environment (lipids and water) as a function of the distance to the membrane center. The protein particles are denoted by their common abbreviations and are put into their respective interaction groups: apolar, polar, aromatic and neutral, see the text.

The phantom particle is then pulled at a constant velocity and, through the spring, the true particle is pulled forward. The velocity at which this induced pulling occurs depends on the spring's force constant, the pulling velocity for the phantom bead, and the interaction of the particle with the environment. Thus, the protein particles are pulled from the water phase through the membrane, to the other side of the membrane, back into the water again. It is important to pull slowly in order not to disturb the equilibrium of the system too much. Therefore, in all of these simulations the phantom particles are pulled at 0.002 nm/ps, which is two orders of magnitude below average thermal fluctuations. The force constant of the harmonic spring is set at 100 kJ/mol/nm^2. The MD simulations are performed at constant pressure (1 atm) and constant temperature (324

K), using a Berendsen barostat and thermostat, respectively, and the size of the time step in the leapfrog integration scheme equals 12 fs. During the simulations the non-bonded interaction energy between the protein particle and the lipid-water environment is recorded as a function of the distance to the center of the membrane, see Figure 5. A closer look at these potential energy graphs shows that with the current model all particles behave similar with respect to the potentials of mean force presented by Ulmschneider. This indicates that protein-environment interactions in the CG protein model are as expected.

More recently, others have also performed similar experiments to obtain potential energy graphs with their CG protein model [33,59,60]. Comparing these profiles with ours shows different behavior for several residues, especially for the neutral residues in our model. However, whereas their models aim to include as much chemical detail as possible in the CG model, our model aims to model the general behavior, and, thus, using the implicit potentials based on the occurrence of residues with respect to the membrane is appropriate.

As mentioned previously, in order to analyze the behavior of the CG protein model at a larger level, five different large protein complexes are simulated for tens of nanoseconds. These five proteins are selected to cover a wide variety of size, shape and function to serve as test cases, and, therefore, are members of two different classes of proteins. Three of the selected proteins are transmembrane proteins (rhodopsin, β_2AR, and the KCSA potassium channel) and two are water-soluble (cytochrome P450-CAM and fasculin 1). The crystal structures for each of these proteins are obtained from the protein data bank and converted into their respective CG representations. The water-soluble proteins are solvated in water only, whereas the transmembrane proteins are inserted in a periodic lipid membrane of appropriate thickness, depending on the length of the transmembrane region of the protein, and thereafter solvated in water. Hence, rhodopsin, the β_2AR and the KCSA potassium channel are inserted in a membrane consisting of H4T4T4-molecules, having an approximate area of 19×19 nm^2. Two important remarks have to be made. First, the four independent subunits of the KCSA potassium channel are not connected by any bonds, and are thus treated as independent molecules in the simulations. Secondly, some sections of the proteins are not included in the crystal structures and are, therefore, not included in the

CG model. In particular for rhodopsin many of the non-helical intra- and extracellular regions are left out. In Figure 6 the starting conformations for rhodopsin and KCSA potassium channel are shown.

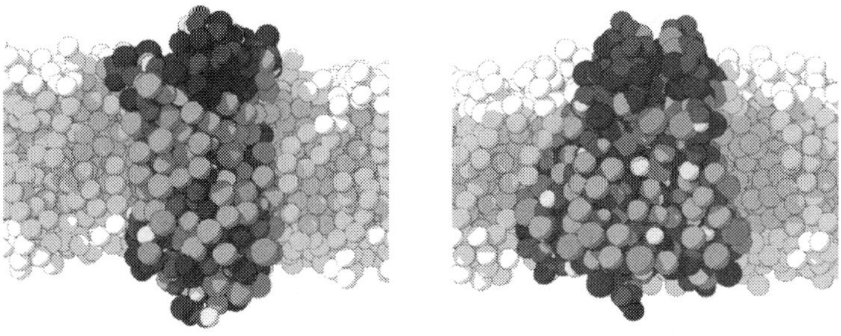

Figure 6: Initial configuration for the CG MD simulations for rhodopsin (on the left) and the KCSA potassium channel (on the right). Both proteins are embedded in an H4T4T4 lipid membrane, for both configurations only the part surrounding the proteins is shown and water and some lipids have been removed for clarity.

In all of the five MD simulations a timestep of 12 fs is being used, allowing for sufficient sampling of the fastest oscillations in the CG model (those in the bonded interactions), for a total simulation time of approximately 60 ns. Both the temperature and the pressure are kept constant (at 324 K and 1 atm respectively).

Whilst taking into account the different nature of each protein, all five simulations are analyzed in a similar fashion. For every protein the coordinate root mean square difference of the C_α-atoms ($CRMS_\alpha$) with respect to the initial structure, and the change in radius of gyration (R_g) are computed. Moreover, for each of the proteins the bond and angle distributions of the backbone particles are determined.

In Table 1 the $CRMS_\alpha$ is shown for each of the simulations. This $CRMS_\alpha$ is an average over the last 240 ps of the 60 ns simulation, and the crystal structure is used as a reference point. For KCSA two values are shown, one for the individual subunits and one for the entire complex, which takes the internal motions of the subunits into account. When the $CRMS_\alpha$ is compared to the characteristic length scale of the model (being the backbone bond length of 0.384 nm), four out of five proteins appear not to change their conformations considerably. Only the entire KCSA-complex shows somewhat more conformational change. However, since the independent subunits of the KCSA potassium channel give a much lower $CRMS_{\alpha}$ value than the total complex, it is apparent that the subunits have moved with respect to each other, leading to a higher $CRMS_{\alpha}$ value. The results for the $CRMS_{\alpha}$ suggest that the overall structure of these proteins is preserved reasonably well in the CG model.

Table 1: Analysis results for five different protein CG MD simulations. Both the $CRMS_{\alpha}$ and R_g (for the initial and final conformations) are given. All numbers are an average over 240 ps of simulation time. For the KCSA potassium channel the results are given for the total complex as well as for an average of the separate subunits

Protein	PDB-code	$CRMS_{\alpha}$ (nm)	R_g^{init} (nm)	R_g^{fin} (nm)
rhodopsin	1U19	0.323 ± 0.004	2.23 ± 0.005	2.19 ± 0.005
β_2AR	2RH1	0.382 ± 0.004	2.02 ± 0.005	1.97 ± 0.005
KCSA (total)	1K4C	0.511 ± 0.002	2.12 ± 0.005	2.03 ± 0.005
KCSA (individual)	1K4C	0.393 ± 0.003	-	-
cytochrome P450-CAM	2CPP	0.323 ± 0.002	2.13 ± 0.005	1.97 ± 0.005
fasculin 1	1FAS	0.196 ± 0.003	1.12 ± 0.005	1.02 ± 0.005

Besides the $CRMS_{\alpha}$ also the radius of gyration, R_g, is shown in Table 1 for both the initial and the final conformation of the proteins. The radius of gyration is computed by taking the root mean square distance for all individual protein particles with respect to the geometric center of the protein [61]. From the results it can be seen that in general the radius of gyration decreases during the simulations. However, the amount of decrease is not the same for all proteins and, apparently, depends on the nature of the protein. The water-soluble protein fasculin 1 (a small randomly coiled protein) shows a small decrease (7%) in its radius of gyration. The only protein that has a relatively large decrease of its radius of gyration is cytochrome P450-CAM. However, this protein contains α-helices as well as β-sheets and random coils, which can explain the reduction in the radius of gyration. The other three transmembrane proteins (rhodopsin, β_2AR and KCSA potassium channel) have a decrease in their radius of gyration of less than 5%. Probably these proteins are less compressible than the other two.

For further evaluation of the performance of the CG protein model, the results of the CG simulations are compared to those from all-atom MD simulations for two proteins: rhodopsin and β_2AR. The simulations of these two are already available (and have been used before [8,41]). For each of the two all-atom simulations a trajectory of 1 ns is used for analysis, with snapshots taken every several picoseconds.

¿From each of the CG simulations, as well as from each of the all-atom simulations, the bond distributions for the bonds between backbone particles and the angle distributions within a triplet of backbone particles are computed, see Figure 7. With respect to the bond distributions it is clear that the distributions for the CG simulations are slightly wider than for the all-atom simulations, but still acceptable. For the angle distributions along the backbone it can be seen that the all-atom simulations have a high peak around 91° and a shoulder around 123°, which coincides with the expected locations for the α-helices on one and the loops and random coils on the other side. In the CG model the backbone angle interaction is governed by the double angle potential. The parameters for this potential are chosen to mimic the α-helices of transmembrane proteins best. ¿From Figure 7(b) it can be seen that the angle distribution for all CG simulations are similar to the all-atom angle distributions for the transmembrane proteins rhodopsin and the β_2AR.

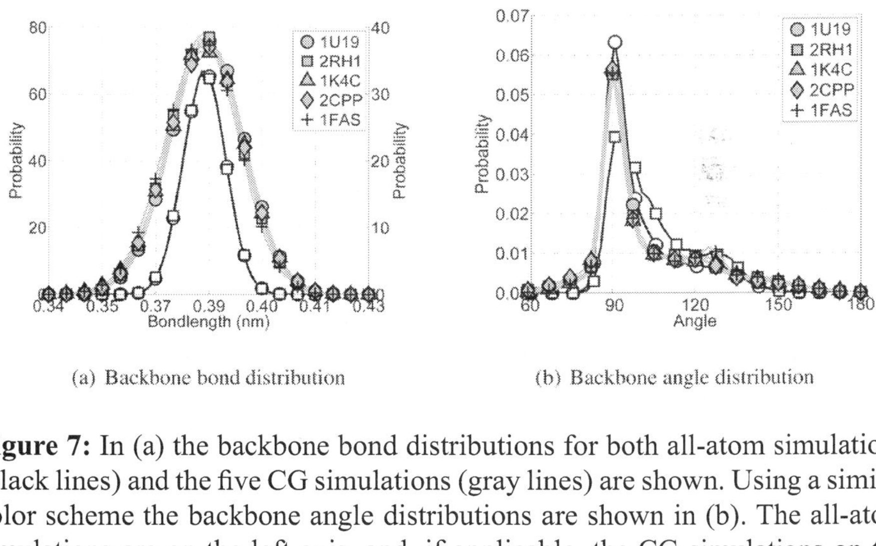

(a) Backbone bond distribution

(b) Backbone angle distribution

Figure 7: In (a) the backbone bond distributions for both all-atom simulations (black lines) and the five CG simulations (gray lines) are shown. Using a similar color scheme the backbone angle distributions are shown in (b). The all-atom simulations are on the left axis, and, if applicable, the CG simulations on the right axis. The protein data bank codes for the proteins are used in the legend, see Table 1.

When developing the CG model special attention has been paid in recovering the secondary structure in the model, without imposing too much knowledge about the secondary structure upon the model. Thus, the performance of the model can partially be measured by the correct representation of the protein's secondary structure. Since the model is aimed at reproducing α-helices in transmembrane proteins, several non-bonded backbone particle distances (or C_α distances in the all-atom simulations) along the backbone chain are investigated for the rhodopsin and β_2AR simulations. Because a regular α-helix is formed when there is an H-bond between particle i and its neighbor on position i + 4, this distance is tracked throughout the simulations. Also the distances are monitored between particles which can form H-bonds leading to tighter and wider helices (i to i + 3 and i to i + 5).

All possible H-bond distributions are shown in Figure 8. ¿From the all-atom distributions (Figure 8(a)) it can be seen that rhodopsin and the β_2AR have similar H-bond distributions for all possible H-bond forming pairs (i + 3 to i + 5). The only difference arises from the fact that in the all-atom model of the β_2AR more intra- and extracellular loops are present, which arises in the figure as longer tails for each of the three distributions. Turning the attention to the CG model for these two proteins and their respective potential H-bond forming pairs (see Figure 8(b)) a clear peak can be seen around 0.61 nm, which coincides with the distance μ_{ij} in the Lennard-Jones H-bond potential. However, while the peak for the i to i + 5 pairs is still present, albeit with a much lower probability, the peak for the i to i + 3 pairs has moved to 0.61 nm as well. Although the Lennard-Jones hydrogen bond potential is not active between backbone particles that are bonded or connected directly with an angle, it is active between the i to i + 3 pairs, favoring these pairs to be at the ideal distance for the H-bond part of the Lennard–Jones H-bond potential. On the other hand visual inspection of both trajectories showed that the structure of these CG proteins is still helical, although the helices are slightly tighter than true α-helices. However, this does not significantly affect the lengths of the helices, nor does it cause the sidechain particles to point inward toward the backbone helix. Therefore, the current representation of the CG α-helix is considered acceptable.

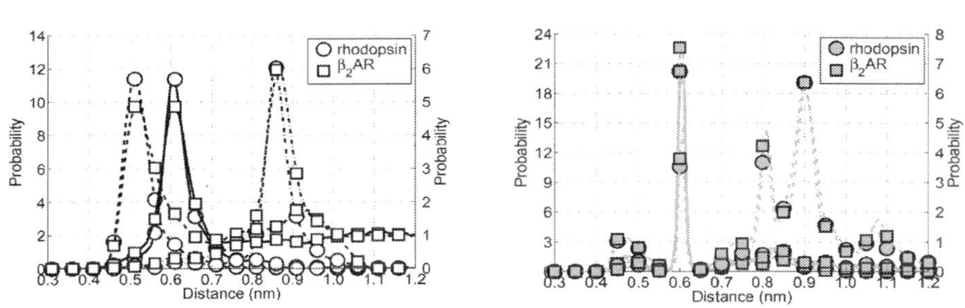

(a) All-atom distribution (b) CG distribution

Figure 8: Distributions of potential H-bond forming pairs in the all-atom simulations (on the left) and in the CG simulations (on the right). For the possible H-bond forming pairs of rhodopsin and the β_2AR backbone particles along the chain are considered, where the dashed lines indicate i to i + 3 pairs, solid lines i to i + 4 pairs, and dash-dotted lines i to i + 5 pairs. In the left figure the distribution for β_2AR is put on the right y-axis, whereas in the right figure the distributions for the i to i + 5 pairs are on the right y-axis.

Concluding the investigation of the performance of the CG model, it has been shown that proteins with α-helices can be modeled quite accurately. Furthermore, the protein-environment interaction energies are reproduced satisfactorily, as are the bond and angle distributions.

CASE STUDY: WALP-PEPTIDES

With respect to the structure and function of biological membranes the interaction between proteins and lipids is crucial. An important factor of this interaction is matching the hydrophobic thickness of the membrane with the hydrophobic length of the transmembrane protein segment. Processes that can be influenced by this hydrophobic matching include protein activity, membrane domain formation and protein sorting [62,63]. Therefore, in order to understand the effects of hydrophobic matching it is important to have detailed information at the molecular level.

Because large protein complexes embedded in lipid membranes are difficult to study both experimentally and theoretically, small model systems to understand the effect of hydrophobic matching have been introduced which mimic the transmembrane segments of membrane proteins. The best known example of these model systems is a family of synthetic α-helical transmembrane model peptides, having a repeated alanine-leucine-sequence flanked on both sides by tryptophan residues, called WALP-peptides [63–67]. The flanking tryptophan residues have been chosen because these residues are frequently found in membrane proteins near the membrane interfacial region and because they orient with the lipid headgroups [68]. At room temperature the helical axis of these WALP-peptides is nearly perpendicular to the plane of the membrane and, the peptide being membrane embedded, strongly protected from the solvent [69]. Furthermore, because these hydrophobic peptides form well-defined α-helices, as has been proven by several experiments [63,66,67,70–72], they are suitable to investigate the packing of α-helices in transmembrane proteins, such as G-protein coupled receptors [73].

The previously mentioned experimental studies have been complemented by all-atom MD simulations investigating the properties of these peptides and their folding into the membrane [74–77]. More recently, MD simulation studies of the insertion of WALP-peptides into the lipid membrane have been performed using an adapted version of Marrink's CG model [37].

To show the applicability of the currently presented CG model for transmembrane proteins, MD simulations of WALP-peptides of different length embedded in lipid membranes of different thickness are performed. Fourteen peptides of systematically increasing length are used, denoted as WALPx, where x is the number of residues in the peptide, ranging from 16 to 41. For example, the amino acid sequence of WALP16 is GWWLALALALALAWWA. For longer peptides the LA-sequence is extended.

To create the starting conformations for each of the peptides, first the all-atom α-helical representation has been built. Subsequently, these all-atom representations have been converted into their CG counterparts. The peptides are then combined with two different lipid membranes, either dilauroylphosphatidylcholine (DLPC) or DPPC, represented in the CG model by H4T3T3- and H4T4T4-molecules respectively. The two different membranes are chosen to expose all WALP-peptides to different membrane thicknesses. Finally, the entire system is solvated in water. The area for the lipid membrane is approximately $10 \times 10\ nm^2$. An example of the CG representation of WALP27 (with water omitted for clarity) is shown in Figure 9(a). In this figure also the definition of the hydrophobic mismatch is depicted, which is the difference between the length of the hydrophobic part of the peptide (dP) and the thickness of the hydrophobic part of the membrane (dL). A negative hydrophobic mismatch means that the hydrophobic part of the peptide is shorter than the hydrophobic part of the membrane, and a positive hydrophobic mismatch the opposite.

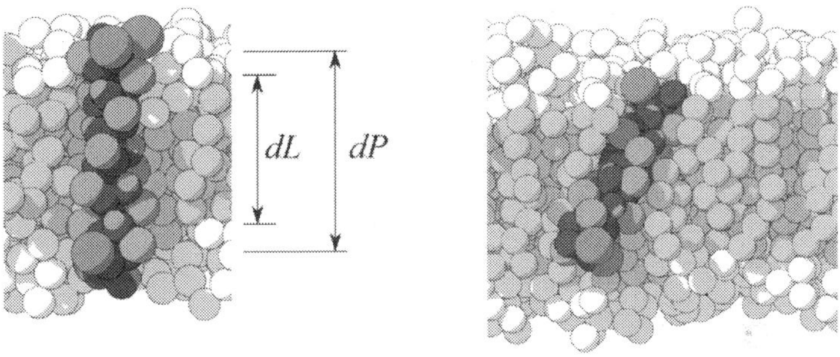

(a) Hydrophobic mismatch (b) Final WALP35

Figure 9: Configurations for different WALPs. In (a) the initial configuration for WALP27 embedded in an H4T3T3 lipid membrane is shown, with the thickness of the hydrophobic part of the membrane (dL) and the hydrophobic part of the peptide (dP) indicated. In (b) the final configuration for WALP35 in an H4T4T4 lipid membrane are depicted. In each figure water and some lipids are removed for clarity.

With fourteen different WALP-peptides in two different lipid membranes, at least 28 CG MD simulations are to be performed. However, to improve statistical accuracy, all simulations are performed three times. In each of these three repeated simulations the initial conformations are the same, but the initial velocities are changed. Furthermore, all simulations are performed for 24 ns, with a time step size of 12 fs. Both the temperature and pressure are kept constant around atmospheric conditions, again using the Berendsen coupling schemes. All simulations are performed 10 K above the melting temperature of the lipids in the membrane, which equals 293 K for H4T3T3 and 324 K for H4T4T4. Unless mentioned otherwise all presented results are the averages of the three repeated simulations.

To verify whether the structure of the WALP-peptides remains intact after the CG MD simulations, the $CRMS_{\alpha}$ is computed for each peptide with respect to its initial configuration. In Table 2 these values are shown for each peptide in both type of membranes. For most simulations the $CRMS_{\alpha}$ is very low, with values below 0.2 nm for 22 out of 28 peptides. Moreover, all peptides have a $CRMS_{\alpha}$ which is below the previously used threshold of 0.384 nm (the characteristic length scale of the model). Visual inspection of the few peptides that have a somewhat larger $CRMS_{\alpha}$ revealed that each of the peptides still maintained its α-helical character. In Figure 9(b) the final configuration of the peptide with the highest $CRMS_{\alpha}$(WALP35 in H4T4T4) is depicted, showing indeed the α-helix being maintained. The most probable cause for its higher $CRMS_{\alpha}$ is a small bend that appeared in the α-helix.

Table 2: The $CRMS_{\alpha}$ (in nm) after 24 ns of coarse grained molecular dynamics simulations for WALP-peptides of different length embedded in two different lipid membranes. The presented values are an average of the final 240 ps of the simulations, and for each of the simulations the standard deviation for the $CRMS_{\alpha}$ equals 0.005 nm

	WALPx													
Lipid	16	17	19	21	23	25	27	29	31	33	35	37	39	41
H4T3T3	0.092	0.054	0.045	0.038	0.061	0.143	0.109	0.045	0.080	0.131	0.191	0.179	0.151	0.126
H4T4T4	0.052	0.188	0.184	0.208	0.048	0.084	0.209	0.241	0.162	0.146	0.323	0.174	0.266	0.261

The stress introduced in a lipid membrane by a hydrophobic mismatch due to an inserted peptide can be relieved by at least two mechanisms [78,79]. The first mechanism involves the membrane adapting its local thickness to match the length of the hydrophobic part of the peptide, while the second mechanism is the tilting of the peptides to decrease any positive hydrophobic mismatch. Obviously a negative hydrophobic mismatch cannot be removed by tilting, because tilting would decrease the hydrophobic mismatch even further. Using the CG MD simulations for the WALP-peptides it is possible to investigate the occurrence of both mechanisms.

In order to compute the local membrane thickness the entire membrane is first divided into a 17 × 17 grid of equally sized sections. For each of these 289 sections the membrane thickness is computed separately by measuring the distance of the tail particles with respect to the central membrane plane of that section. For each distance computation of each lipid only the tail particles connected to the head particles are used.

Multiplying the obtained distance for each tail particle by two gives the local membrane thickness of that lipid. Averaging for all lipids in a section gives the membrane thickness of that specific part. For both lipid membranes (H4T3T3 and H4T4T4) the same procedure is used. Because the membrane thickness is now known throughout the entire 17 × 17 grid, it is possible to compute the membrane adaptation due to the insertion of the peptide. This adaptation is defined as the difference between the membrane thickness near the peptide and far away from the peptide. It is observed that for smaller peptides the membrane adaptation is negative, it increases as the peptide length increases, and eventually becomes positive.

Subsequently, the hydrophobic mismatch is computed based on the thickness of the membrane far away from the peptide (dL) and the hydrophobic length of the peptide (dP). Because in each WALP-peptide the outer amino acids are non-lipophilic, the hydrophobic length of the peptide is assumed to be the distance between the average positions of the four N-terminal and the four C-terminal backbone particles, which should more or less coincide with the beginning of the hydrophobic part of the peptide.

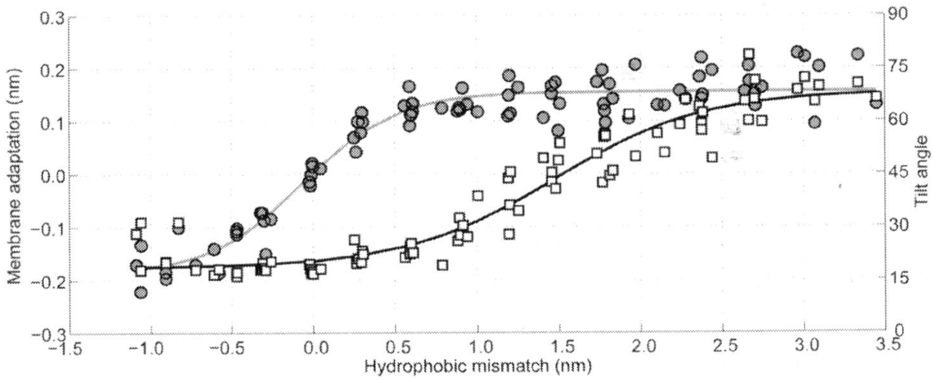

Figure 10: The effect of hydrophobic mismatch on peptide tilt and membrane adaptation. The membrane adaptation is shown by the circles and indicated on the left vertical axis, while the peptide tilt is on the right vertical axis and shown with squares. The lines are sigmoid curves fitted to the data and are only to guide the eye; both lines have a regression coefficient of 0.93.

In Figure 10 the membrane adaptation due to the peptide insertion is shown as a function of the hydrophobic mismatch (circles, left axis). As can be observed there is no membrane adaptation when the hydrophobic mismatch is zero. Moreover, when the hydrophobic mismatch decreases the membrane becomes thinner around the peptide, whereas when the hydrophobic mismatch increases the membrane thickens, until a plateau is reached (approximately 0.15 nm of membrane adaptation). Hence, the first proposed mechanism of relieving the stress introduced by the hydrophobic mismatch is confirmed.

The tilting of the WALP-peptides can be computed by determining the angle between the helical axis and the normal to the membrane, where an angle of 0° coincides with a perfect transmembrane alignment. Because all membranes are located in the xy-plane, the membrane normal is chosen to be the z-axis of the system, which is accurate enough as long as the local oscillations of the membrane remain small. The helical axis is computed using the same terminal residues as used for the computation of the hydrophobic length of the peptide. This definition is expected to be accurate for α-helical peptides, such as WALP-peptides. In Figure 10 the tilt angle of the WALP-peptides is also shown, again as a function of the hydrophobic mismatch (squares, right axis). For the peptides with a negative hydrophobic mismatch the tilting varies between 15° and 30°, which can be assumed to be due to diffusion-driven stochastic movements. When the hydrophobic mismatch increases the peptide tilt also increases gradually, up to a tilt of approximately 70°. The tilt angles computed for the WALP-peptides of intermediate length (WALP19, 21, 23, 25) correspond well with previously reported tilt angles from other simulations [37,79–81]. However, experimental tilt angles are much lower than those computed from MD simulations, due to an averaging effect caused by dynamic variations of the tilt and rotation of the peptide [81,82]. It can be shown that when applying the same averaging procedures, which occur within the experiments, to the MD simulations, the same tilt angles are recovered [82]. Thus, it is shown in the simulations that peptide tilting is a mechanism of relieving stress from a hydrophobic mismatched membrane.

One of the most interesting observations that can be made from Figure 10 is that the membrane adaptation occurs at a lower hydrophobic mismatch than the peptide tilting. Moreover, it seems that first the membrane adaptation is used to decrease tension from the hydrophobic mismatch, followed by tilting of the peptides when the membrane can no longer adapt its thickness easily. Although both mechanisms for stress relaxation have been proven previously by others [79], from the current work it can be seen that both mechanisms occur sequentially and not in parallel, with the membrane adaptation coming first.

Although the membrane adaptation has been shown to be the first step in dealing with the introduced stress in the membrane due to the hydrophobic mismatch, this could be an artifact caused by the force field when the lipids are not modeled rigid enough. Therefore, four different WALP-peptides (WALP19, 23, 27 and 31) are put in an H4T3T3-membrane where the force constant belonging to the angles in the lipid tails is doubled (from 5.41 to 10.82 kJ/mol/rad^2). By choosing these peptides and this membrane a range for the hydrophobic mismatch from 0 to 1.8 nm is used for further investigation. The consequence of the increased force constant is that the lipids become much more rigid and, hence, deformations and fluctuations in the membrane are less likely to occur, and, also, the membrane thickens.

In Figure 11(a) the results for the membrane adaptation and peptide tilt angle as a function of the hydrophobic mismatch for these four WALP-peptides are shown. In this figure also the progression of the membrane adaptation and peptide tilt angle for the original simulations is indicated using the (dashed) lines. From this figure it can be seen that the peptide tilt (indicated by the squares) is not significantly different, and also the membrane adaptation is in good agreement with the previous simulations. Only in the case of the higher hydrophobic mismatch the membrane adaptation shows some deviation. However, although the found value lies outside the bandwidth region (plus or minus the standard deviation), from Figure 10 it can be seen that at similar hydrophobic mismatches membrane adaptations ranging from 0.1 to 0.2 nm are found, and, thus, the value for the stiffer membrane is believed to be at least similar to the case with the normal membrane. Consequently, the stiffer membrane has no apparent effect on the order of occurrence of the two mechanisms to reduce stress in the membrane due to peptide insertion, and, thus, these mechanisms are thought to be a genuine property of peptide-lipid interactions.

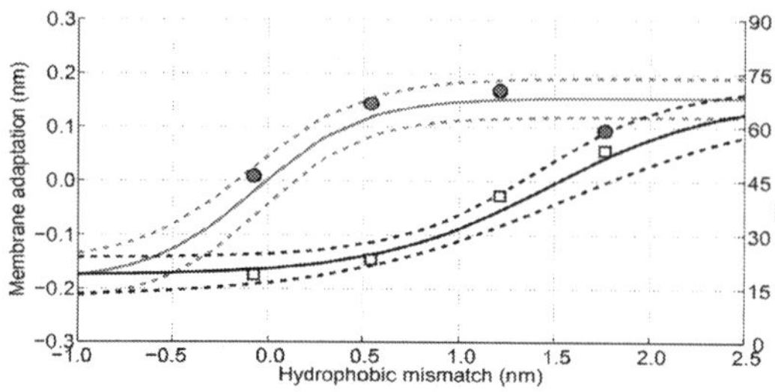

(a) Stiffer *vs.* original membrane

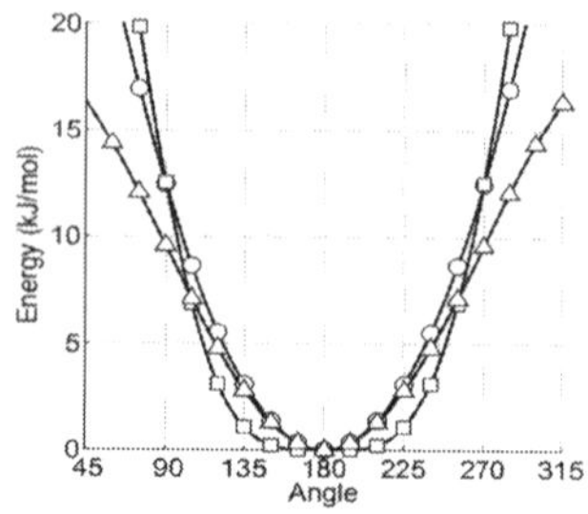

(b) Angle potentials

Figure 11: In (a) the data shown in Figure 10 is repeated using only the sigmoid curves (solid lines). The dotted lines are used to indicate the bandwidth (plus or minus the standard deviation) for both the membrane adaptation (left axis) and the peptide tilt (right axis) as a function of the hydrophobic mismatch. The circles and squares are the values computed from the stiffer membrane. In (b) three lipid tail angle potentials from different CG models are shown (a square denotes Marrink's model, a triangle Klein's, and a circle the current model).

Comparing the angle potential for the lipid tails (with the original force parameter of 5.41 kJ/mol/rad^2) with other force fields, such as Marrink's or Klein's [20,24], it can be seen that all of these are in good agreement, see Figure 11(b). Therefore, changing the force constant by as much as 100% does not seem plausible, and confidence in the observed two-step mechanism in dealing with the hydrophobic mismatch is strengthened once more.

In this case study the newly developed CG model for transmembrane proteins showed that it is applicable to investigate model peptide systems, such as WALP-peptides. Moreover, the model showed that when dealing with a positive hydrophobic mismatch, the peptide-membrane system first thickens the membrane around the peptide before the peptide is tilted in order to reduce the hydrophobic mismatch.

CONCLUSIONS

In this paper the first version of a CG protein model aimed at transmembrane α-helical proteins is described. The main philosophy of coarse graining in this model is based on existing models, for instance the size and mass of the CG particles, but also new interaction potentials are introduced, for example the double angle or the Lennard-Jones H-bond potential. Compared to other CG protein models the main difference with the presented model lies in the number of CG particles used for each side chain particle, with the omission of torsional interactions, and the introduction of new potentials. This last step is a different approach to CG modeling, but in our view a route worthwhile following.

Before showing the applicability of the model, first a performance analysis is conducted, in order to show that the model behaves as expected. To that end it is shown that for each of the CG sidechain particles its preferred location in a lipid membrane is consistent with previously published results, showing that the non-bonded interactions that govern this preference are correctly modeled.

Furthermore, five very different protein complexes have been investigated using MD simulations for several tens of nanoseconds. In general the $CRMS_{\alpha}$ for these proteins are within an acceptable range, indicating the CG model performs well. A closer look at the radius of gyration of the protein complexes confirms this.

For each of the five protein complexes the bond and angle distributions are compared to all-atom MD simulations. Besides the intrinsic property of coarse graining to give rise to softer potentials compared to all-atom potentials, which leads to broader distributions, both all-atom and CG distributions agree very well. However, the possible H-bond forming pairs in both the all-atom and CG representations of two out of five protein complexes shows that the CG model strongly prefers an H-bond distance of 0.61 nm, even for preferred H-bond distances in the all-atom model of around 0.5 nm. The investigation of the performance of the CG model shows that the new model works reasonably well for proteins that are rich in α-helices, especially when keeping in mind that the aim has been the development of a CG model for transmembrane α-helical proteins.

The applicability of the model is shown by a case study for WALP-peptides. The WALP-peptides are used both in experiments and theory as a model system for the interaction between transmembrane peptides and lipid membranes. Using the CG protein model with MD simulations on WALP-peptides of different length and embedded in lipid membranes of different thickness, it is shown that the apparent hydrophobic mismatch between peptide and membrane can be resolved by two mechanisms. In the first mechanism the membrane adapts its thickness to accommodate the peptide in its transmembrane orientation. However, when it is no longer feasible for the membrane to adapt its thickness, the second mechanism forces the peptide to tilt with respect to the membrane normal. It is observed that both mechanisms occur sequentially and not in parallel.

In the future the currently presented first version of the CG protein model needs to be further improved, for instance with respect to protein aggregation, because a possible imbalance in the cross-type non-bonded interaction parameters could cause unwanted protein aggregation behavior. Furthermore, the H-bonding aspects of the current model could probably be improved by adding directionality to the Lennard–Jones H-bond potential, for instance by looking at the angle between H-bond forming pairs.

REFERENCES

1. Lodish, H; Berk, A; Zipursky, SL; Matsudaira, P; Baltimore, D; Darnell, JE. Molecular Cell Biology, 4th ed; W. H. Freeman & Co: New York, NY, USA, 1999.
2. Peleg, G; Ghanouni, P; Kobilka, BK; Zare, RN. Single-molecule spectroscopy of the β2 adrenergic receptor: Observation of conformational substates in a membrane protein. Proc. Natl. Acad. Sci. USA 2001, 98, 8469–8474.
3. Kobilka, BK. G protein coupled receptor structure and activation. Biochim. Biophys. Acta 2007, 1768, 794–807.
4. Kobilka, BK. Agonist-induced conformational changes in the β2 adrenergic receptor. J. Peptide Res 2002, 60, 317–321.
5. Vaidehi, N; Floriano, WB; Trabanino, R; Hall, SE; Freddolino, P; Choi, EJ; Zamanakos, G; Goddard, WA, III. Prediction of structure and function of G protein-coupled receptors. Proc. Natl. Acad. Sci. USA 2002, 99, 12622–12627.
6. Freddolino, PL; Kalani, MYS; Vaidehi, N; Floriano, WB; Hall, SE; Trabanino, RJ; Kam, VWT; Goddard, WA, III. Predicted 3D structure for the human β2 adrenergic receptor and its binding site for agonists and antagonists. Proc. Natl. Acad. Sci. USA 2004, 101, 2736–2741.

7. Kalani, MYS; Vaidehi, N; Freddolino, PL; Hall, SE; Trabanino, RJ; Floriano, WB; Kam, VWT; Kalani, MA; Goddard, WA, III. The predicted 3D structure of the human D2 dopamine receptor and the binding site and binding affinities for agonists and antagonists. Proc. Natl. Acad. Sci. USA 2004, 101, 3815–3820.
8. Spijker, P; Vaidehi, N; Freddolino, PL; Hilbers, PAJ; Goddard, WA, III. Dynamic behavior of fully solvated β2-adrenergic receptor, embedded in the membrane with bound agonist or antagonist. Proc. Natl. Acad. Sci. USA 2006, 103, 4882–4887.
9. Li, Y; Zhu, F; Vaidehi, N; Goddard, WA, III; Sheinerman, F; Reiling, S; Morize, I; Mu, L; Harris, K; Ardati, A; Laoui, A. Prediction of the 3D structure and dynamics of human DP G-protein coupled receptor bound to an agonist and an antagonist. J. Am. Chem. Soc 2007, 129, 10720–10731.
10. Soliman, W; Bhattacharjee, S; Kaur, K. Molecular dynamics simulation study of interaction between a class IIa bacteriocin and its immunity protein. Biochim. Biophys. Acta 2007, 1774, 1002–1013.
11. Heo, J; Han, SK; Vaidehi, N; Wendel, J; Kekenes-Huskey, P; Goddard, WA, III. Prediction of the 3D structure of FMRF-amide neuropeptides bound to the mouse MrgC11 GPCR and experimental validation. Chem. Biol. Chem 2007, 8, 1527–1539.
12. Smit, B; Hilbers, PAJ; Esselink, K; Rupert, LAM; van Os, NM; Schlijper, AG. Computer simulations of a water/oil interface in the presence of micelles. Nature 1990, 348, 624–625.
13. Bolhuis, PG; Louis, AA; Hansen, JP; Meijer, EJ. Accurate effective pair potentials for polymer solutions. J. Chem. Phys2001, 114, 4296–4311.
14. Müller-Plathe, F. Coarse-graining in polymer simulation: From the atomistic to the mesoscopic scale and back. Chem. Phys. Chem 2002, 3, 754–769.
15. Nielsen, SO; Lopez, CF; Srinivas, G; Klein, ML. A coarse grain model for n-alkanes parameterized from surface tension data. J. Phys. Chem. B 2003, 119, 7043–7049.
16. Reith, D; Pütz, M; Müller-Plathe, F. Deriving effective mesoscale potentials from atomistic simulations. J. Comp. Chem2003, 24,

1624–1636.

17. Srinivas, G; Shelley, JC; Nielsen, SO; Discher, DE; Klein, ML. Simulation of diblock copolymer self-assembly, using a coarse-grain model. J. Phys. Chem. B 2004, 108, 8153–8160.
18. Li, X; Ma, X; Huang, L; Liang, H. Developing coarse-grained force fiels for cis-poly(1,4-butadiene) from the atomistic simulation. Polymer 2005, 46, 6507–6512.
19. Li, X; Kou, D; Rao, S; Liang, H. Developing a coarse-grained force field for the diblock copolymer poly(styrene-b-butadiene) from atomistic simulation. J. Chem. Phys 2006, 124, 204909.
20. Shelley, JC; Shelley, MY; Reeder, RC; Bandyopadhyay, S; Klein, ML. A coarse grain model for phospholipid simulations. J. Phys. Chem. B 2001, 105, 4464–4470.
21. Lopez, CF; Moore, PB; Shelley, JC; Shelley, MY; Klein, ML. Computer simulation studies of biomembranes using a coarse grain model. Comp Phys Comm 2002, 147.
22. Marrink, SJ; de Vries, AH; Mark, AE. Coarse grained model for semiquantitative lipid simulations. J. Phys. Chem. B2004, 108, 750–760.
23. Markvoort, AJ; Pieterse, K; Steijaert, MN; Spijker, P; Hilbers, PAJ. The bilayer-vesicle transition is entropy-driven. J. Phys. Chem. B 2005, 109, 22649–22654.
24. Marrink, SJ; Risselada, HJ; Yefimov, S; Tieleman, DP; de Vries, AH. The MARTINI force field: Coarse grained model for biomolecular simulations. J. Phys. Chem. B 2007, 111, 7812–7824.
25. Smeijers, AF; Markvoort, AJ; Pieterse, K; Hilbers, PAJ. A detailed look at vesicle fusion. J. Phys. Chem. B 2006, 110, 13212–13219.
26. Smeijers, AF; Pieterse, K; Markvoort, AJ; Hilbers, PAJ. Coarse-grained transmembrane proteins: Hydrophobic matching, aggregation, and their effect on fusion. J. Phys. Chem. B 2006, 110, 13614–13623.
27. Markvoort, AJ; Smeijers, AF; Pieterse, K; van Santen, RA; Hilbers, PAJ. Lipid-based mechanisms for vesicle fission. J. Phys. Chem. B 2007, 111, 5719–5725.
28. Markvoort, AJ; van Santen, RA; Hilbers, PAJ. Vesicle shapes from

molecular dynamics simulations. J. Phys. Chem. B2006, 110, 22780–22785.

29. Markvoort, AJ; Spijker, P; Smeijers, AF; Pieterse, K; van Santen, RA; Hilbers, PAJ. Vesicle deformation by draining: Geometrical and topological shape changes. J. Phys. Chem. B 2009, 113, 8731–8737.
30. Liwo, A; Arlukowicz, P; Oldziej, S; Czaplewski, C; Makowski, M; Scheraga, HA. Optimization of the UNRES force field by hierarchical design of the protein-energy landscape. 1. Tests of the approach using simple lattice protein models. J. Phys. Chem. B 2004, 108, 16918–16933.
31. Maupetit, J; Tuffery, P; Derreumaux, P. A coarse-grained protein force field for folding and structure prediction. Proteins 2007, 69, 394–408.
32. Zhou, J; Thorpe, IF; Izvekov, S; Voth, GA. Coarse-grained peptide modeling using a systematic multiscale approach. Biophys. J 2007, 92, 4289–4303.
33. Monticelli, L; Kandasamy, SK; Periole, X; Larson, RG; Tieleman, DP; Marrink, SJ. The MARTINI coarse grained force field: extension to proteins. J. Chem. Theory Comput 2008, 4, 819–834.
34. Arkhipov, A; Freddolino, PL; Schulten, K. Stability and dynamics of virus capsids described by coarse-grained modeling. Structure 2006, 14, 1767–1777.
35. Gibbons, MM; Klug, WS. Mechanical modeling of viral capsids. J. Matter. Sci 2007, 42, 8995–9004.
36. Shih, AY; Arkhipov, A; Freddolino, PL; Schulten, K. Coarse grained protein-lipid model with application to lipoprotein particles. J. Phys. Chem. B 2006, 110, 3674–3684.
37. Bond, PJ; Holyoake, J; Ivetac, A; Khalid, S; Sansom, MSP. Coarse-grained molecular dynamics simulations of membrane proteins and peptides. J. Struct. Biol 2007, 157, 593–605.
38. Chothia, C. Structural invariants in protein folding. Nature 1975, 254, 304–308.
39. Levitt, M. A simplified representation of protein conformations for rapid simulation of protein folding. J. Mol. Biol1976, 104, 59–107.
40. Liwo, A; Oldziej, S; Pincus, MR; Wawak, RJ; Rackovsky, S;

Scheraga, HA. A united-residue force field for off-lattice protein-structure simulations. I. Functional forms and parameters of long-range side-chain interaction potentials from protein crystal data. J. Comp. Chem 1997, 18, 849–873.

41. Saam, J; Tajkhorshid, E; Hayashi, S; Schulten, K. Molecular dynamics investigation of primar photoinduced events in the activation of rhodopsin. Biophys. J 2002, 83, 3097–3112.
42. Wallqvist, A; Ullner, M. A simplified amino acid potential for use in structure predictions of proteins. Proteins 1994, 18, 267–280.
43. Liwo, A; Pincus, MR; Wawak, RJ; Rackovsky, S; Oldziej, S; Scheraga, HA. A united-residue force field for off-lattice protein-structure simulations. II. Parameterization of short-range interactions and determination of weights of energy terms by Z-score optimization. J. Comp. Chem 1997, 18, 874–887.
44. Leach, AR. Molecular Modelling - Principles and Applications, 2nd Ed ed; Prentice Hall: Harlow, England, 2001.
45. Faller, R. Automatic coarse graining of polymers. Polymer 2004, 45, 3869–3876.
46. Weeks, JD; Chandler, D; Andersen, HC. Role of repulsive forces in determining the equilibrium structure of simple liquids. J. Chem. Phys 1971, 54, 5237–5247.
47. Ulmschneider, MB; Sansom, MSP; Di Nola, A. Properties of integral membrane protein structures: Derivation of an implicit membrane potential. Proteins 2005, 59, 252–265.
48. Cieplak, M; Hoang, TX. Coarse grain description of protein folding. Phys. Rev. E 1998, 58, 3589–3596.
49. Oldziej, S; Liwo, A; Czaplewski, C; Pillardy, J; Scheraga, HA. Optimization of the UNRES force field by hierarchical design of the protein-energy landscape. 2. Off-lattice tests of the method with single proteins. J. Phys. Chem. B 2004, 108, 16934–16949.
50. Izvekov, S; Voth, GA. A multiscale coarse-graining method for biomolecular systems. J. Phys. Chem. B 2005, 109, 2469–2473.
51. Izvekov, S; Voth, GA. Multiscale coarse graining of liquid-state systems. J Chem Phys 2005, 123, 134105:1–134105:13.
52. Ercolessi, F; Adams, JB. Interatomic potentials from first-principles

calculations: The force-matching method. Europhys. Lett 1994, 26, 583–588.

53. Chu, JW; Izvekov, S; Voth, GA. The multiscale challenge for biomolecular systems: Coarse-grained modeling. Mol. Simul 2006, 32, 211–218.
54. Bond, PJ; Sansom, MSP. Insertion and assembly of membrane proteins via simulation. J. Am. Chem. Soc 2006, 128, 2697–2704.
55. Bond, PJ; Sansom, MSP. Bilayer deformation by the Kv channel voltage sensor domain revealed by self-assembly simulations. Proc. Natl. Acad. Sci. USA 2007, 104, 2631–2636.
56. Carpenter, T; Bond, PJ; Khalid, S; Sansom, MSP. Self-assembly of a simple membrane protein: Coarse-grained molecular dynamics simulations of the influenza M2 channel. Biophys. J 2008, 95, 3790–3801.
57. Han, W; Wu, YD. Coarse-grained protein model coupled with a coarse-grained water model: Molecular dynamics study of polyalanine-based peptides. J. Chem. Theory Comput 2007, 3, 2146–2161.
58. Periole, X; Huber, T; Marrink, SJ; Sakmar, TP. G protein-coupled receptors self-assemble in dynamics simulations of model bilayers. J. Am. Chem. Soc 2007, 129, 10126–10132.
59. Bond, PJ; Wee, CL; Sansom, MSP. Coarse-grained molecular dynamics simulations of the energetics of helix insertion into a lipid bilayer. Biochemistry 2008, 47, 11321–11331.
60. MacCallum, JL; Bennet, WFD; Tieleman, DP. Distribution of amino acids in a lipid bilayer from computer Simulations. Biophys. J 2008, 94, 3393–3404.
61. Goldstein, H; Poole, CP; Safko, JL. Classical Mechanics, 3rd ed; Pearse Educational International: Upper Saddle River, NJ, USA, 2002.
62. Rinia, HA; Kik, RA; Demel, RA; Snel, MME; Killian, JA; van der Eerden, JPJM; de Kruiff, B. Visualization of highly ordered striated domains induced by transmembran peptides in supported phophatidylcholine bilayers. Biochemistry2000, 39, 5852–5858.
63. de Planque, MRR; Goormaghtigh, E; Greathouse, DV;

Koeppe, RE, II; Kruijtzer, JAW; Liskamp, RMJ; de Kruijff, B; Killian, JA. Sensitivity of single membrane-spanning α-helical peptides to hydrophobic mismatch with a lipid bilayer: Effects on backbone structure, orientation, and extent of membrane incorporation. Biochemistry 2001, 40, 5000–5010.

64. de Planque, MRR; Greathouse, DV; Koeppe, RE, II; Schäfer, H; Marsh, D; Killian, JA. Influence of lipid/peptide hydrophobic mismatch on the thicknessof diacylohosphatidylcholine bilayers. A ^{2}H NMR and ESR study using designed transmembrane α-helical peptides and gramicidin A. Biochemistry 1998, 37, 9333–9345.
65. Killian, JA. Synthetic peptides as models for intrinsic membrane proteins. FEBS Lett 2003, 555, 134–138.
66. Strandberg, E; Özdirekcan, S; Rijkers, DTS; van der Wel, PCA; Koeppe, RE, II; Liskamp, RMJ; Killian, JA. Tilt angles of transmembrane model peptides in oriented and non-oriented lipid bilayers as determined by ^{2}H solid-state NMR. Biophys. J 2004, 86, 3709–3721.
67. Nymeyer, H; Woolf, TB; Garcia, AE. Folding is not required for bilayer insertion: Replica exchange simulations of an α-helical peptide with an explicit bilayer. Proteins: Struct. Funct. Bioinf 2005, 59, 783–790.
68. Strandberg, E; Morein, S; Rijkers, DTS; Liskamp, RMJ; van der Wel, PCA; Killian, JA. Lipid dependence of membrane anchoring properties and snorkeling behavior of aromatic and charged residues in transmembrane peptides. Biochemistry 2002, 41, 7190–7198.
69. Siegel, DP; Cherezov, V; Greathouse, DV; Koeppe, RE, II; Killian, JA; Caffrey, M. Transmembrane peptides stabilize inverted cubic phases in a biphasic length-dependent manner: Implications for protein-induced membrane fusion. Biophys. J 2006, 90, 200–211.
70. Morein, S; Koeppe, RE, II; Lindblom, G; de Kruiff, B; Killian, JA. The effect of peptide/lipid hydrophobic mismatch on the phase behavior of model membranes mimicking the lipid composition in Escherichia coli membranes. Biophys. J2000, 78, 2475–2485.
71. Ramamoorthy, A; Kandasamy, SK; Lee, DK; Kidambi, S; Larson, RG. Structure, topology, and tilt of cell-signaling peptides containing nuclear localization sequences in membrane bilayers

determined by solid-state NMR and molecular dynamics simulation studies. Biochemistry 2007, 46, 965–975.

72. van der Wel, PCA; Reed, ND; Greathouse, DV; Koeppe, RE, II. Orientation and motion of tryptophan interfacial anchors in membrane-spanning peptides. Biochemistry 2007, 46, 7514–7524.
73. Lemaître, V; de Planque, MRR; Howes, AP; Smith, ME; Dupree, R; Watts, A. Solid state ^{17}O NMR as a probe for structural studies of proteins in biomembranes. J. Am. Chem. Soc 2004, 126, 15320–15321.
74. Petrache, HI; Zuckerman, DM; Sachs, JN; Killian, JA; Koeppe, RE, II; Woolf, TB. Hydrophobic matching mechanism investigated by molecular dynamics simulations. Langmuir 2002, 18, 1340–1351.
75. Im, W; Brooks, CL, III. Interfacial folding and membrane insertion of designed peptides studied by molecular dynamics simulations. Proc. Natl. Acad. Sci. USA 2005, 102, 6771–6776.
76. Estaban-Martín, S; Salgado, J. Self-assembling of peptide/membrane complexes by atomistic molecular dynamics simulations. Biophys. J 2007, 92, 903–912.
77. Ulmschneider, MB; Ulmschneider, JP. Folding peptides into lipid bilayer membranes. J. Chem. Theory Comput 2008, 4, 1807–1809.
78. Morein, S; Killian, JA; Sperotto, MM. Characterization of the thermotropic behavior and lateral organization of lipid-peptide mixtures by a combined experimental and theoretical approach: Effects of hydrophobic mismatch and role of flanking residues. Biophys. J 2002, 82, 1405–1417.
79. Kandasamy, SK; Larson, RG. Molecular dynamics simulations of model trans-membrane peptides in lipid bilayers: A systematic investigation of hydrophobic mismatch. Biophys. J 2006, 90, 2326–2343.
80. Sengupta, D; Meinhold, L; Langosch, D; Ullmann, GM; Smith, JC. Understanding the energetics of helical peptide orientation in membranes. Proteins: Struct. Funct. Bioinf 2005, 58, 913–922.
81. Estaban-Martín, S; Salgado, J. The dynamic orientation of membrane-bound peptides: Bridging simulations and experiments. Biophys. J 2007, 93, 4278–4288.

82. Özdirekcan, S; Etchebest, C; Killian, JA; Fuchs, PFJ. On the orientation of a designed transmembrane peptide: Toward the right tilt angle? J. Am. Chem. Soc 2007, 129, 15174–15181.

CHAPTER 6

TRANSMEMBRANE HELIX DYNAMICS OF BACTERIAL CHEMORECEPTORS SUPPORTS A PISTON MODEL OF SIGNALLING

Benjamin A. Hall, Judith P. Armitage, Mark S. P. Sansom

Oxford Centre for Integrative Systems Biology, Department of Biochemistry, University of Oxford, Oxford, United Kingdom

ABSTRACT

Transmembrane α-helices play a key role in many receptors, transmitting a signal from one side to the other of the lipid bilayer membrane. Bacterial chemoreceptors are one of the best studied such systems, with a wealth of biophysical and mutational data indicating a key role for the TM2 helix in signalling. In particular, aromatic (Trp and Tyr) and basic (Arg) residues help to lock α-helices into a membrane. Mutants in TM2 of *E. coli* Tar and related chemoreceptors involving these residues implicate changes in helix location and/or orientation in signalling. We have investigated the detailed *structural* basis of this via high throughput coarse-grained molecular dynamics (CG-MD) of Tar TM2 and its mutants in lipid bilayers. We focus on the position (shift) and orientation (tilt, rotation) of TM2 relative to the bilayer and how these are perturbed in mutants relative to the wildtype. The simulations reveal a clear correlation between small (ca. 1.5 Å) shift in position of TM2 along the bilayer normal and downstream changes in signalling activity. Weaker correlations are seen with helix tilt, and little/none between signalling and helix twist. This analysis of relatively subtle changes was only possible because the high throughput simulation method allowed us to run large (n=100) ensembles for substantial numbers of different helix sequences, amounting to ca. 2000 simulations in total. Overall, this analysis supports a swinging-piston model of transmembrane signalling by Tar and related chemoreceptors.

INTRODUCTION

Signalling by transmembrane (TM) receptor proteins is central to the biology of both bacterial and eukaryotic cells. For many receptors (other than the GPCRs) the TM domain is relatively simple, consisting of just one or two TM helices per protein subunit. In these receptors changes in TM helix position and orientation relative to the membrane lipid bilayer almost certainly play a key role in the mechanism of signalling across the membrane (for a recent review see [1]). However, because of the difficulties of determining high resolution structures of intact receptor proteins, our understanding of such mechanisms is incomplete. By combining functional data with computational biophysical descriptions of receptor TM helix interactions with membranes we can better understand these complex membrane systems.

Bacterial chemotaxis receptors (chemoreceptors) provide a well-studied example of such systems. Structural studies of *E. coli* chemoreceptors [2], [3], [4], [5], [6] have shown that e.g. the Tar receptor consists of an extracellular (periplasmic) receptor domain and an α-helical TM domain. The latter is linked via a HAMP domain to a long intracellular coiled-coil domain (Fig. 1AB) (HAMP is a linker domain present in Histidine kinases, Adenyl cyclases, Methyl-accepting proteins and Phosphatases which is found in bacterial sensor and chemotaxis proteins and in eukaryotic histidine kinases.) The receptor is dimeric, and the homodimers seem to be arranged in higher order structures based on a hetero-trimer of homodimers unit [6]. Signalling occurs via ligand binding to the extracellular receptor leading to a change in interactions of the intracellular coiled-coil domain with CheW (an adaptor protein) and CheA (a histidine kinase). For example, in Tar the binding of a chemoattractant (aspartate) to the receptor is thought to causes a conformational change which is propagated through the transmembrane (TM) domain to the signal processing HAMP domain. This in turn cause further changes which propagate from individual dimers, altering the packing and orientation in trimer of dimers structures (as observed by cryoelectron microscopy tomography, [6]), and from there to a wider, mostly hexagonal array consisting of MCPs, histidine kinase CheAs and adaptor CheW proteins [7]. This "conformational spread" of information from individual receptors through an array creates a highly sensitive amplification system with a high signal to noise ratio [8], [9].

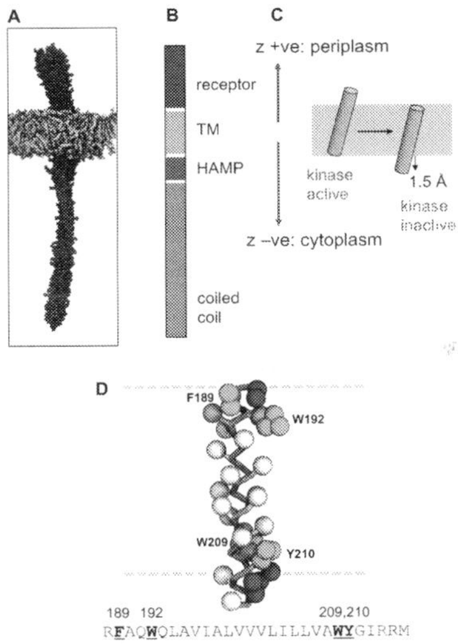

Figure 1: The Tar chemoreceptor and its TM2 helix. **A** Structural model of a chemoreceptor (Tsr) dimer (the two subunits in red and blue) in a phospholipid bilayer (grey). **B** Schematic of the domains of the Tar receptor, showing the periplasmic receptor (blue), transmembrane (TM; green), and cytoplasmic HAMP (red) and coiled-coil (grey) domains. **C** Illustration of the piston model of signalling via displacement of the TM helix (green) relative to the lipid bilayer (grey). The z-axis convention used in subsequent figures is defined, with a positive displacement corresponding to movement in the periplasmic direction and a negative displacement to movement towards the cytoplasm. **D** The Tar

TM2 helix structure in coarse-grained representation. Apolar residue sidechains are represented as white spheres, polar sidechains as cyan spheres, aromatic sidechains as green spheres, and basic sidechains as blue spheres. The peptide backbone is drawn as grey bonds. The approximate location of the lipid bilayer is indicated in by the broken horizontal lines.

The nature of the conformation changes in the receptor in response to ligand binding events are not clearly understood, despite a wealth of structural and genetic information. Attractant binding at the receptor domain causes motions in the second transmembrane helix, TM2, which links the receptor domain to the HAMP domain. The coiled coil region is able to undergo other internal dynamic changes on activation of the receptor. Together, these conformational changes cause a reduction in the activity of the histidine kinase CheA, which alters downstream signalling events [10], [11].

The dynamic behaviour of the TM2 helix in signalling is the subject of on-going debate. Studies of aspartate chemoreceptor Tar suggest that on binding of attractant to the receptor domain TM2 undergoes a 1.5 Å piston-like motion in the membrane, towards the cytoplasm. This was initially proposed from lock-on/lock-off disulphide mutants [12] and identified from the crystal structures of the Tar apo and ligand bound forms of the periplasmic domain [13]. It was subsequently supported by solid state NMR measurements, spin labelling measurements, and data from mutagenesis studies [14], [15], [16]. However, a rotational mechanism has also been invoked on the basis of studies of the HAMP domain [4], [6]. In order to dissect out the elements of the transbilayer signalling mechanism of Tar and related chemoreceptors, it is important to establish how TM2 interacts with a lipid bilayer membrane, and how mutations in TM2, which perturb signalling, alter such interactions. This behaviour has also been studied extensively in the minor chemoreceptor Trg, using genetic and simulation techniques [17], [18], [19], [20], revealing the conservation of the piston model of signalling across related proteins.

Central to a piston-like model (Fig. 1C) is the question of how the TM2 helix is positioned/oriented relative to a lipid bilayer and how changes in this position/orientation may be used to signal across a membrane. This is related to the more general question of what factors 'lock' a TM helix into position in a bilayer. This has been extensively studied using synthetic model TM helices [21], [22], [23], [24], [25], with an emphasis on the roles of tryptophan (W) and arginine (R) sidechains, mirroring their importance in the structures of more complex membrane proteins and their interactions with lipids [26], [27], [28], [29]. Alongside experimental approaches, molecular dynamics simulations of these helices have provided insights into folding processes in membranes and the precise interactions made between individual residues and the membrane [25], [30], [31], [32], [33], [34]. Typically such simulations employ full atomistic detail, which is CPU intensive and therefore limits the statistical significance of the observed behaviour. In contrast, coarse grain (CG) simulations by reducing the level of detail in a system (by a factor of roughly 4:1 in the case of MARTINI [35], [36] and related CG models [37], [38], [39]) yield an effective increase in computational speed of between 1 and 2 orders of magnitude. A number of studies on synthetic peptides have shown the CG-MD can accurately reproduce the experimentally observed positions and orientations of such helices in lipid bilayers [25], [31], [32], [38], [40], [41]. To facilitate the running and analysis of large and multiple ensembles of CG-MD simulations, we have recently developed a high throughput approach [25], [42]. With this approach, we have shown that the improved sampling of helix/bilayer interactions made possible by such ensembles of simulations allows, for example, the helix/bilayer interactions of synthetic model peptides as seen in solid state NMR to be probed [25], [42].

Here we use comparative CG-MD simulations enabled by our high throughput simulation pipeline to understand the structural consequences of several sets of mutations in the TM2 of two chemoreceptors Tar, the high abundance aspartate receptor from *E.coli* and Salmonella, and Trg, the low abundance ribose chemoreceptor from *E.coli*. We build simulations including a roughly 30 residue helix representing TM2 in an explicit membrane with explicit solvent. The outcome of this analysis supports a piston-like model of TM signalling, whilst also highlighting a potential role for helix tilt in Tar.

RESULTS

Tar TM2 C-terminal aromatic scanning mutants

Aromatic sidechains (specifically tryptophan, W, and tyrosine, Y) have been shown to play a key role in anchoring TM helices in a given position in membranes, through H-bonding interactions to the phospholipid carbonyl oxygens in the interfacial region of the bilayer [23], [46]). Thus, by mutating/inserting aromatic residues it should be possible to modulate/ tune the location/behaviour of a TM helix in a membrane [30], [47]. This approach was applied in a set of studies [43], [44] in which the signalling behaviour of Tar from *E. coli* was modulated by performing an asymmetric scan of the cytoplasmic anchoring WY sidechain pair (W209/Y210; see Fig. 2A) in the TM2 helix. By shifting the WY pair up to 3 residues towards either the C-terminal or N-terminal end of the TM2 helix (WY+3 to WY−3 respectively in Fig. 2A), it was shown that downstream signalling and flagellar behaviour correlated with the position of the aromatic anchors.

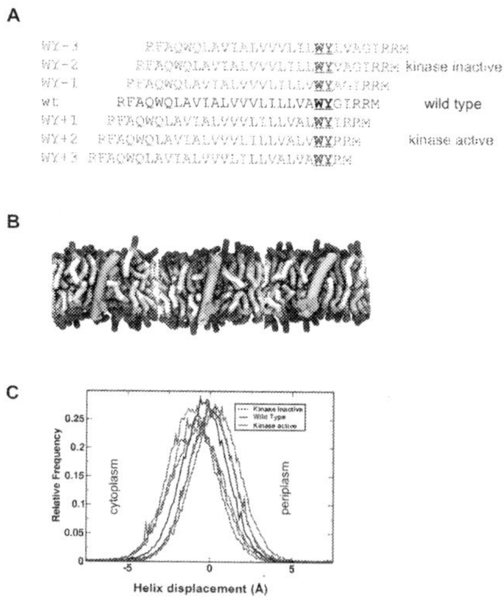

Figure 2: Tar TM2 C-terminal aromatic scanning mutants. **A** Sequences of WY-scanning mutants of the *E. coli* Tar TM2 helix are shown aligned to maintain the position of the C-terminal aromatic motif (WY; underlined). Three mutants with an inactive kinase (determined from flagellar rotation behaviour [44]) which are anticipated to have the TM2 helix translated towards the cytoplasm direction are in red, the wild type sequence in black, and three mu-

tants which are anticipated to translate the TM2 helix in a periplasmic direction are in blue. **B** Tar TM2 WY scanning mutants, show helix reorientation in the membrane. Snapshots of WY-3, wild type and WY+3 helices (left to right), demonstrating interactions of aromatic residues (in green) to the interfacial region. The phospholipid hydrophobic tails are in pale grey, and their phosphate groups in brown. The peptide helix is shown as a yellow tube with the aromatic sidechains in green. **C** Helix displacements with respect to the bilayer centre of the mutant TM2 helices. Kinase inactive mutants (red) are displaced towards the cytoplasm, and kinase active mutants (blue) are displaced towards the periplasm. Histograms shown are for all time points of all members of each ensemble (excluding rare interfacial events).

We mimicked these WY-scan experiments by performing simulations in which the TM2 helix was altered to correspond to the WY-scan mutants. Self-assembly simulations were then performed (ca. 100 per TM2 sequence) and the resultant helix/bilayer systems (Fig. 2B) analysed in terms of helix position and orientation relative to the phospholipid bilayer.

Analysis of the WY scans (Fig. 2C), reveals that the mutations induced changes in position, tilt and rotation of the transmembrane helix (Fig. 2C) relative to the lipid bilayer. Visualisation of the simulation trajectories suggests that the interactions of the WY pair with the lipid headgroups dictates the position and dynamics of the helix in the membrane. Both the wild type and mutants helices adopted a transmembrane orientation, as expected, and so statistical analysis of the simulation ensembles was required to dissect out the relatively subtle impacts of the WY-scan mutations on helix bilayer interactions. From such analysis (Fig. 2C) it was evident that the signalling activity of the mutants correlates quite clearly with changes in the position of the mutant helices relative to the wild type ensemble. In particular reduced kinase activity mutants result in a *cytoplasmic* shift of the TM2 helix in the bilayer relative to the wild type, whereas increased kinase activity mutants resulted in a *periplasmic* shift.

In addition to the correlation between signalling state and helix translation there was some evidence for changes in helix orientation (e.g. two of the reduced kinase activity mutants, WY−2, WY−3 appear to have an increased helix tilt angle relative to the bilayer normal). However no clear patterns emerged relating the activities of the mutants with helix rotation (about the helix axis) suggesting that such rotations may not play a role in signalling. It is noteworthy that the difference in position relative to the bilayer centre of the reduced and increased kinase activity mutant helix populations is ca. 1.5 Å (with a change in helix tilt angle of ca. 5°), which matches the crystallographically determined displacement of the connecting helix of Tar TM2 from a number of sources [12], [13], [15], [16].

Point mutations of the helix anchoring tryptophan residues (W192 and W209; Fig. 3A) to alanine in TM2 of Tar have also been studied [43]. Visualisation of simulations corresponding to these mutations reveals that the helices also remain transmembrane and do not undergo major changes in orientation. Quantitative analysis of single point mutations (i.e. W192A and W209A) showed in each case a ca. 2.5° increase in the tilt of the helix relative to the membrane, and an increase of ca. 5° for the double mutation (W192A, W209A; Fig. 3). These changes in tilt correlate with changes in signalling activity. In contrast, shifts relative to bilayer of these tryptophan to alanine mutations are minimal. This suggests that there may be a role for *both* shift and tilt of helices in the signal transduction process.

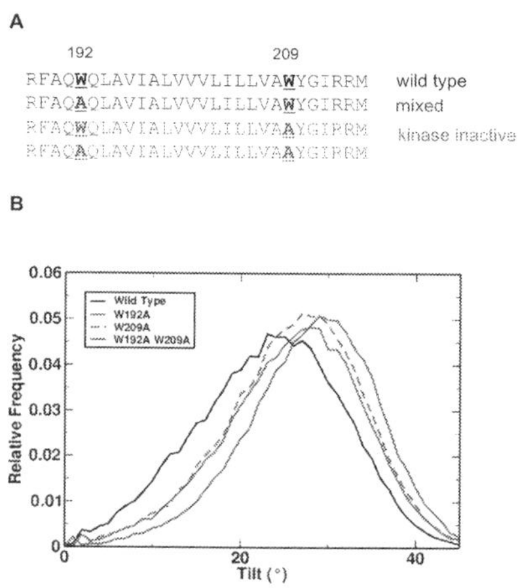

Figure 3: Position and orientation of Tar TM2 tryptophan to alanine mutations [43]. **A** Sequences and phenotypes of mutants. **B** Tilt distributions of mutant helices relative to the bilayer normal, revealing an additive effect of mutating individual tryptophan residues to alanine, with individual mutants causing a ~2.5° increase in tilt and the double mutant a ~5° increase.

Tar TM2 arginine scanning mutants

Arginine and lysine sidechains may also play important roles in anchoring transmembrane helices, via 'snorkelling' interactions with the phosphate moiety of lipid headgroups [24], [25], [47], [48], [49]. This interaction might be anticipated to be somewhat stronger than that of tryptophan and tyrosine. As such, replacement of anchoring aromatic residues with arginine residues would be expected to draw the helix towards one side of the bilayer. Furthermore, introduction of basic residues into the core of the hydrophobic helix may be expected to alter the rotation of the helix due to the strong barrier to placing cationic groups within the bilayer core [25]. Within this context, we simulated helices which correspond to two sets of studies in which basic residue were introduced into TM2. Thus, a site directed mutagenesis approach has been described in which arginine was introduced into the interfacial region of Tar from *Salmonella* [45], and a genetic screen of *E.coli* Trg identified two cases of a basic residue introduced into the bilayer core [20].

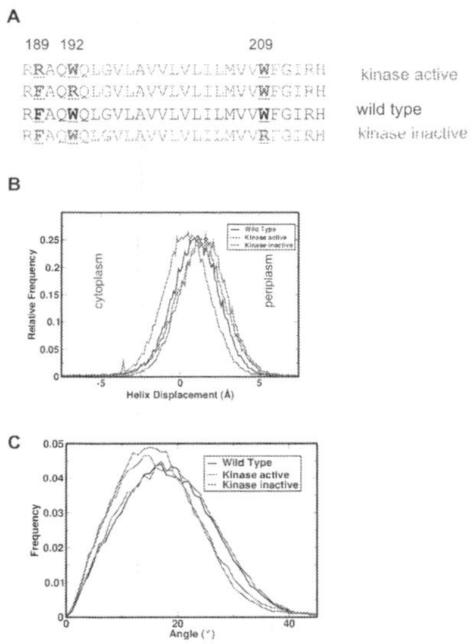

Figure 4: Tar TM2 arginine scanning mutants [45]. **A** Sequences and phenotypes of arginine scanning mutants of the *Salmonella typhimurium* Tar TM2 helix are shown with the inserted arginine residue underlined and aligned to maintain the position of the C-terminal aromatic motif (WY; underlined). Two mutants (F189R and W192R; blue) have an active kinase and are anticipated to have the TM2 helix translated towards the periplasm; the inactive kinase mutant

(W209R; red) is anticipated to translate the TM2 helix in a cytoplasmic direction. **B** Helix displacements with respect to the bilayer centre of the arginine mutant TM2 helices. The kinase inactive mutant (W209R; red) is displaced towards the cytoplasm, and the kinase active mutants (F189R and W189R; blue) are displaced towards the periplasm. Histograms shown are for all time points of all members of each ensemble (excluding rare interfacial events). **C** Helix tilt distribution relative to the bilayer normal, revealing a reduced tilt in the kinase active mutants (F189R and W189R; blue) and an unchanged tilt in the kinase inactive mutant (W209R; red).

Three key arginine-introducing mutations of Tar from *S. typhimurium* have been studied to exemplify reduced/increased kinase activity phenotypes: F189R and W192R (increased kinase activity) and W209R (reduced kinase activity; see Fig. 4). In the simulations, W209R modified-helices are displaced towards the cytoplasm relative to wild type, and have a minimally altered tilt angle in the membrane. In contrast, W192R and F189R both show slight displacement towards the periplasm, and show a reduced tilt angle, consistent with the observed changes in tryptophan to alanine point mutants (where increased inhibition is associated with increased tilt). Examination of rotation angles shows no clear link between rotation and kinase activity, though W192R has a distinctly reduced angular preference.

A genetic screen of Trg revealed three basic mutations in the bilayer core: T215K and T215R (increased kinase activity) and L216R (reduced kinase activity) [20]. Simulations of the corresponding TM2 helices (Fig. 4) indicate that in all three cases the helix is shifted towards the cytoplasm. Significantly, the reduced kinase activity mutant (L216R) shows an increased tilt in the membrane relative to wild type. Furthermore the two increased kinase activity mutants are rotated (i.e. a change in helix twist) by ca. 180°. This is significantly greater than the 26° rotation proposed to confer signals through the membrane [4]. It is likely that this extreme rotation change, coupled with the strong angular dependence in this specific pair of mutations, is leading to an increased kinase activity phenotype through reorienting the signalling helix to a state which is not accessible in the wild type. Additionally, this could also reflect novel interactions created within the transmembrane helix bundle, which cannot be explored in this single helix system.

DISCUSSION

In this study we have shown that a high throughput approach to CG-MD simulation of TM helices and their interactions with lipid bilayers allows the integration and interpretation of a large body of mutational data on bacterial chemoreceptors. Using simulations to examine a wide variety of signalling mutations from the past two decades of chemotaxis research [20], [43], [44], [45] we have found that the majority of the data are consistent with a swinging piston mechanism for signal transduction by Tar-like chemoreceptors. In particular, our results suggest that signalling through the membrane is dominated by translational motions of helix TM2, with some additional contribution from changes in helix tilt. This has been most clearly demonstrated for the WY scanning mutations of Tar from *E. coli* (Fig. 2), but is also supported by data from a range of additional TM2 mutations.

Alongside such mutational data, there exists a substantial body of knowledge on the role of aromatic and basic sidechains in determining the position, orientation and dynamics of synthetic model TM helices in lipid bilayer membranes [23], [50]. Many of the mutations described in this study relate directly to the alteration of the anchoring residues which are directly studied in these approaches, and highlight the importance of understanding interactions of peptides with the membrane through the comparison of TM helices in signalling proteins and synthetic peptides. Here again alongside experimental approaches, simulations of the helices have provided insights into folding processes in membranes and the precise interactions made between individual residues and the membrane [25], [30], [31], [40]. For example, the snorkelling behaviour of arginine in the membrane has been extensively studied in model peptides [24], [25] revealing that the effect of snorkelling by arginine on helix orientation is dependent on location of the arginine in the membrane. Taken together, these results suggest that MD simulations may be used not only to help establish the underlying principles of the interactions of transmembrane helices with their lipid bilayer environment, but also that simulations may be used to probe dynamic shifts in TM helix location and orientation which underlie signalling events in complex receptor systems.

The approach we have described has some limitations. In particular, we neglect possible roles of helix packing interactions within the TM domain and their possible influence on signalling via the TM2 helix. There are some indications in the literature for possible roles of TM1-TM2 and TM2-TM2 interactions. Substitution of a isoleucine with a phenylalanine in *E. coli* Tar TM2 can cause a significant change in receptor activation [51]. Additionally, in the TM2 helices of some chemoreceptors from *R. sphaeroides* there is a (small)XXX(small) helix dimerization sequence motif [52], suggestive of relatively strong interactions between the TM helices of this system. However, mutagenesis of the hydrophobic membrane core of Tar TM2 has revealed a tolerance to a wide range of different amino acid substitutions including large to small, small to large, and non-polar to polar [53]. Future studies will examine the interactions of the helices within the bundle to better define the nature of these interactions, and should include the activity lock mutations from the literature (such as those described in [12] and reviewed in [14]). Furthermore, future studies on larger structures may wish to take into account the restraining effects of the complete receptor on the position and orientation of the helix in the membrane.

Whilst our simulations have indicated that changes in position and, to a lesser extent, orientation of TM2 induced by mutants alter signalling, it remains uncertain how these local helix motions may alter the conformation of the chemoreceptor protein as a whole. In particular, future studies will need to include more complete models of the structure of the entirety of Tar in order to understand e.g. the effect of TM2 helix motions on the adjacent HAMP domain and its interactions [4].

This study highlights the value of simulation of the dynamic structural implications of TM helix mutants to aid interpretation of experimental data on signalling systems. Through the close examination of a wide variety of different datasets it is possible to discern similarities in effect of highly distinct modifications to the wild type signalling helices. Whilst we have developed this approach in the context of bacterial chemosensing, it could be readily adapted to e.g. simulation of helix dimerization by TM helices of mammalian receptor tyrosine kinases [54] for which there is considerable evidence for the role of TM helix mutations in disease [55].

METHODS

CG-MD simulations

Coarse grained molecular dynamics (CG-MD) simulations were performed using the MARTINI forcefield as described previously [36], [56], [57]. This employs an approximately 4:1 mapping of non-H atoms to CG particles was used. Inter-particle interactions were treated with Lennard Jones interactions between 4 classes of particles; polar (P), charged (Q), mixed polar/apolar (N) and hydrophobic apolar (C). These were then split into subtypes to reflect differing hydrogen bonding capabilities or polarity. In MARTINI P and C particle types were subdivided to reflect varying degrees of polarity. Interactions in each case were based on a lookup table, with 9 levels in MARTINI. Short range electrostatic interactions were treated Coulombically, shifted to zero between 0 and 12 Å. Lennard Jones interactions were shifted to zero between 9 and 12 Å. The integrity of the α-helices was maintained via dihedral restraints. The peptide termini were treated as uncharged.

Simulations were performed using Gromacs 3.3 (www.gromacs.org) [58]. Temperature was coupled using a Berendsen thermostat at 323 K (τ_T=1 ps), and pressure was coupled semi-isotropically (across XY/Z) at 1 bar (compressibility=3×10^{-5} bar−1, τ_P=10 ps). The initial simulation timestep was 20 fs.

Protein and lipids

Initial models of the TM α-helices were generated as ideal, atomistically detailed a-helices using standard backbone angles and side- chain conformers. These are then converted to coarse grained as described previously Around 128 DPPC molecules were used in each simulation along with around 3000 CG water particles, giving a final system size of $\sim65\times65\times13$ Å^3.

Mutants and phenotypes

Sequences for characterised mutations were taken from [35], [36], [37], [38],[20], [43], [44]. The first results generated experimentally for Tar were the arginine scanning mutations [45], followed by the tryptophan point mutations [43]. These were later extended to an asymmetric scan of the tryptophan/tyrosine pair at the cytoplasmic side of the membrane [44]. These were preceeded by the genetic screen of Trg which revealed a range of behaviours [20].

Single point and double interfacial tryptophan to alanine mutants (W192A, W209A and WAWA) in Tar from *E. coli* [35] reduced swarm expansion rates in aspartate, maltose and glycerol, with W209A (the cytoplasmic tryptophan) mutation with the most significant reduction in swarm expansion. W209A and WAWA mutants also were found to have reduced CheY phosphate levels, indicating a reduced CheA kinase activity, and an increased methylation state of the receptors (both in the presence and absence of aspartate and nickel).

Tryptophan/tyrosine scanning mutations of *E. coli* Tar [36] were characterised by measurement of flagellar rotation and reversal frequency, methylation states of the mutant receptors, sensitivity of the receptor to aspartate (through cell swarming behaviour and *in vitro* assays), and *in vitro* kinase activation. WY mutants show a correlation between position of the WY pair in the helix and bias in rotation direction; specifically WY mutations which reposition the pair towards the periplasm create a counter clockwise bias, and WY mutations which reposition the pair towards the cytoplasm create a clockwise bias. Furthermore, WY−3, WY+2 and WY+3 have a reduced mean reversal rate relative to the wild type, indicating that methylation may not be compensating for the degree of perturbation of the signalling state induced by the mutations. Deletion of CheR/CheB from the system (responsible for adaptive methylation) increases the flagellar rotational biasing effects of the mutants, indicating that without the methylation apparatus, the effects of the mutations could not be compensated for. WY−1, WY−2, WY−3 mutations all caused increased methylation, and WY+1, WY+2, WY+3 mutations reduced methylation. All mutants were found to be less responsive to aspartate through swarm assays, and WY+1, WY+2, and WY+3 were found to have higher Ki values *in vitro*. *In vitro* kinase assays (in the absence of CheR/CheB) showed that WY−3, WY−2, WY−1 had a strongly reduced kinase activity, and WY+2 and WY+3 unexpectedly had a slightly reduced kinase activity.

Arginine mutations of *S. typhimurium* Tar [37] were characterised in terms of kinase and methylation activity using *in vitro* assays. F189R and W192R were found to have increased kinase activity and reduced methylation activity, whilst W209R was found to have reduced kinase activity and increased methylation activity.

Mutations of *E. coli* Trg [38] were identified using a genetic screen and characterised by examination of the formation of chemotactic rings on semi solid media containing galactose, ribose or tryptone, the ability of the cells to migrate in a spatial gradient, and the methylation status of the receptor (assessed by electrophoretic mobility). L216K was found to have increased methylation (and was classed "induced") and T215K and T215R were found to have reduced methylation (as was classed as "reduced").

Sidekick

The Sidekick high throughput approach is applied as described previously [25], [42]. Briefly, each helix sequence was simulated 100×100 ns. The approach described here is distinct from previous high throughput approaches [41] in several features. The simulation pipeline is entirely automated, with simulations systems built, multiple simulations performed across a cluster, and the resulting simulation trajectories analysed from an input of a sequence and key simulation variables (temperature, forcefield, simulation length etc.). Simulations were performed over a mixed computational grid consisting of a dedicated 56 core MacOS cluster and workstations. Sidekick is written in python using the numpy and matplotlib libraries for calculations and plotting graphics. Xgrid is used to distribute calculations across the grid.

AUTHOR CONTRIBUTIONS

Conceived and designed the experiments: BAH JPA MSPS. Performed the experiments: BAH. Analyzed the data: BAH MSPS. Contributed reagents/materials/analysis tools: BAH. Wrote the paper: BAH JPA MSPS.

REFERENCES

1. Falke JJ, Erbse AH (2009) The piston rises again. Structure 17: 1149–1151.JJ FalkeAH Erbse2009The piston rises again. Structure1711491151

2. Yeh JI, Biemann HP, Prive GG, Pandit J, Koshland DE, et al. (1996) High-resolution structures of the ligand binding domain of the wild-type bacterial aspartate receptor. J Mol Biol 262: 186–201.JI YehHP BiemannGG PriveJ. PanditDE Koshland1996High-resolution structures of the ligand binding domain of the wild-type bacterial aspartate receptor.J Mol Biol262186201
3. Kim KK, Yokota H, Kim SH (1999) Four-helical-bundle structure of the cytoplasmic domain of a serine chemotaxis receptor. Nature 400: 787–792.KK KimH. YokotaSH Kim1999Four-helical-bundle structure of the cytoplasmic domain of a serine chemotaxis receptor. Nature400787792
4. Hulko M, Berndt F, Gruber M, Linder JU, Truffault V, et al. (2006) The HAMP domain structure implies helix rotation in transmembrane signaling. Cell 126: 929–940.M. HulkoF. BerndtM. GruberJU LinderV. Truffault2006The HAMP domain structure implies helix rotation in transmembrane signaling.Cell126929940
5. Swain KE, Falke JJ (2007) Structure of the conserved HAMP domain in an intact, membrane-bound chemoreceptor: A disulfide mapping study. Biochem 46: 13684–13695.KE SwainJJ Falke2007Structure of the conserved HAMP domain in an intact, membrane-bound chemoreceptor: A disulfide mapping study.Biochem461368413695
6. Khursigara CM, Wu X, Zhang P, Lefman J, Subramaniam S (2008) Role of HAMP domains in chemotaxis signaling by bacterial chemoreceptors. Proc Natl Acad Sci U S A 105: 16555–16560.CM KhursigaraX. WuP. ZhangJ. LefmanS. Subramaniam2008Role of HAMP domains in chemotaxis signaling by bacterial chemoreceptors. Proc Natl Acad Sci U S A1051655516560
7. Khursigara CM, Lan G, Neumann S, Wu X, Ravindran S, et al. (2011) Lateral density of receptor arrays in the membrane plane influences sensitivity of the *E. coli* chemotaxis response. EMBO J 30: 1719–1729.CM KhursigaraG. LanS. NeumannX. WuS. Ravindran2011Lateral density of receptor arrays in the membrane plane influences sensitivity of the *E. coli* chemotaxis response. EMBO J3017191729
8. Duke T, Bray D (1999) Heightened sensitivity of a lattice of membrane receptors. Proc Natl Acad Sci USA 96: 10104–10108.T.

DukeD. Bray1999Heightened sensitivity of a lattice of membrane receptors.Proc Natl Acad Sci USA961010410108

9. Sourjik V, Berg HC (2004) Functional interactions between receptors in bacterial chemotaxis. Nature 428: 437–441.V. SourjikHC Berg2004Functional interactions between receptors in bacterial chemotaxis.Nature428437441
10. Erbse AH, Falke JJ (2009) The core signaling proteins of bacterial chemotaxis assemble to form an ultrastable complex. Biochem 48: 6975–6987.AH ErbseJJ Falke2009The core signaling proteins of bacterial chemotaxis assemble to form an ultrastable complex. Biochem4869756987
11. Bhatnagar J, Borbat PP, Pollard AM, Bilwes AM, Freed JH, et al. (2010) Structure of the ternary complex formed by a chemotaxis receptor signaling domain, the CheA histidine kinase, and the coupling protein CheW as determined by pulsed dipolar ESR spectroscopy. Biochem 49: 3824–3841.J. BhatnagarPP BorbatAM PollardAM BilwesJH Freed2010Structure of the ternary complex formed by a chemotaxis receptor signaling domain, the CheA histidine kinase, and the coupling protein CheW as determined by pulsed dipolar ESR spectroscopy.Biochem4938243841
12. Chervitz SA, Lin CM, Falke JJ (1995) Transmembrane signaling by the aspartate receptor: engineered disulfides reveal static regions of the subunit interface. Biochem 34: 9722–9733.SA ChervitzCM LinJJ Falke1995Transmembrane signaling by the aspartate receptor: engineered disulfides reveal static regions of the subunit interface. Biochem3497229733
13. Chervitz SA, Falke JJ (1996) Molecular mechanism of transmembrane signaling by the aspartate receptor: a model. Proc Natl Acad Sci U S A 93: 2545–2550.SA ChervitzJJ Falke1996Molecular mechanism of transmembrane signaling by the aspartate receptor: a model.Proc Natl Acad Sci U S A9325452550
14. Falke JJ, Hazelbauer GL (2001) Transmembrane signaling in bacterial chemoreceptors. Trends Biochem Sci 26: 257–265.JJ FalkeGL Hazelbauer2001Transmembrane signaling in bacterial chemoreceptors.Trends Biochem Sci26257265

15. Murphy OJ, Kovacs FA, Sicard EL, Thompson LK (2001) Site-directed solid-state NMR measurement of a ligand-induced conformational change in the serine bacterial chemoreceptor. Biochem 40: 1358–1366.OJ MurphyFA KovacsEL SicardLK Thompson2001Site-directed solid-state NMR measurement of a ligand-induced conformational change in the serine bacterial chemoreceptor.Biochem4013581366
16. Ottemann KM, Xiao W, Shin YK, Koshland DE (1999) A piston model for transmembrane signaling of the aspartate receptor. Science 285: 1751–1754.KM OttemannW. XiaoYK ShinDE Koshland1999A piston model for transmembrane signaling of the aspartate receptor.Science28517511754
17. Lee GF, Dutton DP, Hazelbauer GL (1995) Identification of functionally important helical faces in transmembrane segments by scanning mutagenesis. Proc Natl Acad Sci U S A 92: 5416–5420. GF LeeDP DuttonGL Hazelbauer1995Identification of functionally important helical faces in transmembrane segments by scanning mutagenesis.Proc Natl Acad Sci U S A9254165420
18. Hughson AG, Hazelbauer GL (1996) Detecting the conformational change of transmembrane signaling in a bacterial chemoreceptor by measuring effects on disulfide cross-linking in vivo. Proc Natl Acad Sci U S A 93: 11546–11551.AG HughsonGL Hazelbauer1996Detecting the conformational change of transmembrane signaling in a bacterial chemoreceptor by measuring effects on disulfide cross-linking in vivo.Proc Natl Acad Sci U S A931154611551
19. Peach ML, Hazelbauer GL, Lybrand TP (2002) Modeling the transmembrane domain of bacterial chemoreceptors. Protein Sci 11: 912–923.ML PeachGL HazelbauerTP Lybrand2002Modeling the transmembrane domain of bacterial chemoreceptors.Protein Sci11912923
20. Baumgartner JW, Hazelbauer GL (1996) Mutational analysis of a transmembrane segment in a bacterial chemoreceptor. J Bacteriol 178: 4651–4660.JW BaumgartnerGL Hazelbauer1996Mutational analysis of a transmembrane segment in a bacterial chemoreceptor.J Bacteriol17846514660
21. de Planque MRR, Kruijtzer JAW, Liskamp RMJ, Marsh D,

Greathouse DV, et al. (1999) Different membrane anchoring positions of tryptophan and lysine in synthetic transmembrane alpha-helical peptides. J Biol Chem 274: 20839–20846.MRR de PlanqueJAW KruijtzerRMJ LiskampD. MarshDV Greathouse1999Different membrane anchoring positions of tryptophan and lysine in synthetic transmembrane alpha-helical peptides.J Biol Chem2742083920846

22. Ridder A, Morein S, Stam JG, Kuhn A, de Kruijff B, et al. (2000) Analysis of the role of interfacial tryptophan residues in controlling the topology of membrane proteins. Biochem 39: 6521–6528.A. RidderS. MoreinJG StamA. KuhnB. de Kruijff2000Analysis of the role of interfacial tryptophan residues in controlling the topology of membrane proteins.Biochem3965216528

23. Killian JA, Nyholm TKM (2006) Peptides in lipid bilayers: the power of simple models. Curr Opin Struct Biol 16: 473–479.JA KillianTKM Nyholm2006Peptides in lipid bilayers: the power of simple models.Curr Opin Struct Biol16473479

24. Vostrikov VV, Daily AE, Greathouse DV, Koeppe RE (2010) Charged or aromatic anchor residue dependence of transmembrane peptide tilt. J Biol Chem 285: 31723–31730.VV VostrikovAE DailyDV GreathouseRE Koeppe2010Charged or aromatic anchor residue dependence of transmembrane peptide tilt.J Biol Chem2853172331730

25. Vostrikov VV, Hall BA, Greathouse DV, Koeppe RE, Sansom MSP (2010) Changes in transmembrane helix alignment by arginine residues revealed by solid-state NMR experiments and coarse-grained MD simulations. J Amer Chem Soc 132: 5803–5811.VV VostrikovBA HallDV GreathouseRE KoeppeMSP Sansom2010Changes in transmembrane helix alignment by arginine residues revealed by solid-state NMR experiments and coarse-grained MD simulations.J Amer Chem Soc13258035811

26. Yau WM, Wimley WC, Gawrisch K, White SH (1998) The preference of tryptophan for membrane interfaces. Biochem 37: 14713–14718. WM YauWC WimleyK. GawrischSH White1998The preference of tryptophan for membrane interfaces.Biochem371471314718

27. Killian JA, von Heijne G (2000) How proteins adapt to a membrane-water interface. Trends Biochem Sci 25: 429–434.JA KillianG. von

Heijne2000How proteins adapt to a membrane-water interface. Trends Biochem Sci25429434

28. Ulmschneider MB, Sansom MSP (2001) Amino acid distributions in integral membrane protein structures. Biochim Biophys Acta 1512: 1–14.MB UlmschneiderMSP Sansom2001Amino acid distributions in integral membrane protein structures.Biochim Biophys Acta1512114
29. Scott KA, Bond PJ, Ivetac A, Chetwynd AP, Khalid S, et al. (2008) Coarse-grained MD simulations of membrane protein-bilayer self-assembly. Structure 16: 621–630.KA ScottPJ BondA. IvetacAP ChetwyndS. Khalid2008Coarse-grained MD simulations of membrane protein-bilayer self-assembly.Structure16621630
30. Petrache HI, Killian JA, Koeppe RE, Woolf TB (2000) Interaction of WALP peptides with lipid bilayers: Molecular dynamics simulations. Biophys J 78: 1917Pos.HI PetracheJA KillianRE KoeppeTB Woolf2000Interaction of WALP peptides with lipid bilayers: Molecular dynamics simulations.Biophys J781917Pos
31. Monticelli L, Kandasamy , Tieleman DP, Fuchs PFJ (2010) Interpretation of ^{2}H-NMR experiments on the orientation of the transmembrane helix WALP23 by computer simulations. Biophys J 99: 1455–1464.L. MonticelliKandasamyDP TielemanPFJ Fuchs2010Interpretation of ^{2}H-NMR experiments on the orientation of the transmembrane helix WALP23 by computer simulations. Biophys J9914551464
32. Gkeka P, Sarkisov L (2010) Interactions of phospholipid bilayers with several classes of amphiphilic α-helical peptides: insights from coarse-grained molecular dynamics simulations. J Phys Chem B 114: 826–839.P. GkekaL. Sarkisov2010Interactions of phospholipid bilayers with several classes of amphiphilic α-helical peptides: insights from coarse-grained molecular dynamics simulations.J Phys Chem B114826839
33. Ulmschneider MB, Doux JPF, Killian JA, Smith JC, Ulmschneider JP (2010) Mechanism and kinetics of peptide partitioning into membranes from all-atom simulations of thermostable peptides. J Amer Chem Soc 132: 3452–3460.MB UlmschneiderJPF DouxJA KillianJC SmithJP Ulmschneider2010Mechanism and kinetics of

peptide partitioning into membranes from all-atom simulations of thermostable peptides.J Amer Chem Soc13234523460

34. Ulmschneider MB, Smith JC, Ulmschneider JP (2010) Peptide partitioning properties from direct insertion studies. Biophys J 98: L60–L62.MB UlmschneiderJC SmithJP Ulmschneider2010Peptide partitioning properties from direct insertion studies.Biophys J98L60L62
35. Marrink SJ, Risselada J, Yefimov S, Tieleman DP, de Vries AH (2007) The MARTINI forcefield: coarse grained model for biomolecular simulations. J Phys Chem B 111: 7812–7824.SJ MarrinkJ. RisseladaS. YefimovDP TielemanAH de Vries2007The MARTINI forcefield: coarse grained model for biomolecular simulations.J Phys Chem B11178127824
36. Monticelli L, Kandasamy SK, Periole X, Larson RG, Tieleman DP, et al. (2008) The MARTINI coarse grained force field: extension to proteins. J Chem Theor Comp 4: 819–834.L. MonticelliSK KandasamyX. PerioleRG LarsonDP Tieleman2008The MARTINI coarse grained force field: extension to proteins.J Chem Theor Comp4819834
37. Bond PJ, Sansom MSP (2006) Insertion and assembly of membrane proteins via simulation. J Amer Chem Soc 128: 2697–2704. PJ BondMSP Sansom2006Insertion and assembly of membrane proteins via simulation.J Amer Chem Soc12826972704
38. Bond PJ, Holyoake J, Ivetac A, Khalid S, Sansom MSP (2007) Coarse-grained molecular dynamics simulations of membrane proteins and peptides. J Struct Biol 157: 593–605.PJ BondJ. HolyoakeA. IvetacS. KhalidMSP Sansom2007Coarse-grained molecular dynamics simulations of membrane proteins and peptides.J Struct Biol157593605
39. Bond PJ, Wee CL, Sansom MSP (2008) Coarse-grained molecular dynamics simulations of the energetics of helix insertion into a lipid bilayer. Biochem 47: 11321–11331.PJ BondCL WeeMSP Sansom2008Coarse-grained molecular dynamics simulations of the energetics of helix insertion into a lipid bilayer. Biochem471132111331
40. Schäfer LV, de Jong DH, Holt A, Rzepiela AJ, de Vries AH, et al.

(2011) Lipid packing drives the segregation of transmembrane helices into disordered lipid domains in model membranes. Proc Natl Acad Sci U S A. LV SchäferDH de JongA. HoltAJ RzepielaAH de Vries2011Lipid packing drives the segregation of transmembrane helices into disordered lipid domains in model membranes.Proc Natl Acad Sci U S Awww.pnas.org/cgi/doi/10.1073/pnas.1009362108. www.pnas.org/cgi/doi/10.1073/pnas.1009362108.

41. Chetwynd AP, Scott KA, Mokrab Y, Sansom MSP (2008) CGDB: a database of membrane protein/lipid interactions by coarse-grained molecular dynamics simulations. Molec Memb Biol 25: 662–669. AP ChetwyndKA ScottY. MokrabMSP Sansom2008CGDB: a database of membrane protein/lipid interactions by coarse-grained molecular dynamics simulations.Molec Memb Biol25662669

42. Hall BA, Chetwynd A, Sansom MSP (2011) Exploring peptide-membrane interactions with coarse grained MD simulations. Biophys J 100: 1940–1948.BA HallA. ChetwyndMSP Sansom2011Exploring peptide-membrane interactions with coarse grained MD simulations. Biophys J10019401948

43. Draheim RR, Bormans AF, Lai R-Z, Manson MD (2005) Tryptophan residues flanking the second transmembrane helix (TM2) set the signaling state of the Tar chemoreceptor. Biochem 44: 1268–1277. RR DraheimAF BormansR-Z LaiMD Manson2005Tryptophan residues flanking the second transmembrane helix (TM2) set the signaling state of the Tar chemoreceptor.Biochem4412681277

44. Draheim RR, Bormans AF, Lai R-Z, Manson MD (2006) Tuning a bacterial chemoreceptor with protein-membrane interactions. Biochem 45: 14655–14664.RR DraheimAF BormansR-Z LaiMD Manson2006Tuning a bacterial chemoreceptor with protein-membrane interactions.Biochem451465514664

45. Miller AS, Falke JJ (2004) Side chains at the membrane-water interface modulate the signaling state of a transmembrane receptor. Biochem 43: 1763–1770.AS MillerJJ Falke2004Side chains at the membrane-water interface modulate the signaling state of a transmembrane receptor.Biochem4317631770

46. Chattopadhyay A, Rawat S, Greathouse D, Kelkar D, Koeppe R (2008) Role of tryptophan residues in gramicidin channel organization

and function. Biophys J 95: 166–175.A. ChattopadhyayS. RawatD. GreathouseD. KelkarR. Koeppe2008Role of tryptophan residues in gramicidin channel organization and function.Biophys J95166175

47. Kandasamy SK, Larson RG (2006) Molecular dynamics simulations of model trans-membrane peptides in lipid bilayers: A systematic investigation of hydrophobic mismatch. Biophys J 90: 2326–2343. SK KandasamyRG Larson2006Molecular dynamics simulations of model trans-membrane peptides in lipid bilayers: A systematic investigation of hydrophobic mismatch.Biophys J9023262343
48. Strandberg E, Killian JA (2003) Snorkeling of lysine side chains in transmembrane helices: how easy can it get? FEBS Lett 544: 69–73.E. StrandbergJA Killian2003Snorkeling of lysine side chains in transmembrane helices: how easy can it get?FEBS Lett5446973
49. Ozdirekcan S, Rijkers DTS, Liskamp RMJ, Killian JA (2005) Influence of flanking residues on tilt and rotation angles of transmembrane peptides in lipid bilayers. A solid-state H-2 NMR study. Biochem 44: 1004–1012.S. OzdirekcanDTS RijkersRMJ LiskampJA Killian2005Influence of flanking residues on tilt and rotation angles of transmembrane peptides in lipid bilayers. A solid-state H-2 NMR study.Biochem4410041012
50. Chung L, Lear J, DeGrado W (1992) Fluorescence studies of the secondary structure and orientation of a model ion channel peptide in phospholipid vesicles. Biochem 31: 6608–6616.L. ChungJ. LearW. DeGrado1992Fluorescence studies of the secondary structure and orientation of a model ion channel peptide in phospholipid vesicles. Biochem3166086616
51. Jeffery CJ, Koshland DE (1994) A single hydrophobic to hydrophobic substitution in the transmembrane domain impairs aspartate receptor function. Biochem 33: 3457–3463.CJ JefferyDE Koshland1994A single hydrophobic to hydrophobic substitution in the transmembrane domain impairs aspartate receptor function. Biochem3334573463
52. Senes A, Engel D, DeGrado WF (2006) Folding of helical membrane proteins: the role of polar, GxxxG-like and proline motifs. Curr Opin Struct Biol 14: 465–479.A. SenesD. EngelWF DeGrado2006Folding of helical membrane proteins: the role of polar, GxxxG-like and proline motifs.Curr Opin Struct Biol14465479

53. Jeffery CJ, Koshland DE (1999) The Escherichia coli aspartate receptor: sequence specificity of a transmembrane helix studied by hydrophobic-biased random mutagenesis. Prot Engng 12: 863–872.CJ JefferyDE Koshland1999The Escherichia coli aspartate receptor: sequence specificity of a transmembrane helix studied by hydrophobic-biased random mutagenesis.Prot Engng12863872
54. Finger C, Escher C, Schneider D (2009) The single transmembrane domain of human tyrosine kinases encode self interactions. Science Signal 2: ra56, 51–58.C. FingerC. EscherD. Schneider2009The single transmembrane domain of human tyrosine kinases encode self interactions.Science Signal2ra56, 5158
55. Li E, Hristova K (2006) Role of receptor tyrosine kinase transmembrane domains in cell signaling and human pathologies. Biochem 45: 6241–6251.E. LiK. Hristova2006Role of receptor tyrosine kinase transmembrane domains in cell signaling and human pathologies.Biochem4562416251
56. Marrink S-J, Risselada HJ, Yefimov S, Tieleman DP, Vries AHd (2007) The MARTINI force field: coarse grained model for biomolecular simulations. J Phys Chem B 111: 7812–7824.S-J MarrinkHJ RisseladaS. YefimovDP TielemanAHd Vries2007The MARTINI force field: coarse grained model for biomolecular simulations.J Phys Chem B11178127824
57. Marrink S-J, Risselada J, Mark A (2005) Simulation of gel phase formation and melting in lipid bilayers using a coarse grained model. Chem Phys Lipids 135: 223–244.S-J MarrinkJ. RisseladaA. Mark2005Simulation of gel phase formation and melting in lipid bilayers using a coarse grained model.Chem Phys Lipids135223244
58. Lindahl E, Hess B, van der Spoel D (2001) GROMACS 3.0: a package for molecular simulation and trajectory analysis. J Molec Model 7: 306–317.E. LindahlB. HessD. van der Spoel2001GROMACS 3.0: a package for molecular simulation and trajectory analysis.J Molec Model7306317

PART III: COARSE-GRAINED MODELS FOR DNA AND RNA

CHAPTER 7

A COARSE-GRAINED DNA MODEL PARAMETERIZED FROM ATOMISTIC SIMULATIONS BY INVERSE MONTE CARLO

Nikolay Korolev [1], Di Luo [1], Alexander P. Lyubartsev [2] and Lars Nordenskiöld [1]

[1]School of Biological Sciences, Nanyang Technological University, 639798, Singapore

[2]Division of Physical Chemistry, Arrhenius Laboratory, Stockholm University, Stockholm, S 10691 Sweden

ABSTRACT

Computer modeling of very large biomolecular systems, such as long DNA polyelectrolytes or protein-DNA complex-like chromatin cannot reach all-atom resolution in a foreseeable future and this necessitates the development of coarse-grained (CG) approximations. DNA is both highly charged and mechanically rigid semi-flexible polymer and adequate DNA modeling requires a correct description of both its structural stiffness and salt-dependent electrostatic forces. Here, we present a novel CG model of DNA that approximates the DNA polymer as a chain of 5-bead units. Each unit represents two DNA base pairs with one central bead for bases and pentose moieties and four others for phosphate groups. Charges, intra- and inter-molecular force field potentials for the CG DNA model were calculated using the inverse Monte Carlo method from all atom molecular dynamic (MD) simulations of 22 bp DNA oligonucleotides. The CG model was tested by performing dielectric continuum Langevin MD simulations of a 200 bp double helix DNA in solutions of monovalent salt with explicit ions. Excellent agreement with experimental data was obtained for the dependence of the DNA persistent length on salt concentration in the range 0.1–100 mM. The new CG DNA model is suitable for modeling various biomolecular systems with adequate description of electrostatic and mechanical properties.

Keywords: DNA; coarse-grained model; molecular dynamics; persistence length

INTRODUCTION

During decades, the DNA molecule has been studied in a vast number of experimental, theoretical, and modeling research investigations. The study of polyelectrolyte properties of DNA takes an important place in these works since electrostatic interactions and salt effects have a profound influence on many other DNA properties and interactions of DNA with its surrounding. An important problem in this context is related to understanding the mechanism of DNA compaction and self-assembly. In the nucleus of eukaryotic cells, the DNA molecule exists as a hierarchically ordered structure formed by the nucleosome core particles consisting of complexes of DNA with histone proteins, and there are strong indications [1,2,3,4] that the stability of chromatin (and, thus, DNA transcription and replication) is modulated by electrostatic interactions. The problem of polyelectrolyte and DNA condensation in solution has a long history of experimental studies [5,6]. Like-charged polyelectrolytes would be expected to repel each other based on a treatment of the electrostatics using mean field theories (e.g., the Poisson-Boltzmann equation). However, dynamic ion-ion correlations give rise to attractive electrostatic force contributions that can explain experimental aggregation of DNA induced by multivalent cations [7,8,9,10,11,12]. In order to model and describe the folding and aggregation of chromatin, Poisson-Boltzmann based models, including the popular Debye-Hückel type approaches [13,14,15], have serious limitations which may be a source of artefacts in predictions of aggregation properties. The reason is that such models do not include effects due to the explicit presence of mobile ions and also do not take into account the molecular nature of the solvent.

An accurate description of molecular properties of condensed matter can in principle be reached by using atomistic molecular dynamics computer simulations. With computer simulations, we are able to follow every detail of the molecular motion and give, within some theoretical model, quantitative answers, directly comparable with experiments, thus, providing a link between the theory and the experiment. This is especially important in studies of macromolecular systems, such as biopolymers, which always have a certain degree of intrinsic disorder. At present, computer simulations have reached a state where simulation times and length scales allow a direct comparison with experiments in well-characterized systems. Molecular simulations are also becoming an intrinsic part of applied research, such as drug design, materials science, and nanotechnologies.

In order to model very large condensed matter systems such as DNA-nucleosome complexes, an atomistic description is not practical as the number of particles would be very large. Therefore, simplified coarse-grained (CG) models are necessary, allowing a coarser description of the molecules and interactions when moving towards description at larger length and time scales. Many of today's coarse-grained models are constructed from empirical parameterization of effective interaction potentials and usually do not contain a rigorous description of the molecular water [16,17]. For example, in the primitive electrolyte model, ions are presented as charged hard spheres interacting in a media with dielectric permittivity of water. Within a multi-scale modeling scheme, the macromolecules can be reduced to a CG description with effective sites, representing many atoms. The interactions between the CG sites are represented by effective potentials that include the effects of the water solvent and which are deduced from the atomistic simulations. This enables the modeling of very large molecular assemblies, with a reduction of the number of particles of several orders of magnitudes, but still using effective forces that implicitly include a detailed description at the atomic level. Simulations of such large systems can then be performed over a range of time scales and hence it is a multi-scale approach. There are two basic approaches for constructing effective coarse-grained potentials from the results of atomistic simulations. One is based on fitting of structural properties, for which the radial distribution functions (RDFs) are typically used (the inverse Monte Carlo, also called the Newton inversion method) [18,19]. Another approach is based on reproduction of forces for specific snapshots of the system (the force matching approach) [20,21]. The former approach is also equivalent to the principle of relative entropy minimization [22], and, together with the force matching methodology, can be described within a general information function concept [23]. The statistical-mechanical equations expressing the canonical properties in terms of parameters of the potential functions can be inverted and solved numerically according to the iterative Newton scheme. In the Inverse Monte Carlo method, RDFs are inverted to reconstruct pair potentials, while in a more general approach the targets can be other canonical averages [18].

A number of theoretical coarse-graining models of DNA [24,25,26,27, 28,29,30,31,32,33,34,35] and chromatin [13,14,15,36,37,38,39,40,41,42 ,43,44,45,46,47,48,49] have been reported recently. In recent works from the Papoian group, a CG model of a single DNA molecule was developed [24,25,26,50]. The model is based on underlying atomistic simulations, using an approach that is similar to the inverse MC method (called molecular renormalization group coarse-graining). DNA was coarse-grained into two chains of sites representing the phosphate groups with bond-only interactions in between them. A recent all-atom simulation of a single NCP applied Boltzmann inversion to obtain effective potentials that were used in a CG NCP simulation [40,51], but as it is based on the Boltzmann inversion, it is of limited value for treating long-range electrostatic interactions [52].

We have previously performed modeling of an NCP solution [38,47] and of a chromatin fiber (a nucleosome array) [53,54,55]. These simplified coarse-grained continuum electrostatic simulation models were successfully applied to qualitatively reproduce histone tail-mediated chromatin folding and nucleosome core particle aggregation [38,47,54] induced by multivalent cations. Long-range electrostatic interactions were included by the presence of explicit mobile ions (representing the salt), but using a simplified structural model of the NCP and of the charge distribution on the array. The effects of the molecular nature of water were not included as the description is based on a continuum dielectric approximation of the solvent.

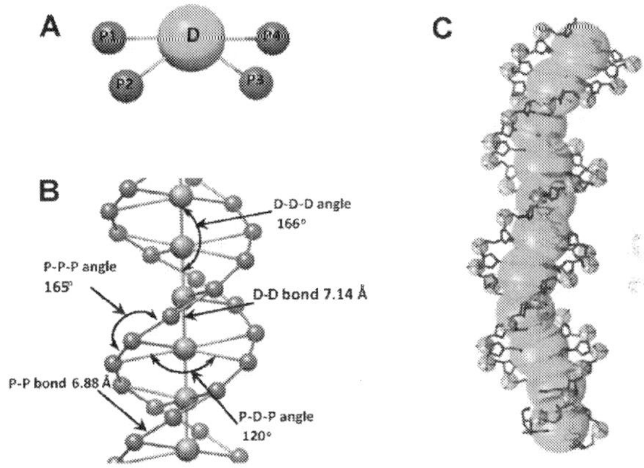

Figure 1: Coarse-Grained DNA model and definition of sites and intramolecular interactions: (**A**) A 5-site two base pair elementary DNA unit. The central particle (green) approximates the pentose and base atoms of a two base pair DNA fragment. The orange beads represent the phosphate groups at the positions of an idealized B-form double helix; (**B**) The geometry of the CG model of double helical DNA with bonds and angles set to maintain B-form conformation. Equilibrium values of bond lengths and angles shown in the figure were calculated by the IMC method from data of all-atom MD simulations (see Table 1); (**C**). Superimposed all-atom and coarse-grained representation of the 22-bp fragment of DNA; All-atom DNA shown in stick representation (blue color); the CG model of DNA is a combination of particles placed at the centers of mass of the phosphate group (orange) and atoms of the pentose and bases from the 2-bp unit (green spheres). Data from all-atom MD simulations of the four 22-bp DNA fragments was used to determine the structure and force parameters for the CG DNA.

An advanced coarse-grained DNA model was recently introduced by us [56] in order to model DNA in nucleosome complexes. It consists of two types of sites: DNA core ("D") beads, each representing two base pairs and "P" beads, representing phosphate groups. The internal DNA structure was maintained by four bond potentials (D–D, D–P, P–P, and P–P across minor grove) and three angular potentials (D–D–D, P–P–P and P–D–P), see Figure 1A,B. All these potentials were of harmonic type with *ad hoc* selected parameters and to maintain the integrity of the NCP, the DNA was connected to the histone octamer by artificial bonds. Numerous observations have shown that in solution the NCP behaves as stable structure with rather fixed attachment of the DNA to the histone core [57,58,59,60,61], therefore the description of the mechanical properties of the DNA when modeling NCP-NCP interactions dominated by electrostatic forces is not critical. However, adequate modeling bending and torsional flexibility of the DNA becomes crucial in the description of chromatin structure and dynamics when a realistic model of the chromatin fiber as a linear array of NCPs connected by stretches of linker DNA is being developed.

In the present work we parameterize the coarse-grained DNA model introduced in [56] (Figure 1A,B) with the aim to reproduce the experimental salt dependence of DNA flexibility. The parameterization is based on atomistic simulations of DNA oligomers in water solution with K^+ as counterions. The internal coarse-grained DNA potentials are then computed by the inverse Monte Carlo method using the MagiC software [62] to reproduce the internal DNA structure and dynamics obtained in atomistic simulations. The Inverse Monte Carlo method with this software has been shown to be robust and rigorous [63] and is implemented with 2-body interactions as well as 3-body angular potentials and the inclusion of these interactions can be done straightforwardly and with high efficiency in computational effort. Our model and parameters are then benchmarked in simulation of a 200 bp long DNA fragment in the presence of explicit salt at varying concentrations. Without any further adjustment of the obtained DNA internal structural parameters, remarkably good agreement with experimental persistent length (L_p) data was obtained for the salt-dependent bending behavior of DNA over a wide range of salt from 0.0001 M to 0.1 M, while the torsional rigidity of the present CG-DNA model requires further optimization. Potentially the proposed CG model could be further developed to include sequence-specific dependence of DNA flexibility. However, this CG model does not include modeling separation of the DNA strands.

METHODOLOGY

Design of the Flexible CG DNA Model and Principles of Its Parameterization

The aim of the present CG approach is to develop a model that is simple enough in its design and yet captures the structural form and properties of double helical B-form DNA, which is suitable for rigorous parameterization based on all atom MD simulations followed by the inverse Monte Carlo method to obtain relevant CG DNA structural and interaction potentials. The model should also pay particular emphasis to electrostatic interactions, which includes explicit mobile ions and explicit charges of DNA phosphate groups and it should also have the potential for refinements, such as inclusion of base sequence specificity (presently not considered).

To satisfy this level of precision, we have chosen to describe the system via a collection of coarse-grained beads representing different fragments of the DNA with intramolecular bonded potentials that maintains the DNA structural integrity while reproducing experimentally observed flexibility of the B-DNA double helix. DNA is modeled as a sequence of units each representing a two base pair fragment of DNA. Each unit consists of five beads, four P units that represent the phosphate groups and one D unit corresponding to the four pentoses with attached base fragments in between (Figure 1A). The double-helix DNA structure was maintained by three bond and three angular potentials between the beads listed in Table 1 (the P–P bond potential across minor groove introduced in [56] was found unnecessary to keep the helical structure and was not included into the current model). The equilibrium geometry of the beads corresponds closely to the B-form of DNA and the whole structure forms a helical DNA with two clearly distinguishable strands of phosphate groups and minor and major grooves between them (Figure 1B). The overall DNA structure thus resembles space-filling grooved DNA models used previously to model the ionic environment of DNA [12,64,65,66].

Table 1: Parameters of the coarse-grained (CG) DNA model developed from the inverse Monte Carlo analysis [62] using data of all atom molecular dynamic (MD) simulations. Force constants were derived from potential functions approximated by harmonic potential

Type of bond/angle	Equilibrium distance/angle	IMC derived force constant ($k_BT \cdot Å^{-2}$ or $k_BT \cdot rad^{-2}$)
D–D bond	7.14 Å	14.8
D–P bond	9.04 Å	9.7
P–P bond	6.88 Å	19.4
D–D–D angle	166.00°	37.0
P–D–P angle	119.50°	100.0
P–P–P angle	165.00°	32.0

Atomistic Molecular Dynamics Simulations

Atomistic MD simulations were run for four 22-bp DNA oligomers (sequence 5'-d(GATGCAGTCACCGCGAATTGGC) × 5'-d(GCCAATTCGCGGTGACTGCATC)) dissolved in 21,200 water molecules with 168 K^+ counterions. The initial structure of the DNA double helix was built using the NAB package [67] with idealized B-DNA structure parameters. The simulation software was the Gromacs package [68,69] with the CHARMM27 force field [70,71] using the simple point charge model of water TIP3P [72,73,74]. The dimension of the cubic simulation cells was set to 90 Å. The system was equilibrated by energy minimization at 50 K keeping DNA fixed followed by release of the DNA constraints and running a series of 10 ps MD simulations under constant volume conditions (NVT) with stepwise heating (50 K, 40 ps) from 50 K to 298 K. Subsequently, the cell size was decreased by about 2 Å in order to reproduce a pressure fluctuating around 1 bar in the NVT ensemble. The temperature was controlled by the Nose-Hoover thermostat applied separately for DNA and solvent, with relaxation time 0.3 ps. All the bonds were constrained using LINKS algorithm allowing for time step 2 fs. The Lennard-Jones interactions were cutoff at 9 Å. The long-range electrostatic interactions were treated by the particle mesh Ewald method [75] of order 4 with Fourier grid spacing 0.8 Å and tolerance 10^{-5}. Simulations were run for 40 ns collecting atom coordinates with 1 ps resolution. Detailed description of the simulations and major results are reported in our earlier work [76].

Coarse-Grained DNA and Computations of Effective Potentials

The trajectories of the atomistic simulations were mapped onto the CG DNA model producing a CG trajectory. The P sites were determined by the positions of the phosphates of the atomistic model and the D cites were computed as center-of-masses of the atoms belonging to the two base pairs involved. Superposition of the atomistic DNA structure on our 5-site CG-model is displayed in Figure 1C. In order to take into account possible end effects, special types "DT" and "PT" were defined for the terminal sites in addition to the two site types for description of all other (not terminal) CG DNA sites. The counterions were also included into CG trajectory. The CG trajectory generated from the atomistic simulations was used to extract effective coarse-grained potentials by the inverse Monte Carlo method [19,62]. For that, radial distribution functions between the sites of the CG model, as well as internal bond and angle distributions between the CG sites were determined from the CG trajectory and were used as input to the inverse Monte Carlo procedure. The CG sites connected by bonds or angles were excluded from the non-bonded interactions, as well as from intermolecular RDF calculations. Since the CG system consists of 5 different site types (four DNA particles: "D", "P", "DT" and "PT" plus DNA counterions, K^+), totally 15 pairs of intermolecular distribution functions were computed from the CG trajectory. Additionally, distributions of bond distances and angles, corresponding to bond and angle internal DNA potentials (see Figure 1B) were computed. Taking into account that the end sites were assigned special types, the total number of bond potentials (and, hence, bond length distributions) became 8 and the total of number of angular potentials (and angular distribution) became 6. As a result of the inverse Monte Carlo computations, effective pair potentials between all CG sites, and internal bond and angle potentials were determined which reproduce the RDFs, as well as bond and angle distributions of the atomistic model. The computations were carried out using the software package MagiC [62].

We also performed inverse Monte Carlo calculations without distinguishing the oligomer terminal groups in order to check their effect. As expected, the effect of the terminal groups was non-negligible and we discuss this in the Results Section below.

Coarse-Grained DNA Model: Large-Scale Simulations

In the present work we have used the results of the inverse MC procedure to obtain the parameters that determines the internal structure and flexibility of the coarse-grained DNA model, the topology of which is illustrated in Figure 1 and that includes harmonic interactions for bonds and angles [56].

The bond and angle potentials for the bound sites were approximated by the equations:

$$V_{\text{bond}}(r) = \frac{1}{2} k_{\text{b}} (r - r_0)^2 \tag{1}$$

$$V_{ang}(\varphi) = \frac{1}{2} k_a (\varphi - \varphi_0)^2 \tag{2}$$

where k_b, k_a, and r_0, φ_0 are respectively bond and angle force constants and equilibrium values for bond length and angle. The final parameters for the bonded interactions are given in Table 1 and a detailed discussion on this fitting is given below.

During this first test of the model, only internal DNA rigidity parameters were fitted to the results of the inverse Monte Carlo computations, while the standard Coulombic potential Equation (3), for charges (q_i, q_j) in a dielectric continuum with permittivity $\epsilon = 78$, and short-range potential defined by Equation (4) were used to describe non-bonded interactions:

$$V_{\text{el}}(r_{\text{ij}}) = \frac{q_{\text{i}} q_{\text{j}}}{4\pi\epsilon_0 \epsilon r_{\text{ij}}} \tag{3}$$

$$V^{\text{SR}}(r) = \begin{cases} \infty & \text{for} \quad r < R_{\text{ij}} \\ 4\varepsilon_{\text{ij}}^{\text{SR}} \left(\left(\frac{\sigma_{\text{ij}}}{r\text{-}R_{\text{ij}}} \right)^{12} - \left(\frac{\sigma_{\text{ij}}}{r\text{-}R_{\text{ij}}} \right)^{6} + 0.25 \right) & \text{for} \quad R_{\text{ij}} < r < 2^{1/6} R_{\text{ij}} \\ 0 & \text{for} \quad r > 2^{1/6} R_{\text{ij}} \end{cases} \tag{4}$$

Here (i, j) are bead numbers. In Equation (4) the hard core particle contact distance $R_{ij} = R_i + R_j$ is determined by the values of the hard core radii of both beads. Parameters σ_{ij} and ε_{ij}^{SR} were set to $\sigma_{ij} = 4$ Å and

$\varepsilon_{ij}^{SR} = 1$ ($k_B T$ units) for all interactions. Thus, the effective radius of each particle was determined by a sum of the soft radius of $\sigma_{ij}/2 = 2$ Å and the hard radius R_i which was defined separately for each particle type. The hard radii R_i as well as charges q_i are given in Table 2. The charges were obtained by summation of partial charges of the corresponding CG groups of the CHARMM27 force field. Thus, the total interaction potential (potential energy) of the system is:

$$V_{tot} = \sum_{i,j}\left(V_{el}\left(r_{ij}\right) + V^{SR}\left(r_{ij}\right)\right) + \sum_{bonds} V_{bond}(r) + \sum_{angles} V_{ang}(\varphi) \quad (5)$$

In the present CG application, which aims to parameterize the DNA internal parameters with the purpose of reproducing its salt dependent bending flexibility, we consider the solvent as a dielectric medium. These interaction potentials represent an approximation to the solvent-mediated effective potential between the charged particles in the solvent medium. Thus, ε is a parameter of the effective ion-ion potential. Many studies at an all atom level have shown that the effective potential of mean force of ions in water usually have one or two oscillations around the Coulomb potential with $\epsilon = 78$ for small (within 8 Å) distances between the ions [77]). These oscillations reflect the molecular nature of the solvent. Since the global conformational DNA properties on the scale of a few tens of nm are determined mainly by the long-range electrostatic interactions, and the IMC-derived effective potentials essentially coincide with the Coulombic potential scaled by the appropriate dielectric constant for distances larger than about 8 Å, we do not expect major effects of neglecting the solvent-induced oscillations in the effective potentials on calculation of the monovalent salt dependence of the DNA persistence length (though this question is definitely worth discussing, which will be a subject of future work). For many applications, including strong polyelectrolytes and high (physiological) salt concentrations, continuum dielectric models with constant dielectric permittivity have been very successful [78,79].

In a dielectric continuum model the ion interactions play an important role. It is important to assign appropriate radii, reflecting the effective hydration of the ions. Here, the following values of the hard radii were used: $R(K^+) = R(Cl^-) = 0$, which corresponds to effective radii of 2.0 Å for these monovalent ions. The phosphate group of DNA was approximated as a bead with radius R (P^-) = 1.0 Å, which gives an effective radius of 3.0 Å. The choice of sizes for K^+ and Cl^- is justified by our previous extensive comparison of experimental as well as computer modeling results of ion-DNA, ion-NCP and ion-chromatin interactions and ion-ion competitions [4,12,53,64,65,80,81,82,83,84]. These values are typical for hydrated ion radii used in a number of models and well reproduce thermodynamic properties and the features of ion distributions compared to all-atom simulations [12,65,66,82,85]. They also fit well the radius of the repulsive core region of the effective potentials derived within the IMC procedure.

Table 2: Non-bonded parameters of the coarse-grained DNA model

Site	Charge	Radius R_i (Å)
D (DNA)	+0.8	4.0
P (DNA)	−1.2	1.0
K^+ (ion)	+1.0	0.0
Cl^- (ion)	−1.0	0.0

The CG DNA model defined by the parameters given in Table 1 and Table 2, was used in Langevin dynamics simulations of a 200 bp DNA fragment in presence of different concentration of monovalent salt. The box sizes as well as the number of ions in each simulation are given in Table 3. The friction coefficient and random forces, mimicking the effect of solvent, were set by the dimensionless parameter $\gamma = 0.1$. The time step was set to 0.01 in the chosen (reduced) units, and we made a total of 10^8 MD steps in each simulation run where the first 5×10^7 steps were considered as equilibration. The Langevin dynamics simulations were performed by the ESPResSo package [58]. Convergence of the simulations was checked by comparing the analysis of various parts of the trajectories, which produced essentially the same persistence length values as for the full set of data (data not shown).

We also confirmed that the local structure of DNA remains preserved in comparison to the all-atom simulations from which its structural parameters were derived. The distribution of distances for some pairs of DNA beads were computed (data not shown). The D–D distance showed a somewhat extended average value (7.3 Å, compared to 7.14 Å in the all-atom simulations), corresponding to an increase of about 0.1 Å per base pair. On the other hand, the distribution of P–P bond distances is the same as in the all-atom simulations (with an average of 6.9 Å). These average distances show that the DNA model is robust, while still allowing for fluctuations and that the main features of the double helical DNA are maintained and close to that of the B form structure.

Table 3: Parameters of Langevin simulations (box size and the number of ions) of the coarse-grained DNA model. DNA length was 200 base pairs in all cases

Ionic strength	Box size (Å)	Number of cations (K^+)	Number of anions (Cl^-)
Uncharged [1]	800	300	300
100.0 mM	500	7900	7500
30.0 mM	500	2650	2250
10.0 mM	800	3400	3000
3.0 mM	800	1300	900
1.0 mM	800	700	300
0.3 mM	1400	900	500
0.1 mM	1420	580	180

[1] Note: charges of the D and P sites were set to zero.

RESULTS AND DISCUSSION

Harmonic Approximation for Internal Potentials

The effective coarse-grained potentials were originally obtained in a tabulated form. In order to facilitate their use in general purpose molecular dynamics software, we approximated the internal bond and angle potentials of DNA by harmonic functions. The results of original potentials for non-terminal DNA sites (output from the inverse MC calculations), their harmonic approximation, as well as the distribution of the corresponding bond lengths and angles (calculated from the all atom MD simulations), are shown in Figure 2. Fitted harmonic parameters are also presented in Table 1.

Merging terminal "P" and "D" particles "contaminates" the statistics, which uses only the central 9 (18 base pairs) DNA units and some of the potential functions lose their parabolic shape. The reason for these deviations is that the terminal base pairs are more flexible and in one case (of total eight base pairs at the DNA ends) the hydrogen bonds between bases were destroyed and these two nucleotides fluctuated considerably during the simulation run. These fluctuations were reflected by appearance of shoulders in the RDFs and with corresponding changes in the potential functions.

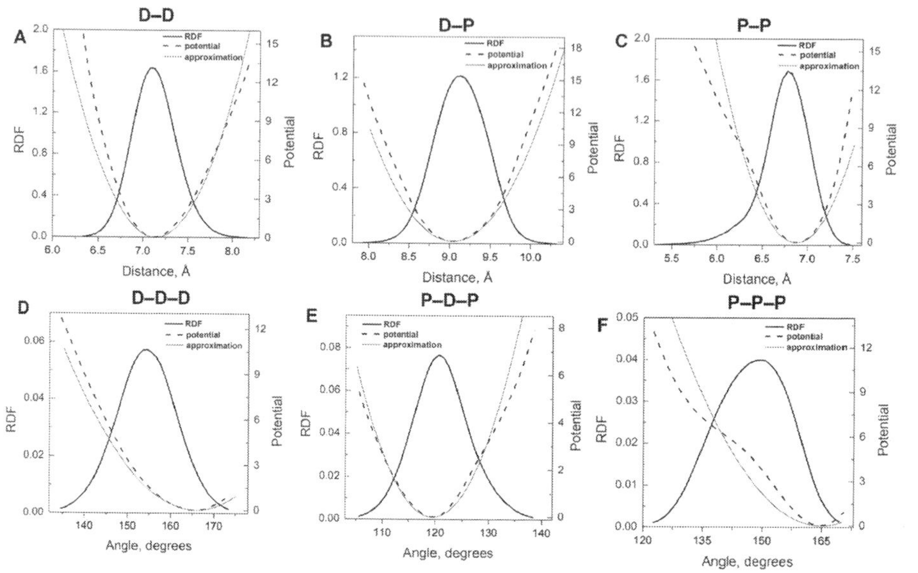

Figure 2: Radial distribution (RDF, solid lines) and force potential (dashed curves) functions for the internal bonds (**A–C**) and angles (**D–F**) of the coarse-grained model of DNA. Types of bonds and angles are indicated at the top of the graphs (see Figure 1B). RDFs were calculated from the data of all-atom MD simulations; potential functions were determined using inverse Monte Carlo method [62]. In calculations of RDFs and potentials 22-bp DNA fragment was coarse-grained on two types ("D" and "P") particles and particles in the middle of the DNA were treated separately from the particles at the DNA ends. Thin red lines are harmonic approximations of the potential curves. See text for details.

In addition to the internal bond and angle distributions of the DNA molecules, the combination of the all-atom MD simulations and the inverse MC method generated a set of intermolecular distributions and potentials. Here, we did not use these potentials in the calculations described below aiming at validation of the salt-dependent internal bending flexibility behavior of the CG DNA. Tabulated potentials of the intermolecular interactions might be used in further tests to compare approaches using the complete output from the IMC calculations (with all potentials expressed in tabulated form) with modeling based on analytical description of the intra- and inter-molecular interactions. It is worth mentioning that the intermolecular potentials are expected to be sensitive to the particular details of the all-atom simulation setup used for their generation while intramolecular potentials are expected to be "universal" and transferable since they are generated from all-atom MD data where only one DNA molecule is present in the simulation cell or when DNA molecules do not interact.

Validation of the CG-Model: Salt Dependence of DNA Persistence Length

To validate the CG-model of DNA we carried out a series of simulations for a 200 bp fragment of DNA in solution with varying concentration of monovalent salt. The purpose of the simulations was to determine the dependence of the mechanical properties of the CG-DNA, mainly bending on salt concentration and to compare these properties with experimental data. Details of the simulations are given above (Section 2.4 of Computational Methodology). This test of the CG-DNA model is interesting, not only for validation purpose, but also in the light of recent renewed discussion about the relative contributions of electrostatic and mechanical forces to the DNA flexibility [25,86,87]. The persistence length, L_p, was determined from the well-known relationship [88] used for treatment of simulation or statistical data for polymers of limited length:

$$L_P = -\frac{L_C}{\ln(\langle\cos\alpha\rangle)} \tag{6}$$

where $\langle\cos\alpha\rangle$ is the average of the cosine of the angle between two neighbouring segments over the simulation and L_C is the contour length between the segments. There is a technical problem preventing a straightforward calculation of the contour length and assignment of a vector to a DNA segment of the present CG-model of the DNA. The positions of the "D" beads (defined as the center of mass (COM) of the pentose and base atoms of the two DNA base pairs) are not aligned along the axis of the double stranded DNA and form a thin helical structure (see Figure 1C). Therefore, to calculate the contour length, the coordinates of the DNA axis should be determined from a limited number of beads. There are different approaches used to solve this sort of problem [89,90,91]. Here, we used the following algorithm: For each central bead (D) we consider 4 phosphates (P1–P4) attached to it. Then, the COM of the P1–P4 beads was defined. Next, a line connecting the phosphates COM with the central bead (D) was built. On the continuation of this line we put a point at distance "shift" from the D bead. This point will be considered as a point defining DNA axis. The value of "shift" can be obtained empirically and we found that shift value of 1.1 Å gives the best approximation for the DNA axis.

For an ideal worm-like chain Equation (6) is exact for a given value of L_P, but for a real DNA model the L_P determined by Equation (6) depends on the chosen contour length L_C as well as on the precise definition of the segment. In practice, the $L_P$value was calculated from the slope of the initial 20 points of the dependence L_C *versus* In(⟨cosα⟩) as shown in Figure 3. An indication of convergence of the simulations was that analysis of only part of the trajectories (excluding 30%–50% of the data collected at the end of the trajectories) produced essentially the same L_C values as for the full set of data (data not shown).

Inspection of snapshots showed that the DNA molecules had a visible difference in degree of bending when comparing the system with different influence of electrostatic forces. The DNA with charged P and D particles and in low 0.1 mM salt (Figure 4A) was considerably more rigid and stretched than the CG DNA with charges removed (Figure 4C); such an uncharged model contributes only the DNA internal mechanical properties to the persistence length. Figure 4B and Table 4summarize the results of the L_P calculations for a range of salt concentrations. Our simulation data showed a remarkably good agreement with experimental L_P values (Figure 4B). This fact indicates that the CHARMM27 force field that was used in all-atom MD runs adequately reproduced the contributions of structural and electrostatic components of the DNA rigidity over a wide range of monovalent salt concentration. In addition to validation of our DNA CG-model, this result demonstrates that the contribution of electrostatic forces to the DNA flexibility is significant. It appears that screening effect of added salt makes an essential influence to the DNA bending at concentrations lower than 10 mM. In agreement with experiment, the persistent length of DNA shows little change at moderate ("physiological") and high salt concentrations being quite close to the value determined for the CG model with electrostatic forces turned off by setting to zero charges of the P and D particles.

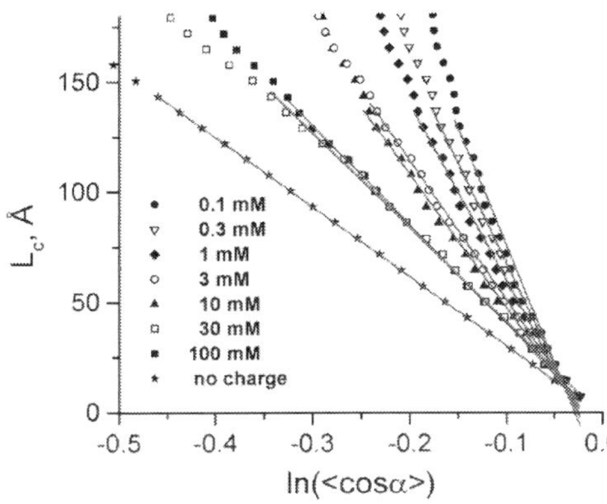

Figure 3: Determination of the persistence length, L_p, from the slope of dependence of DNA contour length, L_C, on ln(⟨cosα⟩). Data calculated for different concentrations of monovalent salt and for the uncharged CG-DNA are displayed as points; thin red lines show linear fitting.

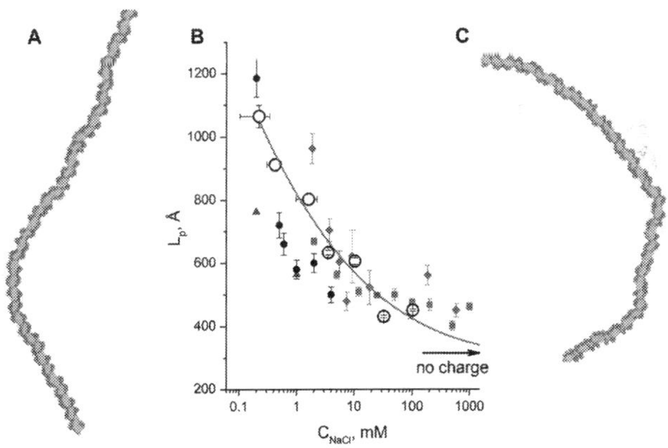

Figure 4: Dependence of DNA persistence length, L_p, on monovalent salt concentration. (**A**) and (**C**) show snapshots from the simulations at low (0.1 mM) salt concentration, (**A**), and for the uncharged DNA, (**C**). (**B**) Comparison of the simulation results with experimental data: circles with red line to guide the eye are the data from simulations using the CG model of DNA with force constants calculated by IMC software [64] from the data of all-atom MD simu-

lations (CHARMM27 force field). Experimental data are also shown to illustrate the good agreement between theory and experiment. An arrow marked "no charge" indicates L_P value calculated for the CG DNA model with charges on the phosphate group beads set to zero. Experimental data are taken from: black circles [92]; khaki rhombi [93]; green squares [94]; blue triangles [95].

Table 4. Flexibility of the CG-DNA as a function of concentration of monovalent salt

Cation concentration [1], (mM)	Persistent length, L_C (Å)	Torsional persistence length, L_T (Å)
0.104–0.340	1065 ± 35	1446
0.303–0.545	912 ± 20	1461
0.970–2.270	803 ± 17	1415
2.910–4.210	633 ± 7	1463
9.730–11.030	606 ± 10	1409
29.800–35.200	431 ± 4	1424
99.600–105.000	451 ± 3	1399
No charge	316 ± 1	1349

[1] Note: it is incorrect to define the NaCl concentration (C_{NaCl}) alone as the parameter characterizing the ionic environment since the dissociated DNA counterions contribute to the ion concentration in the simulation cell. Therefore a range of concentrations are used where lower value is C_{NaCl} and the higher one is $C_{NaCl} + C_P$ (where C_P is total concentration of DNA counterions).

Recently, the dependence of DNA elastic properties on ionic environment and analysis of the relative contributions of electrostatic (due to phosphate-phosphate repulsion) and structural (mostly due to base-base stacking) factors to the DNA stiffness has been investigated using all-atom MD simulations followed by so-called group renormalization calculation of effective potentials used in a CG model of DNA [25,86,87]. The CG model of DNA developed in the Papoian laboratory uses different coarse graining of the DNA (without central beads) and employs explicit modeling of the ionic environment but the method used for derivation of intramolecular potentials from all-atom MD simulations is similar to the approach used in the present work. It was found that agreement between the experimental and calculated dependence of the persistent length on concentration of monovalent salt could be achieved if the force constants derived from the all-atom DNA MD simulations, using the Amber10 force field were scaled by factor 0.7 [25]. Here, we obtained excellent agreement with experiment without adjustment of the potentials derived by the IMC procedure from all-atom DNA simulations using the CHARMM27 force field. This discrepancy might be due the different all-atom force fields used in our work (CHARM27) compared to the

Amber10 force field used in the Papoian group work [25,86,87]. Another difference is the detailed coarse-graining of DNA, which employed somewhat different strategies including the use of explicit bending angle potentials in the present work. Additionally, in the present work the electrostatic interactions were described within a continuum model while Papoian and co-workers used ionic interaction potentials derived from the all-atom simulations. Despite these differences the results on the salt dependence of the persistence length from the approaches are similar. Specifically, in the present work we obtained the same estimation for the value of persistent length of the uncharged CG DNA as that obtained by Papoian and co-workers (about 300 Å) [86]). Correspondingly, our results support an analysis of the role of electrostatic forces on DNA stiffness with the conclusion that electrostatic and structural contributions are of similar scale at physiological salt concentrations [25,86,87] which is in contrast to predictions of both the Manning theory (supporting a major role of electrostatics) [96] and the Odjik, Skolnick, and Fixman model (suggesting a major contribution to structural stiffness) [97,98].

Calculation of Torsion Persistence Length of the DNA CG-Model

The torsion persistence length is determined from the following procedure. A torsion angle is defined between all neighboring monomers using vectors from the central bead position to COMs of adjacent phosphates. A torsion angle between more separated monomers (like n and $n + j, j > 1$ at a contour distance L_C) is determined as the sum of torsion angles between all intermediate neighbors. Then the torsional persistence length is defined from [99]:

$$L_T = \frac{L_C}{\delta\varphi^2} \tag{7}$$

where $\delta\varphi^2 = \langle\varphi^2\rangle - \langle\varphi\rangle^2$ is variation of the torsion angle φ between monomers separated by L_C distance.

In all MD simulations of the 22 bp CG DNA, the dependence L_C *versus* $\delta\varphi^2$ shows excellent linearity for full range of L_C giving values of L_T in the range 1350–1500 Å (Table 4). The value obtained here for the torsional persistence length is larger than what is known from the limited available experimental data which give a value around 700 Å [41,100]. Our results with L_T values around 1400 Å thus show that the torsional resistance in the CG model is exaggerated. However, the salt-dependence displays minor effect on the torsional rigidity of DNA. To the best of our knowledge such data on the salt dependence of the torsional DNA rigidity has not been published but as the phosphate-phosphate repulsion is not expected to vary much with torsion, this is the expected result (that indeed has been confirmed, Jan Lipfert, personal communication). The optimization of this feature in the CG model would require fine tuning of the force constants in order to give a shorter torsional persistence length, while still preserving the excellent behavior of the bending persistence length as a function of salt concentration. This will pursued in future work, introducing further development of the CG model.

CONCLUSIONS

We present a CG DNA model that consist of two types of beads representing a set of two base pairs, one P bead for each phosphate group and one central D bead for the pentose and bases of a two-bp unit (Figure 1). The internal structural and dynamic properties of the DNA chain were defined by a set of harmonic bead-bead bond and angular force constants. Using all-atom MD simulations of DNA oligonucleotides in solution, we applied the inverse MC method to extract the force constant parameters defining the internal properties of the DNA polymer chain. By applying this model in large scale CG simulations of a 200 bp DNA molecule with explicit ions at varying salt conditions within a continuum description of the electrostatic interactions, the DNA flexibility as a function of monovalent salt was analyzed by calculation of the DNA bending persistence length, L_P. The CG model combined with the parameters obtained directly from the all-atom simulation without further adjustment or optimization, resulted in excellent agreement for the dependence of L_P on monovalent salt concentration when compared with available experimental data. The results are in agreement with a recently developed two-bead (per bp) CG DNA model [24,25,86,87] in which the internal force parameters of the DNA chain was extracted from all-atom simulations in a manner similar to that employed in the present work. However, one difference between the results of the present model and those obtained from the model developed by Papoian and co-workers, is that we obtained agreement with experiments without further optimization or scaling of the all-atom extracted internal DNA force parameters. This could be due to the different force field used in the all-atom simulations of DNA (CHARMM27 in our case *versus* Amber10 in the Papoian group case) and to the differences in the coarse-graining of DNA or may be related to different form of ionic potentials used.

We also made preliminary analysis of the torsional persistence length of our CG DNA model. While the model produce the expected independence of this parameter on salt, the absolute values for torsional persistence length seem too large and will require further optimization in future work.

The conclusions from our analysis regarding the relative contributions of the electrostatic and elastic forces that defines the bending properties of DNA under physiological conditions is also in agreement with the work of Papoian, Savelyev, and co-workers [25,86,87] and suggest that these two contributions are of equal importance in determining the stiffness of DNA. Our model should represent an important step towards modeling of DNA bending properties in CG multi-scale modeling of a chromatin fiber, for which the description of the stiffness of linker DNA is highly important in order to reproduce chromatin folding induced by multivalent cations [81]. It may also be noted that, although the present approach is based on a CG DNA model without base sequence identity, this can be introduced in a rather straightforward way by assigning different types of D beads for variable two-bp sequence units. The CG DNA internal force field parameters defining such sequence effects can be obtained from all-atom MD simulations of oligonucleotides of varying base sequences.

ACKNOWLEDGMENTS

This work has been supported by the Singapore Agency for Science Technology and Research (A*STAR) through the Biomedical Research Council (LN), by the Singapore Ministry of Education Academic Research Fund (AcRF) through a Tier 1 grant (LN) and by the Swedish Research Council (to APL).

AUTHOR CONTRIBUTIONS

N.K., A.P.L. and L.N. designed and planned simulations and analysis and wrote the paper; N.K., D.L. and A.L. performed simulations and analysis of data.

REFERENCES

1. Widom, J. Physicochemical studies of the folding of the 100 A nucleosome filament into the 300 A filament. Cation dependence. *J. Mol. Biol.* 1986, *190*, 411–424.
2. Clark, D.J.; Kimura, T. Electrostatic mechanism of chromatin folding. *J. Mol. Biol.* 1990, *211*, 883–896.
3. Korolev, N.; Vorontsova, O.V.; Nordenskiöld, L. Physicochemical analysis of electrostatic foundation for DNA-protein interactions in chromatin transformations. *Prog. Biophys. Mol. Biol.* 2007, *95*, 23–49.
4. Korolev, N.; Berezhnoy, N.V.; Eom, K.D.; Tam, J.P.; Nordenskiöld, L. A universal description for the experimental behavior of salt-(in) dependent oligocation-induced DNA condensation. *Nucleic Acids Res.* 2009, *37*, 7137–7150.
5. Bloomfield, V.A. DNA condensation. *Curr. Opin. Struct. Biol.* 1996, *6*, 334–341.
6. Iwaki, T.; Saito, T.; Yoshikawa, K. How are small ions involved in the compaction of DNA molecules? *Surf. B Biointerf.*2007, *56*, 126–133.
7. Ray, J.; Manning, G.S. Effect of counterion valence and polymer charge density on the pair potential of two polyions. *Macromol.* 1997, *30*, 5739–5744.
8. Lyubartsev, A.P.; Nordenskiöld, L. Monte Carlo simulation study of ion distribution and osmotic pressure in hexagonally oriented DNA. *J. Phys. Chem.* 1995, *99*, 10373–10382.
9. Gronbech-Jensen, N.; Mashl, R.J.; Bruinsma, R.F.; Gelbart, W.M. Counterion-induced attraction between rigid polyelectrolytes. *Phys. Rev. Lett.* 1997, *78*, 2477–2480.
10. Guldbrand, L.; Nilsson, L.G.; Nordenskiöld, L. A Monte Carlo simulation study of electrostatic forces between hexagonally packed

DNA double helices. *J. Chem. Phys.* 1986, *85*, 6686–6698.

11. Nilsson, L.G.; Guldbrand, L.; Nordenskiöld, L. Evaluation of the electrostatic osmotic pressure in an infinite system of hexagonally oriented DNA molecules. A Monte Carlo simulation study. *Mol. Phys.* 1991, *72*, 177–192.
12. Lyubartsev, A.P.; Nordenskiöld, L. Monte Carlo simulation study of DNA polyelectrolyte properties in the presence of multivalent polyamine ions. *J. Phys. Chem. B* 1997, *101*, 4335–4342.
13. Arya, G.; Schlick, T. Role of histone tails in chromatin folding revealed by a mesoscopic oligonucleosome model. *Proc. Natl. Acad. Sci. USA.* 2006, *103*, 16236–16241.
14. Arya, G.; Schlick, T. A tale of tails: How histone tails mediate chromatin compaction in different salt and linker histone environments. *J. Phys. Chem. A* 2009, *113*, 4045–4059.
15. Muhlbacher, F.; Schiessel, H.; Holm, C. Tail-induced attraction between nucleosome core particles. *Phys. Rev. E* 2006, *74*.
16. Saunders, M.G.; Voth, G.A. Coarse-graining of multiprotein assemblies. *Curr. Opin. Struct. Biol.* 2012, *22*, 144–150.
17. Takada, S. Coarse-grained molecular simulations of large biomolecules. *Curr. Opin. Struct. Biol.* 2012, *22*, 130–137.
18. Lyubartsev, A.P.; Laaksonen, A. Calculation of effective interaction potentials from radial distribution functions: A reverse Monte Carlo approach. *Phys. Rev. E* 1995, *52*, 3730–3737.
19. Lyubartsev, A.P.; Mirzoev, A.; Chen, L.J.; Laaksonen, A. Systematic coarse-graining of molecular models by the Newton inversion method. *Faraday Discuss.* 2010, *144*, 43–56.
20. Izvekov, S.; Voth, G.A. Multiscale coarse-graining method for biomolecular systems. *J. Phys. Chem. B* 2005, *109*, 2469–2473.
21. Ayton, G.S.; Noid, W.G.; Voth, G.A. Multiscale modeling of biomolecular systems: In serial and in parallel. *Curr. Opin. Struct. Biol.* 2007, *17*, 192–198.
22. Chaimovich, A.; Shell, M.S. Coarse-graining errors and numerical optimization using a relative entropy framework. *J. Chem. Phys.* 2011, *134*.

23. Rudzinski, J.F.; Noid, W.G. Coarse-graining entropy, forces, and structures. *J. Chem. Phys.* 2011, *135*.
24. Savelyev, A.; Papoian, G.A. Molecular renormalization group coarse-graining of polymer chains: application to double-stranded DNA. *Biophys. J.* 2009, *96*, 4044–4052.
25. Savelyev, A.; Papoian, G.A. Chemically accurate coarse graining of double-stranded DNA. *Proc. Natl. Acad. Sci. USA*2010, *107*, 20340–20345.
26. Potoyan, D.A.; Savelyev, A.; Papoian, G.A. Recent successes in coarse-grained modeling of DNA. *WIREs Comput. Mol. Sci.* 2013, *3*, 69–83.
27. Dans, P.D.; Zeida, A.; MacHado, M.R.; Pantano, S. A coarse grained model for atomic-detailed DNA simulations with explicit electrostatics. *J. Chem. Theory Comput.* 2010, *6*, 1711–1725.
28. Demille, R.C.; Cheatham, T.E.; Molinero, V. A coarse-grained model of DNA with explicit solvation by water and ions. *J. Phys. Chem. B* 2011, *115*, 132–142.
29. Cao, Q.; Zuo, C.; Ma, Y.; Li, L.; Zhang, Z. Interaction of double-stranded DNA with a nanosphere: A coarse-grained molecular dynamics simulation study. *Soft Matter* 2011, *7*, 506–514.
30. Freeman, G.S.; Hinckley, D.M.; de Pablo, J.J. A coarse-grain three-site-per-nucleotide model for DNA with explicit ions. *J. Chem. Phys.* 2011, *135*.
31. Hinckley, D.M.; Freeman, G.S.; Whitmer, J.K.; de Pablo, J.J. An experimentally-informed coarse-grained 3-site-per-nucleotide model of DNA: Structure, thermodynamics, and dynamics of hybridization. *J. Chem. Phys.* 2013, *139*.
32. Cragnolini, T.; Derreumaux, P.; Pasquali, S. Coarse-grained simulations RNA and DNA duplexes. *J. Phys. Chem. B*2013, *117*, 8047–8060.
33. Doye, J.P.; Ouldridge, T.E.; Louis, A.A.; Romano, F.; Šulc, P.; Matek, C.; Snodin, B.E.; Rovigatti, L.; Schreck, J.S.; Harrison, R.M.; *et al.* Coarse-graining DNA for simulations of DNA nanotechnology. *Phys. Chem. Chem. Phys.* 2013, *15*, 20395–20414.
34. Ouldridge, T.E.; Louis, A.A.; Doye, J.P.K. DNA nanotweezers studied

with a coarse-grained model of DNA. *Phys. Rev. Lett.* 2010, *104*.

35. Ouldridge, T.E.; Louis, A.A.; Doye, J.P.K. Structural, mechanical, and thermodynamic properties of a coarse-grained DNA model. *J. Chem. Phys.* 2011, *134*.
36. Muhlbacher, F.; Holm, C.; Schiessel, H. Controlled DNA compaction within chromatin: the tail-bridging effect. *Europhys. Lett.* 2006, *73*, 135–141.
37. Wedemann, G.; Langowski, J. Computer simulation of the 30-nanometer chromatin fiber. *Biophys. J.* 2002, *82*, 2847–2859.
38. Korolev, N.; Lyubartsev, A.P.; Nordenskiöld, L. Computer modeling demonstrates that electrostatic attraction of nucleosomal DNA is mediated by histone tails. *Biophys. J.* 2006, *90*, 4305–4316.
39. Langowski, J. Polymer chain models of DNA and chromatin. *Eur. Phys. J. E* 2006, *19*, 241–249.
40. Voltz, K.; Trylska, J.; Tozzini, V.; Kurkal-Siebert, V.; Langowski, J.; Smith, J. Coarse-grained force field for the nucleosome from self-consistent multiscaling. *J. Comp. Chem.* 2008, *29*, 1429–1439.
41. Langowski, J.; Schiessel, H. Chromatin Simulations. from DNA to Chromatin Fibers. In *Computational Studies of RNA and DNA*; Šponer, J., Lankaš, F., Eds.; Springer: Dordrecht, the Netherlands, 2006; pp. 605–634.
42. Arya, G.; Schlick, T. Efficient global biopolymer sampling with end-transfer configurational bias Monte Carlo. *J. Chem. Phys.* 2007, *126*, 044107.
43. Langowski, J.; Heermann, D.W. Computational modeling of the chromatin fiber. *Semin. Cell Dev. Biol.* 2007, *18*, 659–667.
44. Kepper, N.; Foethke, D.; Stehr, R.; Wedemann, G.; Rippe, K. Nucleosome geometry and internucleosomal interactions control the chromatin fiber conformation. *Biophys. J.* 2008, *95*, 3692–3705.
45. Stehr, R.; Kepper, N.; Rippe, K.; Wedemann, G. The effect of internucleosomal interaction on folding of the chromatin fiber. *Biophys. J.* 2008, *95*, 3677–3691.
46. Grigoryev, S.A.; Arya, G.; Correll, S.; Woodcock, C.L.; Schlick, T. Evidence for heteromorphic chromatin fibers from analysis of

nucleosome interactions. *Proc. Natl. Acad. Sci. USA* 2009, *106*, 13317–13322.

47. Yang, Y.; Lyubartsev, A.P.; Korolev, N.; Nordenskiöld, L. Computer modeling reveals that modifications of the histone tail charges define salt-dependent interaction of the nucleosome core particles. *Biophys. J.* 2009, *96*, 2082–2094.
48. Stehr, R.; Schöpflin, R.; Ettig, R.; Kepper, N.; Rippe, K.; Wedemann, G. Exploring the conformational space of chromatin fibers and their stability by numerical dynamic phase diagrams. *Biophys. J.* 2010, *98*, 1028–1037.
49. Kepper, N.; Ettig, R.; Stehr, R.; Marnach, S.; Wedemann, G.; Rippe, K. Force spectroscopy of chromatin fibers: Extracting energetics and structural information from Monte Carlo simulations. *Biopolymers* 2011, *95*, 435–447.
50. Materese, C.K.; Savelyev, A.; Papoian, G.A. Counterion atmosphere and hydration patterns near a nucleosome core particle. *J. Am. Chem. Soc.* 2009, *131*, 15005–15013.
51. Voltz, K.; Trylska, J.; Calimet, N.; Smith, J.C.; Langowski, J. Unwrapping of nucleosomal DNA ends: A multiscale molecular dynamics study. *Biophys. J.* 2012, *102*, 849–858.
52. Hess, B.; Holm, C.; van der Vegt, N. Modeling multibody effects in ionic solutions with a concentration dependent dielectric permittivity. *Phys. Rev. Lett.* 2006, *96*.
53. Korolev, N.; Lyubartsev, A.P.; Nordenskiöld, L. Cation-induced polyelectrolyte-polyelectrolyte attraction in solutions of DNA and nucleosome core particles. *Adv. Coll. Interf. Sci.* 2010, *158*, 32–47.
54. Korolev, N.; Allahverdi, A.; Yang, Y.; Fan, Y.; Lyubartsev, A.P.; Nordenskiöld, L. Electrostatic origin of salt-induced nucleosome array compaction. *Biophys. J.* 2010, *99*, 1896–1905.
55. Allahverdi, A.; Yang, R.; Korolev, N.; Fan, Y.; Davey, C.A.; Liu, C.F.; Nordenskiöld, L. The effects of histone H4 tail acetylations on cation-induced chromatin folding and self-association. *Nucleic Acids Res.* 2011, *39*, 1680–1691.
56. Fan, Y.; Korolev, N.; Lyubartsev, A.P.; Nordenskiöld, L. An advanced coarse-grained nucleosome core particle model for computer

simulations of nucleosome-nucleosome interactions under varying ionic conditions. *PLoS One* 2013, *8*, e54228.

57. Bertin, A.; Mangenot, S.; Renouard, M.; Durand, D.; Livolant, F. Structure and phase diagram of nucleosome core particles aggregated by multivalent cations. *Biophys. J.* 2007, *93*, 3652–3663.
58. Leforestier, A.; Dubochet, J.; Livolant, F. Bilayers of nucleosome core particles. *Biophys. J.* 2001, *81*, 2114–2421.
59. Mangenot, S.; Leforestier, A.; Durand, D.; Livolant, F. Phase diagram of nucleosome core particles. *J. Mol. Biol.* 2003, *333*, 907–916.
60. Li, G.; Levitus, M.; Bustamante, C.; Widom, J. Rapid spontaneous accessibility of nucleosomal DNA. *Nat. Struct. Mol. Biol.* 2005, *12*, 46–53.
61. Böhm, V.; Hieb, A.R.; Andrews, A.J.; Gansen, A.; Rocker, A.; Tóth, K.; Luger, K.; Langowski, J. Nucleosome accessibility governed by the dimer/tetramer interface. *Nucleic Acids Res.* 2011, *39*, 3093–3102.
62. Mirzoev, A.; Lyubartsev, A.P. MagiC: Software package for multiscale modeling. *J. Chem. Theory Comput.* 2013, *9*, 1512–1520.
63. Mirzoev, A.; Lyubartsev, A.P. Systematic implicit solvent coarse graining of DMPC lipids. *J. Comput. Chem.* 2014. in press.
64. Korolev, N.; Lyubartsev, A.P.; Rupprecht, A.; Nordenskiöld, L. Experimental and Monte Carlo simulation studies on the competitive binding of Li^+, Na^+, and K^+ ions to DNA in oriented DNA fibers. *J. Phys. Chem. B* 1999, *103*, 9008–9019.
65. Korolev, N.; Lyubartsev, A.P.; Rupprecht, A.; Nordenskiöld, L. Competitive binding of Mg^{2+}, Ca^{2+}, Na^+, and K^+ to DNA in oriented DNA fibers: Experimental and Monte Carlo simulation results. *Biophys. J.* 1999, *77*, 2736–2749.
66. Montoro, J.C.G.; Abascal, J.L.F. Ionic distribution around simple DNA models. I. Cylindrically averaged properties. *J. Chem. Phys.* 1995, *103*, 8273–8284.
67. Macke, T.; Case, D.A. Modeling unusual nucleic acid structures. In *Molecular Modeling of Nucleic Acids*; Leontes, N.B., SantaLucia, J., Eds.; ACS Publications: Washington, DC, USA, 1998; pp. 379–393.

68. van der Spoel, D.; Lindahl, E.; Hess, B.; Groenhof, G.; Mark, A.E.; Berendsen, H.J.C. GROMACS: Fast, flexible and free. *J. Comp. Chem.* 2005, *26*, 1701–1718.

69. Lindahl, E.; Hess, B.; van der Spoel, D. Gromacs 3.0: A package for molecular simulations and trajectory analysis. *J. Mol. Mod.* 2001, *7*, 306–317.

70. Foloppe, N.; MacKerell, A.D., Jr. All-atom empirical force field for nucleic acids: I. Parameter optimization based on small molecule and condensed phase macromolecular target data. *J. Comp. Chem.* 2000, *21*, 86–104.

71. MacKerell, A.D., Jr.; Banavali, N. All-atom empirical force field for nucleic acids: II. Application to molecular dynamics simulations of DNA and RNA in solution. *J. Comp. Chem.* 2000, *21*, 105–120.

72. Hart, K.; Foloppe, N.; Baker, C.M.; Denning, E.J.; Nilsson, L.; MacKerell, A.D., Jr. Optimization of the CHARMM additive force field for DNA: Improved treatment of the BI/BII conformational equilibrium. *J. Chem. Theory Comput.*2012, *8*, 348–362.

73. Mark, P.; Nilsson, L. Structure and dynamics of liquid water with different long range interaction truncation and temperature control methods in molecular dynamics simulations. *J. Comp. Chem.* 2002, *23*, 1211–1219.

74. Mark, P.; Nilsson, L. Structure and dynamics of the TIP3P, SPC, and SPC/E water models at 298 K. *J. Phys. Chem. A*2001, *105*, 9954–9960.

75. Darden, T.; York, D.M.; Pedersen, L.G. Particle mesh Ewald: An N-log(N) method for Ewald sums in large systems. *J. Chem. Phys.* 1993, *98*, 10089–10092.

76. Korolev, N.; Yu, H.; Lyubartsev, A.P.; Nordenskiöld, L. Molecular dynamics simulations demonstrate the regulation of DNA-DNA attraction by H4 histone tail acetylations and mutations. *Biopolymers* 2014. in press.

77. Lyubartsev, A.P.; Laaksonen, A. Osmotic and activity coefficient from effective potentials for hydrated ions. *Phys. Rev. E* 1997, *55*, 5689–5696.

78. Rashin, A.A.; Bukatin, M.A. A view of thermodynamics of hydration emerging from continuum studies. *Biophys. Chem.* 1994, *51*, 167–190.
79. Koehl, P. Electrostatics calculations: Latest methodological advance. *Curr. Opin. Struct. Biol.* 2006, *16*, 142–151.
80. Korolev, N.; Allahverdi, A.; Lyubartsev, A.P.; Nordenskiöld, L. The polyelectrolyte properties of chromatin. *Soft Matter* 2012, *8*, 9322–9333.
81. Korolev, N.; Fan, Y.; Lyubartsev, A.P.; Nordenskiöld, L. Modelling chromatin structure and dynamics: status and prospects. *Curr. Opin. Struct. Biol.* 2012, *22*, 151–159.
82. Korolev, N.; Lyubartsev, A.P.; Rupprecht, A.; Nordenskiöld, L. Competitive substitution of hexammine cobalt(III) for Na^+ and K^+ ions in oriented DNA fibers. *Biopolymers* 2001, *58*, 268–278.
83. Korolev, N.; Lyubartsev, A.P.; Nordenskiöld, L.; Laaksonen, A. Spermine: an "invisible" component in the crystals of *B*-DNA. A grand canonical Monte Carlo and molecular dynamics simulation study. *J. Mol. Biol.* 2001, *308*, 907–917.
84. Nordenskiöld, L.; Lyubartsev, A.P.; Korolev, N. DNA-DNA interaction. In *DNA Interactions with Polymers and Surfactants*; Dias, R.S., Lindman, B., Eds.; John Wiley & Sons, Inc.: London, UK, 2008; pp. 209–237.
85. Montoro, J.C.G.; Abascal, J.L.F. Ionic distribution around simple B-DNA models. II. Deviations from cylindrical symmetry. *J. Chem. Phys.* 1998, *109*, 6200–6210.
86. Savelyev, A.; Materese, C.K.; Papoian, G.A. Is DNA's rigidity dominated by electrostatic or nonelectrostatic interactions? *J. Am. Chem. Soc.* 2011, *133*, 19290–19293.
87. Savelyev, A. Do monovalent mobile ions affect DNA's flexibility at high salt content? *Phys. Chem. Chem. Phys.* 2012, *14*, 2250–2254.
88. Grosberg, A.Y.; Khokhlov, A.R. *Statistical Physics of Macromolecules*; AIP Press: New York, NY, USA, 1994; p. 350.
89. Åqvist, J. A simple way to calculate the axis of an α-helix. *Comput. Chem. (Oxford)* 1986, *10*, 97–99.
90. Christopher, J.A.; Swanson, R.; Baldwin, T.O. Algorithms for

finding the axis of a helix: Fast rotational and parametric least-squares methods. *Comput. Chem.* 1996, *20*, 339–345.

91. Enkhbayar, P.; Damdinsuren, S.; Osaki, M.; Matsushima, N. HELFIT: Helix fitting by a total least squares method. *Comput. Biol. Chem.* 2008, *32*, 307–310.
92. Hagerman, P.J. Investigation of the flexibility of DNA using transient electric birefringence. *Biopolymers* 1981, *20*, 1503–1535.
93. Baumann, C.G.; Smith, S.B.; Bloomfield, V.A.; Bustamante, C. Ionic effects on the elasticity of single DNA molecules. *Proc. Natl. Acad. Sci. USA* 1997, *94*, 6185–6190.
94. Rizzo, V.; Schellman, J. Flow dichroism of T7 DNA as a function of salt concentration. *Biopolym.* 1981, *20*, 2143–2163.
95. Lu, Y.; Weers, B.; Stellwagen, N.C. DNA persistence length revisited. *Biopolym.* 2002, *61*, 261–275.
96. Manning, G.S. The persistence length of DNA is reached from the persistence length of its null isomer through an internal electrostatic stretching force. *Biophys. J.* 2006, *91*, 3607–3616.
97. Odijk, T. Polyelectrolytes near the rod limit. *J. Polym. Sci.: Polym. Phys. Ed.* 1977, *15*, 477–483.
98. Skolnick, J.; Fixman, M. Electrostatic persistence length of a wormlike polyelectrolyte. *Macromol.* 1977, *10*, 944–948.
99. Lipfert, J.; Wiggin, M.; Kerssemakers, J.W.; Pedaci, F.; Dekker, N.H. Freely orbiting magnetic tweezers to directly monitor changes in the twist of nucleic acids. *Nat. Commun.* 2011, *2*, 439.
100. Hagerman, P.J. Flexibility of DNA. *Annu. Rev. Biophys. Biophys. Chem.* 1988, *17*, 265–286.

CHAPTER 8

MELTING DYNAMICS OF SHORT DSDNA CHAINS IN SALINE SOLUTIONS

Yichen He[1], Yazhuo Shang[1], Yu Liu[2], Shuangliang Zhao[2] and Honglai Liu[1]

[1] State Key Laboratory of Chemical Engineering and Department of Chemistry, East China University of Science and Technology, Shanghai 200237, China

[2] State Key Laboratory of Chemical Engineering, East China University of Science and Technology, Shanghai 200237, China

ABSTRACT

DNA melting has attracted much attention due to its importance in understanding the life-reproduction and metabolism and in the applications of modern DNA-based technologies. While numerous works have been contributed to the determination of melting profiles in diverse environments, the understanding of DNA melting dynamics is still limited. By employing three-site-per-nucleotide (3SPN) double-stranded DNA (dsDNA) model, we here demonstrate the melting dynamics of an isolated short dsDNA under different conditions (different temperatures, ionic concentrations and DNA chain lengths) can be accessed by coarse-grained simulation studies. We particularly show that at dilute ionic concentration the dsDNA, regardless being symmetric or asymmetric, opens at both ends with roughly equal probabilities, while at high ionic concentration the asymmetric dsDNA chain opens at the A-T-rich end. The comparisons of our simulation results to available data are discussed, and overall good agreements have been found.

Keywords: DNA Melting dynamics Molecular simulation Ionic concentration

BACKGROUND

Denaturation of duplex DNA by heat leading to strand separation is known as DNA melting, and such phenomenon has caused more and more research interests in recent decades due to its importance in understanding the life-reproduction and metabolism. Primarily, a double-stranded DNA (dsDNA) dissociating into two single-stranded DNAs (ssDNA) lays a cornerstone for transcribing genetic information from DNA to RNA (Wang et al. 1998; Rief and Spudich 1999), and this process is indispensable during the replication and transcription of DNA. The DNA denaturation is also utilized in developing modern DNA-based technologies (Mirkin et al. 1996; Jin et al. 2003; Wu et al. 2011; Belozerova and Levicky 2012; Song et al. 2013; Wu et al. 2013). For instances, Jin et al. (2003) showed that the well-defined characteristic of the melting transition enables the detection of single-base mismatches between a probe and target DNAs. By measuring the dissociation rates of dsDNA at different temperatures, Reed and Wittwer proposed a method for tracing food products with High Resolution Melting Analysis (Reed and Wittwer 2004).

DNA melting process is usually characterized with the melting temperature. Melting temperature is conventionally defined as the temperature at which the probabilities of finding a given DNA in the hybridized and dissociated states are the same. The melting temperature is closely related to the environment to which the DNA is subjected

(Sorokin et al. 1997; Mrevlishvili et al. 1998; Goobes et al. 2003; Khimji et al. 2013; Nakano et al. 2014). Mrevlishvili et al. (Mrevlishvili et al. 1998) found that the DNA melting temperature increases with the increase of the ionic concentration, and the same conclusion has been drawn by Sorokin et al. through a study on the influences of ions on the thermo-stability of dsDNA (Sorokin et al. 1997). Goobes et al. (2003) added crowding reagent PEG into DNA solution, and found that inertia polymers are in favor with the thermostability of DNA. However, the added reagents may also bring opposite effect due to their molecular interactions with the basepairs in dsDNA. Nakano et al. (2014) found that the melting temperature decreases in the presence of PEG, especially in the region with high PEG concentrations. Different reagents may have different effects on DNA melting depending on the competition between entropic effect and correlation effect. Indeed, Khimji et al. (2013) concluded that the melting temperature of duplex DNA is much higher in polyanions than in non-ionic polymers with similar ionic strength due to an additional electrostatic contribution beyond the excluded volume effect.

Apart from these experimental studies, many coarse-grained theoretical models have been developed to address the DNA melting profile (Drukker and Schatz 2000; Knotts et al. 2007; Sambriski et al. 2009a, b; Liu et al. 2010; Ouldridge et al. 2011; Freeman et al. 2011; De Pablo 2011; Liu et al. 2011, 2012). Drukker and Schatz (2000) proposed a beads-spring model which allows for the investigation of hybridization by using simulation. In this model one nucleotide is represented by two sites, and one is for skeleton and the other for base. The model by Ouldridge et al. (2011) is specifically targeted to reproduce the thermodynamics of DNA melting. However, the model cannot account for electrostatics and sequence specificity. Recently, Knotts and De Pablo et al. (Knotts et al. 2007; Sambriski et al. 2009a, b; Freeman et al. 2011; De Pablo 2011) proposed an advanced three-site-per-nucleotide (3SPN) model and its extensions. This model faithfully captures the characteristics of major and minor grooves and the elasticity of DNA, and it involves an ion-concentration dependent long-range electrostatic interaction between the phosphates in two strands. This model is parameterized by fitting the thermal melting experimental data in a constant ionic strength, and unsurprisingly it can well predict the behaviors of DNA chain including melting profile, bubble formations and the mechanical properties of the molecule as a function of ionic concentration. Besides the simulation models, thermodynamic models are also developed for studying the melting process. Very recently, Liu et al. (2010) developed an interesting molecular thermodynamic model from the perspective of phase equilibrium between the association and dissociation states of each basepair. By constructing a proper free energy function which involves the melting enthalpy contribution from each basepair, they showed that the melting curves and melting temperatures of dsDNA in ionic and crowded solutions or in confined spaces could be successfully predicted (Liu et al. 2010, 2011, 2012).

Two different mechanisms on DNA melting have been discussed: slithering and unzipping. It's widely accepted that long DNA chains display the melting behavior through slithering, while short DNA chains through unzipping (see Fig. 1b for illustration). In addition, insightful investigations on the melting of DNAs with B-form helical geometries have also been reported (Giudice et al. 2001, 2003; Várnai and Lavery 2002; Banavali and MacKerell 2002; Hagan et al. 2003; Giudice and Lavery 2003; Sambriski et al. 2009a, b). By using transition-path sampling, Hagan et al. (2003) have found that the unzipping of dsDNA proceeds via two qualitatively different pathways: in the first one the flipping base breaks its intramolecular hydrogen bonds before it unstacks, and in the other one it ruptures both sets of interactions simultaneously. Almost in all cases, unzipping occurs following a stepwise process in which one base unstacks significantly before the other. The hydrogen bonds break mainly through twisting motions of the bases. Giudice and Lavery (2003) concluded that the base pair lifetimes for canonical B-DNA directly reflect the difficulty of breaking the intramolecular hydrogen bonds and local stacking interactions, while the unusually long lifetimes for A-T pairs within A-tracts reflect the additional constraints imposed by a narrow and rigid minor groove. Sambriski et al. (2009a, b) proposed that in terms of chain length, the free energy barrier of longer oligonucleotides (30 versus 15 base pairs) is higher and slightly narrower, due to increased sharpness associated with the transition. Low ionic strength tends to decrease free energy barriers.

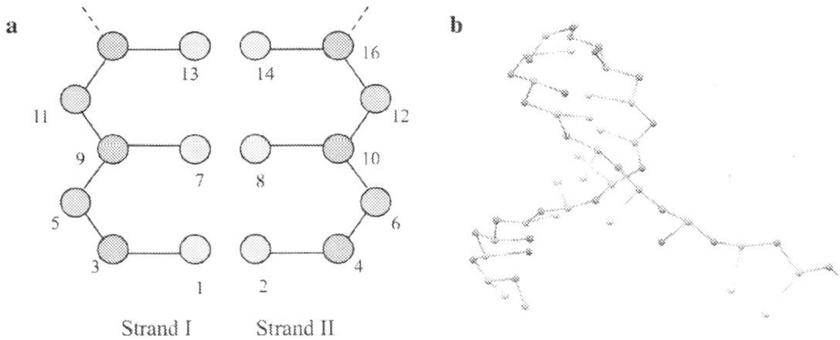

Figure 1: a Illustration of 3SPN model for dsDNA: blue coloured circle denotes base bead, yellow coloured circle for sugar and orange coloured circle for phosphate; **b** illustrative process of DNA unzipping.

While numerous experimental and theoretical works have been contributed to the determination of melting profile of dsDNA in diverse environments, the melting dynamics of dsDNA, especially for those with short chain lengths, is less understood (Drukker et al. 2001; Banerjee and Pal 2007; Wong and Pettitt 2008; Kenward and Dorfman 2009). Generally, the access of intermediate melting transition state is of great interest. Such information is crucial because it's helpful to explore the sequence and end effects and to reveal the role of crowding or confinement environment for DNA melting (Perez and Orozco 2010; Toma et al. 2014).

The aim of present work is to study the dynamics of DNA melting in dilute brine environment. The denaturation of short DNA chains is usually employed in Biosensing (Sendroiu et al. 2011; Toma et al. 2014; Loget and Corn 2014) and grafting or anchor technology (Ma et al. 2009; Seela et al. 2011; Zhao et al. 2012; Woller et al. 2012). Besides, it is straight-forward to explore the mechanism of DNA melting by using short DNA chains than using long DNA chains (Wong and Pettitt 2008; Miyoshi and Sugimoto 2008).

MODELLING AND SIMULATION DETAILS

The 3SPN model developed by Knotts et al. (2007) is employed here to explore the denaturation dynamics of short DNA. The model was developed initially for the investigation of the melting curve. However, as we demonstrate below, it can be utilized to qualitatively explain the melting dynamics. The system under consideration is composed of an isolated dsDNA chain with certain chain sequence. The electrolyte solution is characterized as a continuum with an ion-concentration dependent Debye length scaling the interaction between two charged phosphate beads, i.e.,

$$\kappa_D = \left(\frac{\varepsilon_0 \varepsilon_k RT}{2N_A^2 e_q^2 I} \right)^{0.5}. \tag{1}$$

Here κ_D is Debye length, and I is ionic strength. ε_0 and ε_k are vacuum permittivity and the water relative dielectric constant at room temperature, respectively. T is the absolute temperature of the system and R the universal gas constant; N_A is the Avogadro constant and e_q the elementary charge. As demonstrated by Knotts et al. (2007) and Sambriski et al. (2009a, b), such implicit solvent description allows for capturing the ionic concentration effect on melting profile.

The geometry and force field of 3SPN model are detailed elsewhere (Knotts et al. 2007; Sambriski et al. 2009a, b), and here we only give a brief introduction. As shown in Fig. 1a, the 3SPN dsDNA model is composed of two ssDNA strands, and in each strand three kinds of beads are included which represent sugar, phosphate and nucleotide. Each phosphate bead carries a negative elementary charge, and the nucleotide can be Adenine (A), Guanine (G), Cytosine (C) or Thymine (T). For different basepairs, i.e., A-T and G-C, the hydrogen bonding interactions are distinct. At natural state, the geometry of dsDNA can be characterized by the lengths of the connecting bonds and the values of the triangle and dihedral angles. Besides, the backbones of the second chain should be reproduced by simply rotating the first chain along the axis and then shifting up with a given height for a double helix structure. Under perfect matching condition, the structure of a dsDNA is determined once the sequence of nucleotides in either strand is specified, and in this circumstance we conventionally use a sequence composed of A, G, C and T to represent a dsDNA chain.

The interactions involved in 3SPN model include three sets of contributions. The first one is intra-strand interaction accounting for the structural stability of single strand, and the second one accounts for the interaction between two strands involving the electrostatic repulsion between the phosphates in both strands and the basepairing interaction.

This baseparing interaction is responsible for the hybridization, and provides a key energy barrier from DNA melting. The last contribution accounts for the elasticity of DNA chain, and specifically the electrostatic interaction among the phosphate beads are included, and the strength of this solvent-mediate interaction is dependent on the ion-concentration in solution. The mathematical formats of those interactions and the involved parameters can be found in the original work (Knotts et al. 2007; Sambriski et al. 2009a, b).

In the present work, four samples of short DNA chains are considered for accessing the melting dynamics. The sequences and numbers of corresponding basepairs (bp) are listed in Table 1.

Table 1: Sequences of DNA chains considered in this work

DNA name	Sequence (5′ to 3′)
11 bp	ATC CGT ATG CG
16 bp	ATC CGT ATG CGA TCC G
21 bp	ATC CGT ATG CGA TCC GTA TGC
26 bp	ATC CGT ATG CGA TCC GTA TGC GAT CC

The reasons of choosing above target DNA samples are as follows: firstly, DNA chains of different chain lengths should be considered in order to access the length effect on DNA melting dynamics. Secondly, the sequence effect should be reflected, and thus the other three dsDNA samples are generated by simply repeating, in a whole or half manner, the sequence of the first sample. For simplicity, the four DNA chains are referred to as 11, 16, 21 and 26 bp. The melting temperatures of these dsDNA samples are reported to be within the range of 290–350 K depending on the chain length and the salt concentration (Owczarzy et al. 2004; Freeman et al. 2011).

To implement the force calculation during simulation, we introduce the following method to label the beads in dsDNA chain (see Fig. 1a for illustration): the beads in the first strand are labeled by odd numbers, and the beads in the second strand are labeled by even numbers. In each strand, the beads are labeled from the bottom up and from nucleotide to sugar site.

The Brownian dynamics simulation is carried in NVT ensemble, and the motion of bead i is governed by Langevin equation

$$m_i \frac{d^2\mathbf{r}_i}{dt^2} = -\nabla U_i - \gamma \frac{d\mathbf{r}_i}{dt} + \boldsymbol{W}_i(t), \tag{2}$$

where m_i and $\mathbf{r}_i$ are the mass and position, respectively, and U_i is the total potential on bead i. The random force $W_i(t)$ satisfies $\langle \boldsymbol{W}_i(t)\boldsymbol{W}_i(t')\rangle = 6k_BT\gamma\delta_{ij}\delta(t - t')$. Here γ is the friction coefficient with $\gamma = 0.05\ ps^{-1}$. The masses for different kinds of bead are different. The temperature is controlled by Nosé-Hoover chain thermostat (Martyna et al. 1992). The Nosé-Hoover (NH) and Langevin thermostats are combined mainly due to the following consideration: when a base pair is hybridized it is only weakly coupled to the solvent because the bonding interaction dominates in this circumstance. On the contrary, when the base pair is dissociated, the bases are in close contact with the solvent. The main difference between Nose–Hoover-Langevin (NHL) and Nosé-Hoover thermostats is the addition of a Langevin friction and noise terms. This combination has been initiated by Samoletov et al. (2007) for the investigation of DNA replication and transcription, and thereafter applied by many others (Frank and Gottwald 2011; Xu et al. 2014; Zhang et al. 2015).

The initial configuration of the DNA is generated following the method described by Knotts et al. (2007). The differential equation above is solved by using Velocity-verlet method iteratively with a time step of 0.01 ps. The simulation is started with 40,000 steps (0.4 ns) of steepest descent minimization followed by a 1 ns equilibration process during which the system temperature is set as room temperature. After equilibrium, a temperature jump to the final target temperature is carried out (Qamhieh et al. 2009) for studying the melting dynamics. The target temperatures, though high, are chosen to ensure the occurrence of melting within microsecond timescale. Under a given condition, 100 parallel simulations are run, and then the melting time is averaged and analyzed. Similar to the original work (Knotts et al. 2007), when the distance of basepair G-C is larger than 2.8694 Å or that of basepair A-T is larger than 2.9002 Å, and in addition two nucleotide beads move in opposite direction, the basepair is considered as being dissociated. Three factors that may affect the DNA melting dynamics are investigated including temperature, ionic concentration and chain length.

RESULTS AND DISCUSSIONS

Effect of temperature

Firstly, the temperature effect on the melting speed is investigated. To characterize this effect, the melting duration time is introduced. The melting duration time is defined as the interval of two time points at which the first and the last basepairs in a dsDNA chain are dissociated. As temperature increases, the average kinetic energy per bead becomes greater, and one would expect that the melting duration time is shorter. Figure 2 plots the melting time in terms of the number of dissociated basepairs in 11 bp at 0.05 mol/L ionic concentration at three temperatures. The melting duration time is determined by time difference between the first dissociation and the last one. For the sake of simplicity, we set the time point of the first dissociation as zero. Note that the first nucleotide in each strand is not paired due to the helix DNA structure, and thus only 9 basepairs within the 11 bp DNA. Three temperatures, i.e., 400, 460 and 520 K, are applied, and these temperatures are much beyond the melting temperature and thus ensure the occurrence of dissociation process. It's to be noted that although these temperatures are artificial for a physiological system, they enable the physical description and fast simulation studies for the temperature effect. The curves in Fig. 2 confirm that the melting duration time becomes less and less when the temperature increases from 400 to 460 and to 520 K. In addition, we notice that the dissociation of the second basepair generally takes longer time than the subsequent one. However, as temperature increases, such time difference becomes almost neglectable.

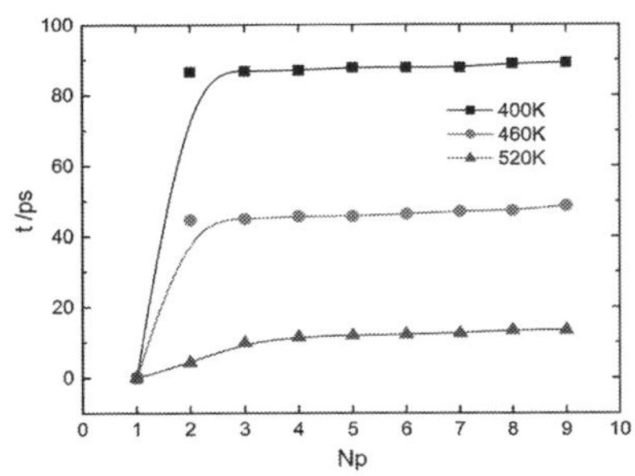

Figure 2: DNA melting time versus the number of dissociated basepairs in 11 bp at different temperatures (0.05 mol/L).

Our result generally accords with other simulation findings (Qamhieh et al. 2009; Perez and Orozco 2010). Qamhieh et al. (2009) performed molecular dynamics simulation to study the DNA melting with an atomistic model (Wong and Pettitt 2004), and they explicitly investigated the melting dynamics of 12-basepair ($A_{12}T_{12}$) B-DNA duplex tethered to a surface through its 5′-amine linker in the solution of 0.1 M NaCl. For comparison, their simulation data are converted by exchanging the X and Y coordinates and displayed in Fig. 3 (red curve with circles). In particular Qamhieh et al. studied the number of separated basepairs in terms of time and demonstrated that at 400 K the first dissociation of a basepair took much longer time than the other ones, and this generally agrees with our result.

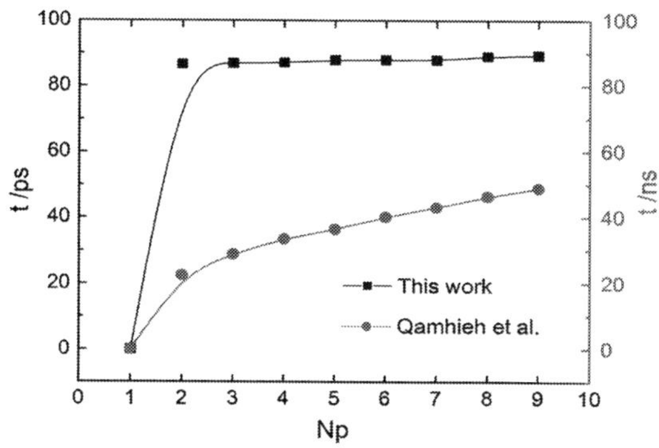

Figure 3: The comparison of DNA melting dynamics at 400 K. The black curve with blocks is from this work for 11 bp with ionic concentration 0.05 M, and the red curve with circles is from (Qamhieh et al. 2009) for a tethered 12 basepairs ($A_{12}{\cdot}T_{12}$) with ionic concentration 0.1 M.

The comparison in Fig. 3 shows that a qualitatively agreement between our simulation and Qamhieh's result can be found, i.e., the speed of basepair dissociation generally becomes larger and larger during the melting process. One may notice that the time scale of melting duration time in our study is much smaller than that in Qamhieh's work, and this is very likely due to the difference in modelling, i.e., in Qamhieh's work the dsDNA chain was described by using an atomistic model while in current work the DNA dissociation process is studied by using a coarse-grained model. Although coarse-grained simulation can give qualitatively or semi-quantitatively similar results as the corresponding all-atom simulation, the simulation time scales in both systems are generally believed to be different. In addition the dsDNA being tethered may present different melting dynamics as in a free space because the tethered dsDNA encounters an extra entropic force from the presence of wall (Zhao et al. 2011). Similar agreement can be obtained by comparing our simulation results to those in Perez's work. Perez and Orozco (2010) performed a simulation study to gain the real-time atomistic description of DNA unzipping, and they investigated the time evolution of the total number of hydrogen bonds for short dsDNA chains (12-mer) at 368 K, which can be easily converted to the time evolution of total number of separated basepairs. This comparison is very similar to Fig. 3 and thus is omitted here. The agreement with the atomistic simulation results rationalizes our calculation and the analysis on the melting dynamics below.

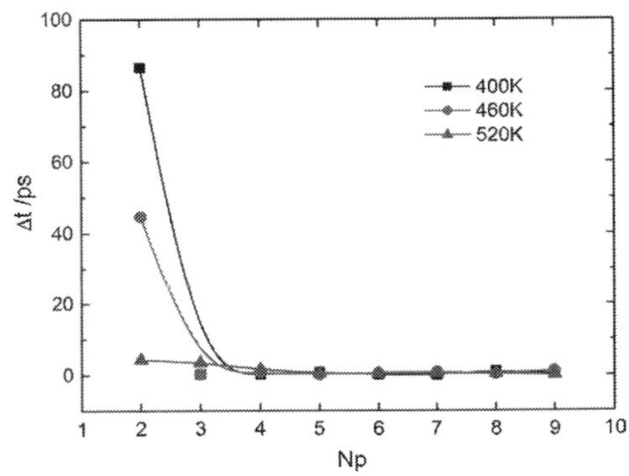

Figure 4: The melting interval time versus the number of dissociated basepairs at different temperatures (0.05 mol/L, 11 bp)

To further analyze the melting dynamics, we plot the time interval between the adjacent dissociations in Fig. 4. It shows that the time interval Δt quickly decreases with the increase in the number of separated base-pairs, and becomes almost a constant after the fourth dissociation occurs. This trend indicates that the posterior dissociations are nearly irrelevant with the sequence during the melting process. This is probably because the temperatures considered in current work are much beyond the melting temperature (usually at 350 K), and then the driving force for dissociation dominates while the energy difference in A-T and G-C hydrogen-bonding interactions becomes neglectable. At a relatively low temperature (400 K), the time interval for the dissociation of the second basepair is much longer than the other dissociation after it. Similar trend has been observed during the melting processes of other dsDNA samples.

Effect of sodium ion concentration

The concentration of salt plays an important role in DNA melting profile (Sorokin et al. 1997; Mrevlishvili et al. 1998), and one can expect the ionic concentration can affect the melting pathway as well. Figure 5 plots the melting time versus the total number of dissociated basepairs in 11 bp at different ionic concentrations. It shows that the DNA melting time becomes shorter in a saline solution with lower ionic concentration, namely a higher melting speed can be achieved by reducing the saline concentration, which elucidates that the stability of DNA can be improved by elevating the ionic concentration, and this trend is likely related to the positive correlation between thermostability of dsDNA and ionic concentration as discovered in experiments (Sorokin et al. 1997; Mrevlishvili et al. 1998). Indeed, the addition of ions can screen the electrostatic interaction between phosphate groups in two strands of dsDNA, leading to the improvement of DNA stability. Similar to the temperature effect as discussed in Fig. 2, the difference in dissociation time of each basepair gradually vanishes as the ionic concentration decreases.

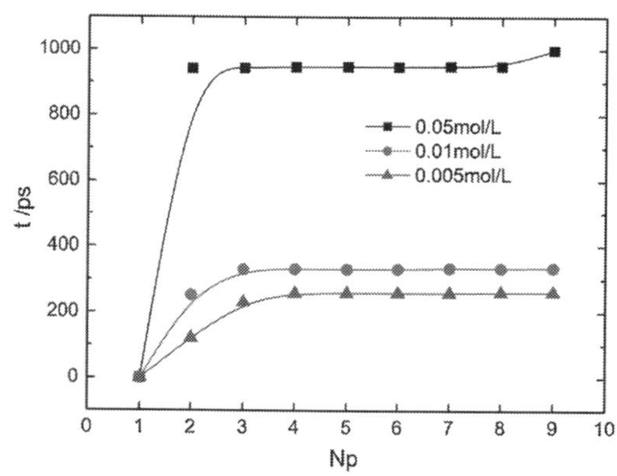

Figure 5: DNA melting time versus the total number of disassociated basepairs (Np) in different ionic concentrations (360 K, 11 bp).

It has been reported that the nature of the transition is controlled by end effects and sequence design (Montrichok et al. 2003). With the help of UV spectroscopy combined with a quenching method, Montrichok demonstrated that the presence of intermediate states could be quantified (Montrichok et al. 2003). By measuring the average fraction of open basepairs and the fraction of completely open molecules, they reported that the symmetric DNA chain (i.e., L42V1, a dsDNA chain with G-C-rich regions at both ends and an A-T-rich region in the middle) in 50 mM ionic concentration opens at both ends during transition, while the asymmetric DNA chain (i.e., L48AS, a dsDNA chain with a G-C-rich region at one end, and an A-T-rich region at the other) opens at one end. Those DNA transitions were characterized during the process of slowly increasing the temperature. The transition details are investigated in our simulation. Figure 6 plots the dissociating time point of the n-th basepair in 11 bp at different ionic concentrations, and it gives the time difference between the dissociations of n-th basepair and the first one and thus reflects the dissociation sequence of the basepairs. We find that the dsDNA opens statistically at one end at high ionic concentration. 11 bp is an asymmetric dsDNA with an A-T-rich region at one end and a G-C-rich region at the other. This melting behavior at high ionic concentration (50 mM) accords with the experimental findings (Montrichok et al. 2003).

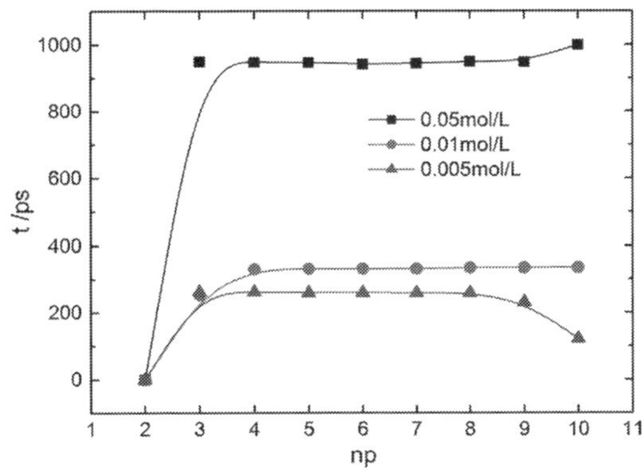

Figure 6: The dissociating time point of the n-th basepair (np) in 11 bp at different ionic concentrations at 360 K.

While at a low ionic concentration, we find the dsDNA opens at both ends during melting transition. This unusual phenomenon can be explained with the help of interaction energy analysis. As the ionic concentration increases, the Debye length κ_D in Eq. (1) decreases, and then the Coulombic repulsion between the phosphates in two strands, i.e., V_{qq}, greatly reduces as stated in Table 2. V_{qq}, together with the thermal fluctuation force, overcomes the hydrogen bonding attraction for transition. The total driving force that accounts for the dissociation dominates during the melting in our study. When V_{qq} decreases, the driving force decreases, and the net force for DNA transition becomes small. In this circumstance, energy difference on hydrogen-bonding interaction at the G-C-rich end and A-T-rich end becomes significant. This finally leads to a less symmetric probability for opening both ends. Such feature is similar to the reported substrate effect on DNA melting transition, i.e., Qamhieh et al. (2009) found that the melting transition of a tethered dsDNA on a surface was less symmetric than in the solution case where for dsDNA both ends melted with roughly equal probability.

Table 2: κ_D and V_{qq} in different ionic concentrations (360 K, 11 bp)

Ionic concentration/mol L^{-1}	0.005	0.01	0.05
κ_D/Å	47.203	33.378	14.927
V_{qq}/× 10^{16} J	3.772	3.164	1.509

Effect of chain length

Finally, the chain length effect on the DNA melting dynamics is investigated. Figure 7 plots the DNA melting duration time of different DNA samples at 520 K and at ionic concentration of 0.05 mol/L. As illustrated, the melting duration time is nearly proportional to the chain length. This trend is generally expected. At a high temperature, the melting time interval for each basepair is nearly a constant as discussed in Fig. 4. Because more basepairs in a long DNA chain are to be melted, the melting duration time is supposed to be proportional to the numbers of basepairs. At a low temperature (i.e., 400 K), the melting duration time is less dependent on chain length (not shown here) since the melting duration time at this condition is mainly dominated by the first dissociation regardless of the chain length.

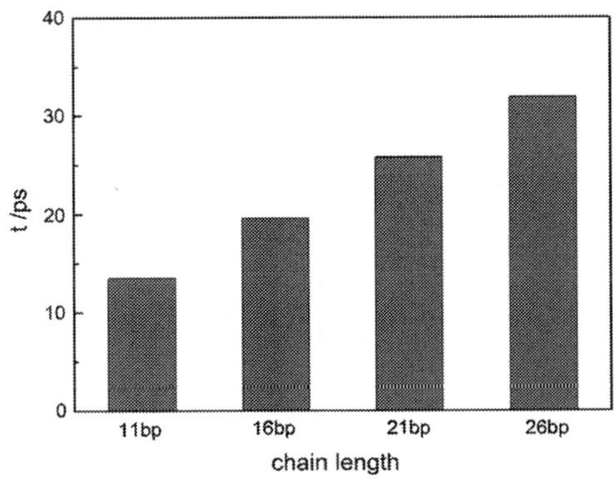

Figure 7: The melting duration time of DNA in different chain lengths (0.05 mol/L, 520 K).

CONCLUSIONS

In the present work, a series of coarse-grained molecular dynamics simulations have been performed for investigating the melting dynamics of short dsDNA chains. By accessing the temperature effect, ionic concentration effect, and chain length and sequence effects, we find that the melting dynamics of short dsDNA chain is no trivial. The addition of ion can improve the stability of DNA. Moreover, the DNA melting has two different pathways: opens at one end, and opens at two ends. At dilute ionic concentration, the dsDNA, regardless of being symmetric or asymmetric, opens at both ends with roughly equal probabilities; while at high ionic concentration, the asymmetric dsDNA chain opens at the A-T-rich end. The comparisons of our simulation results with available data are discussed, and overall good agreements have been found.

The current work reveals the melting dynamics of an isolated short dsDNA chain in saline solution. The majority of DNA in experimental study is long DNA chains and the physiological environment is crowded and geometrically confined. The melting dynamics in practical may be significantly distinct due to the additional steric effect and correlations between DNA chains. Toward that realistic study, a more complex modelling and simulation is required which represents the further direction of this work.

DECLARATIONS

Authors' contributions

SZ and HL conceived this study. YH carried calculation, and all authors performed data analysis. SZ and YH wrote the manuscript. All authors read and approved the final manuscript.

Acknowledgements

This work is supported by the National Key Basic Research Program of China (2014CB748500), National Natural Science Foundation of China (Nos. 21173079, 21206036), and the 111 Project of China (No. B08021) and the Fundamental Research Funds for the Central Universities of China. SZ also acknowledges the support of the Shanghai Science and Technology Committee Rising-Star Program (No. 14QA1401300).

REFERENCES

1. Banavali NK, MacKerell AD Jr (2002) Free energy and structural pathways of base flipping in a DNA GCGC containing sequence. J Mol Biol 319:141–160
2. Banerjee D, Pal SK (2007) Direct observation of essential DNA dynamics: melting and reformation of the DNA minor groove. J Phys Chem B 111:10833–10838
3. Belozerova I, Levicky R (2012) Melting thermodynamics of reversible DNA/ligand complexes at interfaces. J Am Chem Soc 134:18667–18676
4. De Pablo JJ (2011) Coarse-grained simulations of macromolecules: from DNA to nanocomposites. Annu Rev Phys Chem 62(1):555–574
5. Drukker K, Schatz GC (2000) A model for simulating dynamics of DNA denaturation. J Phys Chem B 104(26):6108–6111
6. Drukker K, Wu GS, Schatz GC (2001) Model simulations of DNA denaturation dynamics. J Chem Phys 114:579–590
7. Frank J, Gottwald GA (2011) The langevin limit of the Nose-Hoover-Langevin thermostat. J Stat Phys 143(4):715–724

8. Freeman GS, Hinckley DM, De Pablo JJ (2011) A coarse-grain three-site-per-nucleotide model for DNA with explicit ions. J Chem Phys 135:165104
9. Giudice E, Lavery R (2003) Nucleic acid base pair dynamics: the impact of sequence and structure using free-energy calculations. J Am Chem Soc 125:4998–4999
10. Giudice E, Va´rnai P, Lavery R (2001) Energetic and conformational aspects of A: T base-pair opening within the DNA double helix. Chem Phys Chem 2:673–677
11. Giudice E, Va´rnai P, Lavery R (2003) Base pair opening within B-DNA: free energy pathways for GC and AT pairs from umbrella sampling simulations. Nucleic Acid Res 31:1434–1443
12. Goobes R, Kahana N, Cohen O, Minsky A (2003) Metabolic buffering exerted by macromolecular crowding on DNA–DNA interactions: origin and physiological significance. Biochem 42(8):2431–2440
13. Hagan MF, Dinner AR, Chandler D, Chakraborty AK (2003) Atomistic understanding of kinetic pathways for single base-pair binding and unbinding in DNA. Proc Natl Acad Sci USA 100:13922–13927
14. Jin RC, Wu GS, Li Z, Mirkin CA, Schatz GC (2003) What controls the melting properties of DNA-linked gold nanoparticle assemblies? J Am Chem Soc 125:1643
15. Kenward M, Dorfman KD (2009) Brownian dynamics simulations of single-stranded DNA hairpins. J Chem Phys 130:095101
16. Khimji I, Shin J, Liu JW (2013) DNA duplex stabilization in crowded polyanion solutions. Chem Commun 49:1306–1308
17. Knotts TA, Rathore N, Schwartz DC, De Pablo JJ (2007) A coarse grain model for DNA. J Chem Phys 126(8):084901
18. Liu Y, Kermanpour F, Liu HL, Hu Y, Shang YZ, Sandler SI, Jiang JW (2010) Molecular thermodynamic model for DNA melting in ionic and crowded solutions. J Phys Chem B 114:9905–9911
19. Liu Y, Shang YZ, Liu HL, Hu Y, Jiang JW (2011) DNA melting in slit pores: a reaction density functional theory. J Phys Chem B 115:1848–1855

20. Liu Y, Shang YZ, Liu HL, Hu Y, Jiang JW (2012) Crowding effect on DNA melting: a molecular thermodynamic model with explicit solvent. Phys Chem Chem Phys 14:15400–15405
21. Loget G, Corn RM (2014) Silica nanowire arrays for diffraction-based bioaffinity sensing. Chem Euro J 20:10802–10810
22. Ma YZ, Yin HJ, Wang KZ (2009) A β-d-allopyranoside-grafted Ru(II) complex: synthesis and acid-base and DNA-binding properties. J Phys Chem B 113:11039–11047
23. Martyna GJ, Klein ML, Tuckerman M (1992) Nose-Hoover chains: the canonical ensemble via continuous dynamics. J Chem Phys 97(4):2635–2643
24. Mirkin CA, Letsinger RL, Mucic RC, Storhoff JJ (1996) A DNA-based method for rationally assembling nanoparticles into macroscopic materials. Nature 382:607–609
25. Miyoshi D, Sugimoto N (2008) Molecular crowding effects on structure and stability of DNA. Biochimie 90:1040–1051
26. Montrichok A, Gruner G, Zocchi G (2003) Trapping intermediates in the melting transition of DNA oligomers. Europhys Lett 62(3):452–458
27. Mrevlishvili GM, Metreveli NO, Razmadze GZ, Mdzinarashvili TD, Kakabadze GR, Khvedelidze MM (1998) Partial heat capacity change-fundamental characteristic of the process of thermal denaturation of biological macromolecules (proteins and nucleic acids). Thermochim Acta 308(1–2):41–48
28. Nakano S, Miyoshi D, Sugimoto N (2014) Effects of molecular crowding on the structures, interactions, and functions of nucleic acids. Chem Rev 114(5):2733–2758
29. Ouldridge TE, Louis AA, Doye JPK (2011) Structural, mechanical, and thermodynamic properties of a coarse-grained DNA model. J Chem Phys 134:085101–085121
30. Owczarzy R, You Y, Moreira BG, Manthey JA, Huang L, Behlke MA, Walder JA (2004) Effects of sodium ions on DNA duplex oligomers: improved predictions of melting temperatures. Biochemistry 43(12):3537–3554
31. Perez A, Orozco M (2010) Real-time atomistic description of DNA

unfolding. Angew Chem Int Ed 49:4805–4808

32. Qamhieh K, Wong KY, Lynch GC, Pettitt BM (2009) The melting mechanism of DNA tethered to a surface. Int J Numer Anal Model 6(3):474–488
33. Reed GH, Wittwer CT (2004) Sensitivity and specificity of single-nucleotide polymorphism scanning by high-resolution melting analysis. Clin Chem 50(10):1748–1754
34. Rief MD, Spudich JA (1999) Single-molecule biomechanies with optical methods. Science 283:1689–1695
35. Sambriski EJ, Ortiz V, de Pablo JJ (2009a) Sequence effects in the melting and renaturation of short DNA oligonucleotides: structure and mechanistic pathways. J Phys: Condens Matter 21:034105
36. Sambriski EJ, Schwartz DC, De Pablo JJ (2009b) A mesoscale model of dna and its renaturation. Biophys J 96(5):1675–1690
37. Samoletov AA, Dettmann CP, Chaplain MAJ (2007) Thermostats for “slow” configurational modes. J Stat Phys 128:1321–1336
38. Seela F, Ding P, Budow S (2011) DNA gold nanoparticle conjugates incorporating thiooxonucleosides: 7-deaza-6-thio-20-deoxyguanosine as gold surface anchor. Bioconjugate Chem 22:794–807
39. Sendroiu IE, Gifford LK, Lupták A, Corn RM (2011) Ultrasensitive DNA microarray biosensing via in situ RNA transcription-based amplification and nanoparticle-enhanced SPR imaging. J Am Chem Soc 133:4271–4273
40. Song L, Shan D, Zhao M, Pink BA, Minnehan KA, York L, Gardel M, Sullivan S, Phillips AF, Hayman RB, Walt DR, Duffy DC (2013) Direct detection of bacterial genomic DNA at sub-femtomolar concentrations using single molecule arrays. Anal Chem 85:1932–1939
41. Sorokin VA, Gladchenko GO, Valeev VA, Sysa IV, Petrova LG, Blagoi YP (1997) Effect of salt and organic solvents on DNA thermal stability and structure. J Mol Struct 408–409:237–240
42. Toma M, Cho K, Wood JB, Corn RM (2014) Gold nanoring arrays for near infrared plasmonic biosensing. Plasmonics 9:765–772
43. Várnai P, Lavery R (2002) Base flipping in DNA: pathways and

energetics studied with molecular dynamic simulations. J Am Chem Soc 124:7272–7273

44. Wang MD, Schnitzer MJ, Yin H, Landick R, Gelles J, Block SM (1998) Force and velocity measured for single molecules of rna polymerase. Science 282(5390):902–907
45. Woller JG, Börjesson K, Svedhem S, Albinsson B (2012) Reversible hybridization of DNA anchored to a lipid membrane via porphyrin. Langmuir 28(4):1944–1953
46. Wong KY, Pettitt BM (2004) Orientation of DNA on a surface from simulation. Biopolymers 73:570–578
47. Wong KY, Pettitt BM (2008) The pathway of oligomeric DNA melting investigated by molecular dynamics simulations. Biophys J 95(12):5618–5626
48. Wu J, Zhao SL, Gao L, Wu J, Gao D (2011) Separation of single-stranded DNA fragments at a 10-nucleotide resolution by stretching in microfluidic channels. Lab Chip 11:4036–4040
49. Wu J, Zhao SL, Gao L, Wu J, Gao D (2013) Sorting short fragments of single-stranded DNA with an evolving electric double layer. J Phys Chem B 117(8):2267–2272
50. Xu Y, Liu YL, Gao S, Jiang ZW, Su D, Liu GS (2014) Monolayer adsorption of dodecylamine surfactants at the mica/water interface. Chem Eng Sci 114:58–69
51. Zhang Y, Yildirim E, Antila HS, Valenzuela LD, Sammalkorpi M, Lutkenhaus J (2015) The influence of ionic strength and mixing ratio on the colloidal stability of PDAC/PSS polyelectrolyte complexes. Soft Matter 11:7392–7401
52. Zhao SL, Wu JM, Gao D, Wu JZ (2011) Gaussian fluctuations in tethered DNA chains. J Chem Phys 134(6):065103
53. Zhao XL, Ma YZ, Wang KZ (2012) Synthesis, pH-induced "on-off-on" luminescence switching, and partially intercalative DNA-binding and DNA photocleavage properties of an β-d-allopyranoside-grafted ruthenium(II) complex. J Inorg Biochem 113:66–76

CHAPTER 9

PREDICTION OF SOLUTION PROPERTIES AND DYNAMICS OF RNAS BY MEANS OF BROWNIAN DYNAMICS SIMULATION OF COARSE-GRAINED MODELS: RIBOSOMAL 5S RNA AND PHENYLALANINE TRANSFER RNA

Aarón Ayllón Benítez, José Ginés Hernández Cifre, Francisco Guillermo Díaz Baños and José García de la Torre

Departamento de Química Física, Universidad de Murcia, 30100 Murcia, Spain

ABSTRACT

Background

The possibility of validating biological macromolecules with locally disordered domains like RNA against solution properties is helpful to understand their function. In this work, we present a computational scheme for predicting global properties and mimicking the internal dynamics of RNA molecules in solution. A simple coarse-grained model with one bead per nucleotide and two types of intra-molecular interactions (elastic interactions and excluded volume interactions) is used to represent the RNA chain. The elastic interactions are modeled by a set of Hooke springs that form a minimalist elastic network. The Brownian dynamics technique is employed to simulate the time evolution of the RNA conformations.

Results

That scheme is applied to the 5S ribosomal RNA of E. Coli and the yeast phenylalanine transfer RNA. From the Brownian trajectory, several solution properties (radius of gyration, translational diffusion coefficient, and a rotational relaxation time) are calculated. For the case of yeast phenylalanine transfer RNA, the time evolution and the probability distribution of the inter-arm angle is also computed.

Conclusions

The general good agreement between our results and some experimental data indicates that the model is able to capture the tertiary structure of RNA in solution. Our simulation results also compare quite well with other numerical data. An advantage of the scheme described here is the possibility of visualizing the real time macromolecular dynamics.

Keywords: Brownian dynamics Coarse-grained model Hydrodynamics Transfer RNA Ribosomal RNA Diffusion coefficients Internal dynamics

BACKGROUND

Ribonucleic acid, or RNA, is a key molecule for the synthesis of proteins in living cells. Transfer RNA (tRNA) carries an amino acid to the protein synthetic machinery of a cell, the ribosome, which is the particle responsible for translation of the genetic code. Ribosomes are mainly formed by the so-called ribosomal RNA (rRNA) along with a variety of protein subunits. As many biological macromolecules, the RNA functionality is related with its structure which, in turn, determines the RNA solution properties. Therefore, the latter are essential sources of information on RNA structure and dynamics [1]. However, the structural determination of even small RNA molecules like tRNAs (70 to 80 nucleotides long) or the 5S rRNA of E. Coli (120 nucleotides long [2]) is not an easy task, being most RNAs much larger molecules. For instance, besides 5S rRNA, the main classes of rRNA are 16S and 23S in prokaryotic cells, and 17S, 18S, 26S, and 28S in eukaryotic cells [2] (S stands for Svedberg, the unit of sedimentation coefficient; it is noteworthy that the notation for ribosomal RNAs, as for some other biomacromolecules, is made in terms of a hydrodynamic property).

It is well-known that tRNA and rRNA are single-stranded chains that fold unto themselves giving rise to a secondary structure that contains both ordered (essentially rigid) double-helical regions and disordered zones so-called loops (see Fig. 1). Double helices are formed when complementary base pairs belonging to two pieces of the strand interact with each other – after strand folding – through either Watson-Crick or Hoogsteen (non-Watson-Crick) interactions, whereas loops appear in regions with unpaired bases. The overall view of such structure can be described as some number of arms, or stems, joined by the loop regions, which may act as flexible hinges. Thus, RNA molecules have an intrinsic flexibility that allows them to adopt a wide range of three-dimensional conformations. Although valuable insights about the tertiary structure can be obtained from atomic coordinates coming from x-ray crystallography measurements – available at the Protein Data Bank (PDB) (www.rcsb.org) – [3], experiments where the RNA flexibility can be revealed are most useful to characterize its conformational variability. In that sense, as above mentioned, dilute solution properties (hydrodynamic coefficients, scattering-related quantities, etc.) provide key information on the overall three-dimensional structure of RNA when combined with theoretical predictive models. Clearly, those models should include flexibility in order to make adequate predictions.

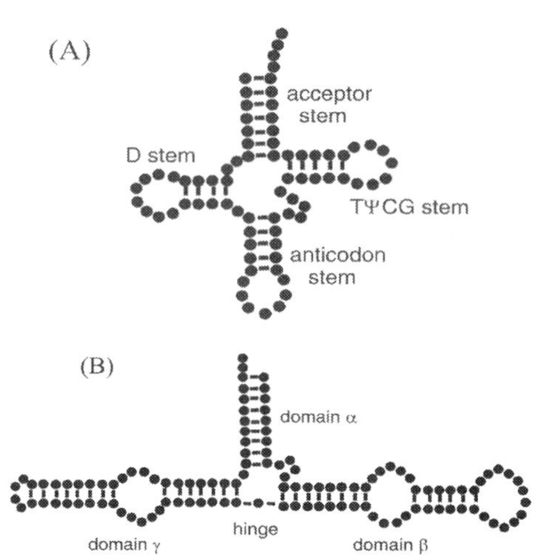

Figure 1: Secondary structure of RNAs. **a** Sketch of the secondary structure of yeast tRNA ^{phe} [45] (note the presence of four helices, three loops and a hinge). **b** Sketch of the secondary structure of the 5S rRNA [20] (note the presence of five double helices, four loops, and a hinge). In each case nucleotides are the black points, double helices are the regions with connections between opposite nucleotides and loops are the "circular" regions without connections between opposite nucleotides.

The tRNA is a particularly simple case as it presents four short stems as sketched in Fig. 1 a. In usual conditions, those stems appear grouped forming two arms joined by a disordered region so that the crystal structure has the typical "L-letter" shape [4]. Thus, tRNA has been presented as a typical case among segmentally flexible macromolecules, composed by rigid parts connected by flexible hinges. Considerable effort has been devoted to the characterization of the structure and degree of flexibility of tRNA in dilute solution [5], by means of either measurements of solution properties [6, 7, 8, 9, 10] or atomistic molecular dynamics simulation [11] (particularly noteworthy are the pioneering works of Harvey, McCammon and coworkers [12, 13, 14, 15]). Recent works and reviews on tRNA dynamics are those of Alexander et al. [16] and Agirrezabala and Valle [17]. Ribosomal RNAs are structurally quite more complex than tRNAs. For example, the secondary structure of the majority of the 5S rRNAs has five double helices, two hairpin loops, two internal loops, and a hinge region organized into three big domains (α, β, and γ) [18, 19, 20] as sketched in Fig. 1 b. Combining results from dilute solution experiments with numerical calculations, some models for the tertiary structure of 5S rRNA have been proposed [19, 20].

The level of detail, or coarse-graining, of a macromolecular model depends on the type of behavior that is intended to capture. Thus, a large variety of coarse-grained models and computational methods are used for biological macromolecules [21]. Since long, our group has developed procedures and devised computational tools to build coarse-grained models and predict solution properties of both rigid [22, 23, 24] and flexible macromolecules [25, 26, 27]. The latter are represented by bead-and-spring chains where, in the classical concept, each bead corresponds to a large piece of the macromolecular chain. Nonetheless, finer coarse-grained (bead-and-spring) models can be used whose elements represent, for instance, the repeating units – amino acid or nucleotide residues – of the biomacromolecules [28, 29]. Indeed, one of the present authors proposed forty years ago [30] to obtain hydrodynamic properties for double-helical DNA using what is nowadays regarded as a coarse-grained model with one bead per nucleotide and the approximate Kirkwood formula for friction coefficients. The problem was revisited, in further work [28], using modern bead-model hydrodynamics [22] for rigid DNA models. Subsequently, bending and twisting internal flexibility was included in the model by means of stiff, elastic springs [31]. The latter simulation of flexible, double-helical DNA, which is precursor of the present work, is in the spirit of a popular family of coarse-grained models for biomacromolecules and supramacromolecular structures, the so-called "elastic network model" [32], which has been applied to RNA structures [33, 34].

Certainly, the preferred simulation method for short time scales has been, following the above mentioned pioneering works [11, 12, 15], the classical molecular dynamics technique. There are many commercial and public domain tools that implement that technique [21] which has been applied to a variety of RNA structures [35, 36, 37]. However, the large size of the macromolecules and the typical long time scales of their global dynamics are not usually accessible to conventional (atomic-level) molecular dynamics simulations. An alternative is to apply the Brownian dynamics simulation technique to some type of coarse-grained model. Thus, in this work, we have attempted to devise a scheme for predicting global solution properties, that would also mimic the internal dynamics, of RNAs using the simple coarse-grained model with one bead per nucleotide, with intra-molecular interactions represented by a minimalist elastic network, along with the Brownian dynamics technique.

METHODS

Coarse-grained model

The RNA molecules (tRNA and rRNA) were modeled by bead-and-spring chains with one bead per nucleotide. In that model, nucleotides along the RNA strand form a string of touching beads all of them with the same hydrodynamic radius, σ=3.15 Å. That value is determined according to the features of the double-helical regions which are modeled following previous works for double-stranded DNA [28, 31], taking into account that RNA double helices appear usually in the A form whereas DNA is commonly found in B (or Z) form. The A form is a right handed double helix with the following features [38]: major groove width 3.8 Å, minor groove width 10.9 Å, helix radius 10 Å, rise 2.8 Å, inclination 16.1°, number of base pairs per helical pitch 11. On the other hand, the phase angle, φ, which define the gap between the complementary helices and therefore the shape and size of the grooves [28], was set to φ= 200°.

Beads are the model elements where the friction and the intra-molecular interactions take place. There are two types of intra-molecular interactions in the model: elastic interactions (represented by Hooke springs) and excluded volume interactions. Hooke springs are used to keep the connectivity between beads along the strand as well as the secondary structure in the helical regions. Also, they introduce the required degree of flexibility in the model [25] and form a kind of elastic network (see Fig. 2). For the sake of minimizing the amount of interactions but keeping the double-helical shape and the stiffness at short scale, we found it adequate [28] to connect each bead i (within every double-helical region) to:

- its first neighbors along its helix piece (beads i±1), which keeps connectivity and bond equilibrium length.

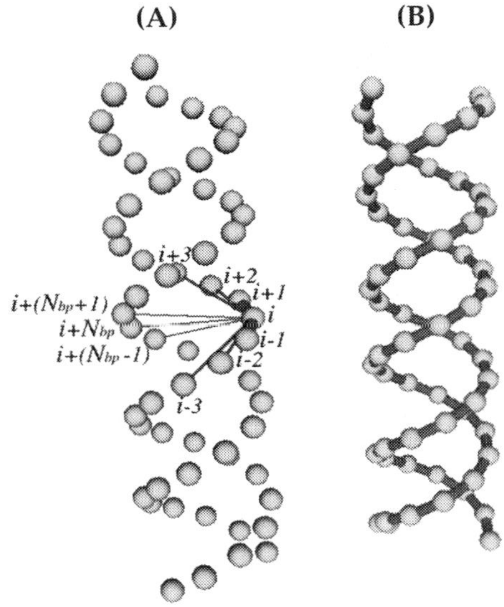

Figure 2: Double-helical model for RNA. **a** All the connectors supported by a given bead i (beads connected to i are labeled using i as reference). **b** All the connectors between first neighbors beads along each helix piece (i.e. connectors between each bead i and beads i±1). Beads appear smaller than in the real model (where they are tangent) for the sake of a better visualization of the connectors.

- its second neighbors along its helix piece (beads i±2), which accounts for bending interactions.
- its third neighbors along its helix piece (beads i±3), which accounts for torsional interactions.
- its counterpart in the complementary helix piece (bead $i+N_{bp}$), which accounts for interactions between nucleotides forming the base pair.
- the first neighbors of its counterpart in the complementary helix piece (beads $i+(N_{bp}\pm1)$), which is necessary in order to keep the double helical structure.

N_{bp} is the number of base pairs within a given double-helical region. In the above description, we use the term "helix piece" instead of "strand" because every nucleotide in the RNA molecule actually belongs to a unique strand. Figure 2 a shows the model for a double helix (in the straight, equilibrium conformation) displaying all the connectors involving one of the innermost beads (the rest of connectors was removed for the sake of clarity). On the other hand, each bead in a loop is only connected to their two first neighbors (beads i±1).

The Hooke spring potential reads

$$V_{ij}^{(conn)} = \frac{1}{2}H(l - l_e), \tag{1}$$

where l is the instantaneous spring length, l_e is the equilibrium spring length, and H is the spring constant. Following previous works [31], we set $H = 200k_BT/l_{e,1}^2$ for all connectors regardless the type of interaction represented. In that expression, k_BT is the Boltzmann factor and $l_{e,1}$ is the equilibrium length of the connector binding bead i to its first neighbor ($l_{e,1}$=2σ=6.30 Å, in our touching-beads model). For the chosen value of H, the root-mean-square fluctuation in the spring length, $\sqrt{\langle l^2\rangle - \langle l_e^2\rangle}$, is 5 % of l_e.

In addition to the elastic interactions, non-connected beads interact with each other through the following excluded volume (EV) Lennard-Jones potential that prevents beads from overlap and mimics good solvent conditions:

$$V_{ij}^{(LJ)} = 4\epsilon_{LJ}\left[\left(\frac{\sigma_{LJ}}{r_{ij}}\right)^{12} - \left(\frac{\sigma_{LJ}}{r_{ij}}\right)^{6}\right], \tag{2}$$

where r_{ij} is the instantaneous distance between the interacting beads, ε_{LJ} =0.1k_BT, and σ_{LJ}=0.8$l_{e,1}$ [39].

Simulation, hydrodynamics, and solution properties

We performed Brownian dynamics (BD) simulations of bead-and-spring models of tRNA and 5S rRNA in order to obtain their solution properties in a solvent with viscosity η_s=0.01 poise at T=293 K. The deterministic forces in the BD algorithm are the derivatives of the potentials defined by Eqs. (1) and (2). Both simulation and analysis were carried out with the software package SIMUFLEX [27] that implements a procedure based on the Ermak-McCammon algorithm [40]. That BD algorithm takes into account the hydrodynamic interaction (HI) effect through the Rotne-Prager-Yamakawa tensor [41]. Although the rigorous inclusion of the HI effect is necessary in order to generate the right macromolecular dynamics, BD simulations without HI (computationally less expensive) can be used as a "smart Monte Carlo" procedure [42] because they sample correctly the equilibrium conformational space. In such an approach, an ensemble of equilibrium conformations is generated and properties are calculated by means of the so-called rigid body treatment [43]. In the rigid body treatment, conformations are considered as instantaneously rigid and the hydrodynamic theory for rigid bead models is applied [23]. According to that procedure, conformational properties (like the radius of gyration) can be rigorously calculated whereas some overall hydrodynamic entities (like the translational diffusion and the intrinsic viscosity) can be acceptably estimated [44].

The initial conformations of the RNA models (including helices and loops) were built using an in-house program. After an equilibration period, long BD simulations with a duration about 100 times the relaxation time of the molecules were carried out. The resulting trajectories were divided into five pieces in order to calculate averages and standard deviations. We computed three properties: the radius of gyration R_g, the translational diffusion coefficient D_t, and the harmonic relaxation time τ_h.

The macroscopic translational diffusion of a Brownian particle (consequence of its Brownian motion) is characterized by the time correlation function of the center of mass or Einstein equation $\langle d^2 \rangle = 6D_t \Delta t$, where $\langle d^2 \rangle$ is the mean squared displacement of the center of mass, Δt is the displacement duration, and D_t is the translational diffusion coefficient of the particle. Long BD-HI simulations were carried out to obtain D_t using that equation. On the other hand, BD-noHI simulations were used as a "smart Monte Carlo" procedure to generate equilibrium conformations and apply the rigid body treatment to obtain τ_h and R_g. The harmonic relaxation time τ_h characterizes the overall tumbling of the molecule and can be compared to experimental rotational relaxation times. In case of a rigid body with arbitrary shape, the harmonic mean of its five characteristic rotational relaxation times is calculated as $\tau_h = 1/(6D_r)$ [23], where D_r is a rotational diffusion coefficient obtained from the eigenvalues of the rotational diffusion tensor. Although the rigid body hydrodynamics is not strictly applicable to the reorientational dynamics of flexible structures (like tRNA and rRNA), it allows for the calculation of an ensemble average $\langle D_r \rangle$ from which to calculate a value $\tau_h = 1/$

(6 $\langle D_r \rangle$). Finally, since R_g is a conformational property, it is correctly obtained from a conformational average over a BD-noHI trajectory.

RESULTS AND DISCUSSION

A bead-and-spring model for the yeast phenylalanine tRNA, yeast tRNA ^{phe} (PDB code 1EHZ), was built according to the structure sketched in Fig. 1 a [45]. After a BD-noHI simulation for equilibrating the initial conformation, BD simulations with and without HI were performed. Figure 3 a shows two of the many conformations adopted by that molecule along the Brownian trajectory once the equilibrium has been reached. Flexibility is noticed in the variation of the angle between arms (inter-arm angle θ) as better appreciated in Fig. 4 a. There, it is depicted the fluctuation along the Brownian trajectory of the angle θ subtended by the two arm vectors, one defined along the "acceptor stem" and the other one defined along the "anticodon stem". As observed, fluctuations are quick and the value of the angle varies in a relatively wide range, in agreement with some molecular dynamics simulations [11]. Figure 4 b is an histogram that shows the frequency of occurrence of the different θ values as taken at intervals of 10°. Its broadness is indicative of the flexibility of the structure. Clearly, there is a most probable equilibrium value around θ~ 90° (the precise average value being θ= 88°), which is coincident with that found in the crystal structure. Therefore, overall (time-averaged) hydrodynamic properties can be reasonably estimated using rigid bead models built upon the atomic coordinates coming from PDB files [46, 47].

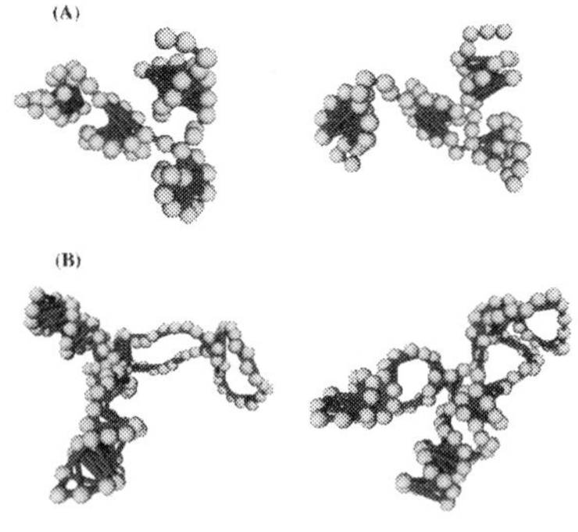

Figure 3: tRNA and rRNA models. Conformations along the Brownian trajectory (after equilibration) for **a** the tRNA model, and **b** the 5S rRNA model.

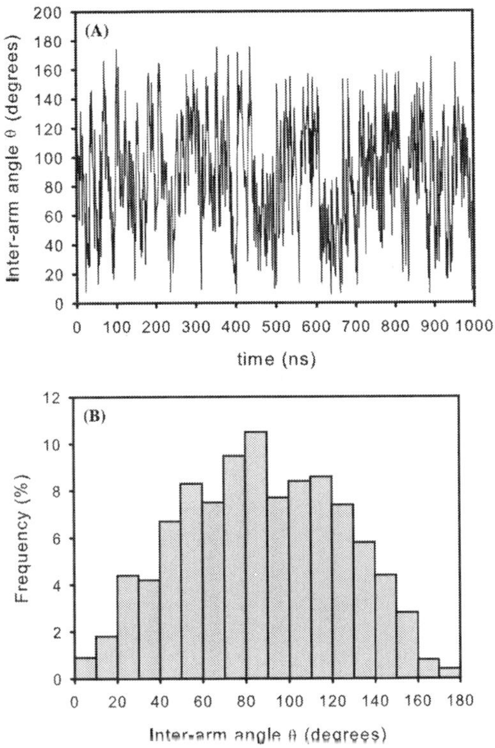

Figure 4: Inter-arm angle (θ). Angle subtended between the "acceptor stem" and the "aticodon stem" of tRNA: **a** fluctuation of θ along the Brownian trajectory; **b** frequency of occurrence of the different θ values.

Table 1 collects our BD results for R_g, D_t and τ_h, along with values calculated previously using rigid body hydrodynamics [47] and experimental results. The rigid body hydrodynamic calculations were carried out with the program HYDROPRO [24] which is able to build a bead model from the atomic (PDB) coordinates. There is a general good agreement among HYDROPRO calculations, BD simulations and experimental results. The advantage of our BD scheme is the possibility of evaluating, in addition to global solution properties, the internal dynamics of the molecule as illustrated by Fig. 4.

Table 1: Experimental and calculated values (via rigid body hydrodynamics and BD simulations) of solution properties of yeast $tRNA^{phe}$

Property	Experimental	Rigid body hydrodynamics[a]	Brownian dynamics
$D_t \times 10^7$ (cm^2/s)	7.3 [45]; 7.9 [6]	7.89 [47]	7.8±0.5
τ_h (ns)	23.4 [7]	24.2 [47]	23.6±0.8
R_g (Å)	23.1 [7]	24.1 [47]	25.1±0.3

[a]Calculated from PDB file 1EHZ

On the other hand, a bead-and-spring model of 5S rRNA of E. Coli (PDB code IC2X:C) was built according to the structure sketched in Fig. 1 b [20]. Again, after equilibrating the initial conformation, BD simulations with and without HI were performed. Figure 3 b illustrates two of the many conformations adopted by our 5S rRNA model along the equilibrium Brownian trajectory. The flexibility is revealed by the variation of the angle between the domains (α, β, and γ) as well as the expansion of the loops whereas the helical regions keep a rigid structure.

Table 2 shows the values for R_g, D_t and τ_h coming from our BD simulations along with some values selected from Skibinska et al. [20]. In that work, authors collect experimental data from several sources and present their own experimental data. Also, they make some rigid body hydrodynamic calculations (partly carried out with the program HYDROPRO [24]) and propose a likely tertiary structure for the 5S rRNA.

Table 2: Experimental and calculated values (via rigid body hydrodynamics and BD simulations) of solution properties of 5S rRNA of E. Coli

Property	Experimental	Rigid body hydrodynamics[a]	Brownian dynamics
$D_t \times 10^7$ (cm^2/s)	6.00 [20]; 6.2 [48]	6.16 [20]	6.2±0.3
τ_h (ns)	107 [20]	79.25 [20]	82±2
R_g (Å)	32.7 [48]; 39 [49]	32.98 [20]	39.2±1.5

[a]Calculated from PDB file IC2X:C

As observed, the results coming from our BD simulations agree relatively well with both experimental data and rigid body hydrodynamic predictions. In particular, our predictions for D_t and R_g are coincident with some of the experimental measurements collected in Table 2 [48, 49]. If we use our predicted D_t value (6.2×10^{-7} cm ^{2}/s) and the Svedberg equation $(s/D_t = M(1-\bar{v}\rho)/RT)$, we obtain a value for the sedimentation coefficient s≃5 S what further supports the validity of the scheme employed in this work. In order to obtain that value of s, we assume for the 5S rRNA a molecular weight M=40000 Da [50] and a partial specific volume v $\bar{\nu} = 0.53\ \mathrm{cm}^3/\mathrm{g}$ [51] as well as a solvent (water) density ρ≃1 g/cm ^{3}.

On the other hand our calculated τ_h is smaller than the experimental value obtained by Skibinska et al. [20], although close to the value predicted by their rigid body hydrodynamic calculations. As already mentioned, the rigid body hydrodynamic treatment does not capture rigorously the reorientational dynamics of flexible molecules and it is not the most adequate procedure to calculate their rotational relaxation times. For the case of 5S rRNA, the rigid body treatment seems to underestimate the rotational relaxation time. In any case, theoretical predictions are reasonably good if one considers the existing discrepancy among the experimental data themselves.

CONCLUSIONS

We have devised a computational scheme to model RNA molecules, which present a partially disordered structure, and calculate their solution properties. According to the model, each nucleotide is replaced by a bead so that the secondary structure is represented in detail. The beads are connected by Hooke springs that form a type of elastic network. The model dynamics can be simulated by any Brownian dynamics algorithm. The algorithm and the analysis procedures employed in this work are implemented in our public-domain software SIMUFLEX available at http://leonardo.inf.um.es/macromol. The previously mentioned program HYDROPRO is also available at that web site.

The good agreement of our simulation results with some experimental data as well as with other theoretical results obtained by alternative numerical approaches ensures the validity of our scheme for the prediction of solution properties of RNAs. It is also indicative that the model is able to capture the tertiary folds of RNA molecules in solution. The advantage of the Brownian dynamics simulation technique (including hydrodynamic interaction) respect to other procedures is the possibility of simulating the long time internal dynamics of big macromolecules in real time.

In a previous work [52], we tested the adequacy of a computational scheme based on a residue level model and the Brownian dynamics simulation to predict the internal dynamics of partially disordered proteins. The present work on the more complex RNA molecule further confirms the success of that type of approach which is expected to be appropriate for the study of other disordered biological macromolecules.

DECLARATIONS

Acknowledgments

Financial support was provided by Ministerio de Economía y Competitividad within project CTQ2012-33717 including FEDER funds and Fundación Séneca de la Región de Murcia (grant 19353/PI/14).

Authors' contributions

AAB carried out the Brownian dynamics simulations and data analysis. JGHC participated in the data analysis and wrote the manuscript. FGB designed the simulations and the analysis. JGT conceived of the study. The authors read and approved the final manuscript.

REFERENCES

1. Bloomfield VA, Crothers DM, Tinoco I. Nucleic Acids. Structures, Properties and Functions. Sausalito California: University Science Books; 2000.
2. Noller HF. Structure of ribosomal RNA. Ann Rev Biochem. 1984; 53:119–62.
3. Hyeon C, Dima RI, Thirumalai D. Size, shape and flexibility of RNA structures. J Chem Phys. 2006; 125:194905.
4. Kim SH, Suddath FL, Quigley GJ, McPherson A, Sussman JL,

Wang AHJ, et al.Three-dimensional tertiary structure of yeast phenylalanine transfer RNA. Science. 1974; 185:435–40.

5. Hagerman P. Flexibility of RNA. Ann Rev Biophys Biomol Str. 1997; 26:139–56.
6. Wang CC, Ford Jr. N C, Fournier MJ. Laser light-scattering analysis of the dimerization of transfer ribonucleic acids with complementary anticodons. Biopolymers. 1981; 20:155–68.
7. Patkowski A, Jen S, Chu B. Intensity-fluctuation spectroscopy and tRNA conformation. II. Changes of size and shape of tRNA in the melting process. Biopolymers. 1978; 17:2643–62.
8. Porschke D, Antosiewicz J. Permanent dipole moment of tRNA's and variation of their structure in solution. Biophys J. 1990; 58:403–11.
9. Müller JJ, Damaschun G, Wilhelm P, Welfle H, Pilz I. Comparison of the structures of the native form of rat liver 5S rRNA and yeast tRNA: small-angle and wide-angle scattering. Int J Biol Macromol. 1982; 4:289–96.
10. Friederich MW, Vacano E, Hagerman PJ. Global flexibility of tertiary structure in RNA: yeast tRNA as a model system. Proc Natl Acad Sci U S A. 1998; 95:3572–7.
11. (McCammon JA, Harvey SC, editors.)1987. Dynamics of Proteins and Nucleic Acids. Cambridge: Cambridge University Press.
12. Harvey SC, McCammon JA. Intramolecular flexibility in phenylalanine transfer RNA. Nature. 1981; 294:286–7.
13. Harvey SC, Prabhakaran M, Mao B, McCammon JA. Phenylalanine transfer RNA: molecular dynamics simulation. Science. 1984; 223:1189–91.
14. Tung C-S, Harvey SC, McCammon JA. Large-amplitude bending motions in phenylalanine transfer RNA. Biopolymers. 1984; 23:2173–93.
15. Harvey SC, Prabhakaran M, McCammon JA. Molecular-dynamics simulation of phenylalanine transfer RNA. I. Methods and general results. Biopolymers. 1985; 24:1169–88.
16. Alexander RW, Eargle J, Luthey-Schulten Z. Experimental and computational determination of tRNA dynamics. FEBS Lett. 2010; 584:376–86.

17. Agirrezabala X, Valle M. Structural insights into tRNA dynamics on the ribosome. Int J Mol Sci. 2015; 16:9866–95.
18. Fox JW, Wong K-P. The hydrodynamic shape, conformation, and molecular model of Escherichia coli ribosomal 5 S RNA. J Biol Chem. 1979; 254:10139–44.
19. Barciszewska MZ, Erdmann VA, Barciszewski J. Ribosomal 5S RNA: tertiary structure and interactions with proteins. Biol Rev. 1996; 71:1–25.
20. Skibinska L, Banachowicz E, Gapiński J, Patkowski A, Barciszewski J. Structural similarity of E. coli 5S rRNA in solution and within the ribosome. Biopolymers. 2004; 73:316–25.
21. Takada S. Coarse-grained molecular simulations of large biomolecules. Current Opinion in Structural Biology. 2012; 22:130–7.
22. García de la Torre J, Navarro S, López Martínez MC, Díaz FG, López Cascales JJ. HYDRO: a computer software for the prediction of hydrodynamic properties of macromolecules. Biophys J. 1994; 67:530–31.
23. Carrasco B, García de la Torre J. Hydrodynamic properties of rigid ps. Comparison of different modelling and computational procedures. Biophys J. 1999; 76:3044–57.
24. García de la Torre J, Huertas ML, Carrasco B. Calculation of hydrodynamic properties of globular proteins from their atomic-level structures. Biophys J. 2000; 78:719–30.
25. García de la Torre J, Pérez Sánchez HE, Ortega A, Hernández Cifre JG, Fernandes MX, Díaz Baños FG, et al.Calculation of the solution properties of flexible macromolecules: methods and applications. Eur Biophys J. 2003; 32:477–86.
26. García de la Torre J, Ortega A, Pérez Sánchez HE, Hernández Cifre JG. MULTIHYDRO and MONTEHYDRO: Conformational search and Monte Carlo calculation of solution properties of rigid and flexible macromolecular models. Biophys Chem. 2005; 116:121–8.
27. García de la Torre J, Hernández Cifre JG, Ortega A, Rodrígez Schmidt R, Fernandes MX, Pérez Sánchez HE, et al.SIMUFLEX : Algorithms and tools for simulation of the conformation and

dynamics of flexible molecules and nanops in solution. J Chem Theor Comput. 2009; 5:2606–18.

28. García de la Torre J, Navarro S, López Martínez MC. Hydrodynamic properties of a double-helical model for DNA. Biophys J. 1994; 66:1573–9.
29. Tepper H, Voth GA. A coarse-grained model for double-helix molecules in solution: Spontaneous helix formation and equilibrium properties. J Chem Phys. 2005; 122:124906.
30. García de la Torre J, Horta A. Sedimentation coefficient and X-ray scattering of double helical model for DNA. J Phys Chem. 1976; 80:2028–35.
31. Huertas ML, Navarro S, López Martínez MC, García de la Torre J. Simulation of the conformation and dynamics of a double-helical model for DNA. Biophys J. 1997; 73:3142–53.
32. Tozzini V. Coarse-grained models for proteins. Curr Opin Struct Biol. 2005; 15:144–50.
33. Xia Z, Gradner DP, Gutell RR, Ren P. Coarse-grained model for simulation of RNA three-dimensional structures. J Phys Chem B. 2010; 114:13497–506.
34. Hyeon C, Denesyuk NA, Thirumali D. Development and applications of coarse grained models for RNA. Isr J Chem. 2014; 54:1358–73.
35. Réblová K, Lankas F, Rázga F, Krasovska MV, Koca J, Sponer J. Structure, dynamics, and elasticity of free 16S rRNA helix studied by molecular dynamics simulations. Biopolymers. 2006; 82:504–20.
36. McDowell SE, Spacková N, Sponer J, Sponer NG. Molecular dynamics simulations of RNA: An in silico single molecule approach. Biopolymers. 2006; 85:169–84.
37. Badu SR, Melnik R, Paliy M, Prabhakar S, Sebetci A, Shapiro BA. Modeling of RNA nanotubes using molecular dynamics simulation. Eur Biophys J. 2014; 43:555–64.
38. Tanaka Y, Fujii S, Hiroaki H, Sakata T, Tanaka T, Uesugi S, et al.A'-form RNA double helix in the single crystal structure of r(UGAGCUUCGGCUC). Nucleic Acids Res. 1999; 27:949–55.

39. Rey A, Freire JJ, García de la Torre J. Monte Carlo calculations for linear and star polymers. 3. Dimensions and hydrodynamic properties in good solvents. Macromolecules. 1987; 20:342–6.
40. Ermak DL, McCammon JA. Brownian dynamics with hydrodynamic interactions. J Chem Phys. 1978; 69:1352–60.
41. Rotne J, Prager S. Variational treatment of hydrodynamic interaction on polymers. J Chem Phys. 1969; 50:4831–7.
42. Rossky PJ, Doll JD, Friedman HL. Brownian dynamics as a smart Monte Carlo method. J Chem Phys. 1978; 69:4628–33.
43. Zimm BH. Chain molecule hydrodynamics by the Monte-Carlo method and the validity of the Kirkwood-Riseman approximation. Macromolecules. 1980; 13:592–602.
44. Fixman M. Variational bounds for polymer transport coefficients. J Chem Phys. 1983; 78:1588–93.
45. Shi H, Moore PB. The crystal structure of yeast phenylalanine tRNA at 1.93 A resolution: A classic structure revisited. RNA. 2000; 6:1091–105.
46. Antosiewicz J, Porschke D. An unusual electrooptical effect observed for DNA fragments and its apparent relation to a permanent electric moment associated with bent DNA. Biophys Chem. 1989; 33:19–30.
47. Fernandes MX, Ortega A, López Martínez MC, García de la Torre J. Calculation of hydrodynamic properties of small nucleic acids from their atomic structures. Nucleic Acids Res. 2002; 30:1782–8.
48. Müller JJ, Zalkova TN, Ziwer D, Misselwitz R, Gast K, Serdyuk IN, et al.Comparison of the structure of ribosomal 5S RNA from E. coli and from rat liver using x-ray scattering and dynamic light scattering. Eur Biophys J. 1986; 13:301–7.
49. Brunel C, Romby P, Westhof E, Ehresmann C, Ehresmann J. Three-dimensional model of escherichia coli ribosomal 5 S RNA as deduced from structure probing in solution and computer modeling. J Mol Biol B. 1991; 221:293–308.
50. Erdmann VA, Barciszewski J, Szymansky M. Barciszewska MZ. 5S ribosomal RNA database. Nucleic Acids Res. 2002; 30:176–8.

51. Funari SS, Rapp G, Perbandt M, Dierks K, Vallazza M, Betzel C, et al.Structure of free thermus flavus 5S rRNA at 1.3 nm resolution from synchrotron solution scattering. J Biol Chem. 2000; 275:31283–8.
52. Amorós D, Ortega A, García de la Torre J. Prediction of hydrodynamic and other solution properties of partially disordered proteins with a simple, coarse-grained model. J Chem Theory Comput. 2013; 9:1678–85.

CHAPTER 10

COARSE-GRAINED PREDICTION OF RNA LOOP STRUCTURES

Liang Liu, Shi-Jie Chen

Department of Physics and Department of Biochemistry, University of Missouri, Columbia, Missouri, United States of America

ABSTRACT

One of the key issues in the theoretical prediction of RNA folding is the prediction of loop structure from the sequence. RNA loop free energies are dependent on the loop sequence content. However, most current models account only for the loop length-dependence. The previously developed "Vfold" model (a coarse-grained RNA folding model) provides an effective method to generate the complete ensemble of coarse-grained RNA loop and junction conformations. However, due to the lack of sequence-dependent scoring parameters, the method is unable to identify the native and near-native structures from the sequence. In this study, using a previously developed iterative method for extracting the knowledge-based potential parameters from the known structures, we derive a set of dinucleotide-based statistical potentials for RNA loops and junctions. A unique advantage of the approach is its ability to go beyond the the (known) native structures by accounting for the full free energy landscape, including all the nonnative folds. The benchmark tests indicate that for given loop/junction sequences, the statistical potentials enable successful predictions for the coarse-grained 3D structures from the complete conformational ensemble generated by the Vfold model. The predicted coarse-grained structures can provide useful initial folds for further detailed structural refinement.

INTRODUCTION

The ability to predict RNA 3D structure is critical for understanding RNA functions. Recent developments in *de novo* prediction of RNA 3D structures have led to highly promising results [1]–[18] (for review, see [19]–[24]). In particular, several *de novo* structure prediction methods have been developed based on the knowledge-based energy function. For example, Dima and co-workers [25] extracted the base-pair stacking parameters from RNA native structures and the extracted parameters agree with the experimental data [26]. Wu et al. [27] explored the correlation between RNA secondary structural motifs and their thermodynamic stability to derive energy parameters for base-pair stackings, and free segments such as hairpin loops, internal loops and bulge loops. Bernauer and co-workers [15] further extracted a set of distance-dependent energy parameters between any two bases, irrespective of the locations of the bases (in a helix or a loop). Das and Baker [5], [9] obtained energy function – made available in the Rosetta software package – based on the base orientations and interactions. Parisien and Major [7] predicted 3D structures with a pipeline of two computer programs: MC-Fold and MC-Sym, developed based on the nucleotide cyclic motifs (NCMs). All of these methods have provided valuable insights into the correlation between loop sequence and their stability. Especially, these methods are particularly useful for selecting the most probable conformation from an ensemble of near-native structures.

A key issue in the predictions of RNA stability is how to compute the loop free energy. For hairpins and RNA secondary structures in general, the nearest neighbor model, which assumes that the total free energy is an additive sum of the free energy of each elements (base-pair stacking, loop), and other models [6], [26], [28]–[32] has enabled successful predictions for RNA structures and stabilities. In most of the existing models, loop stability is often assumed to depend on loop size, the identity of the closing base pair, the interaction of the first mismatch with the closing base pair, and an additional stabilization term for loops with GA or UU first mismatches [33]–[35]. Further detailed sequence-dependence of the loop stability has been ignored. Experimental results suggest that loop stability may be sensitive to the sequence context inside the loop. For example, for unusually stable RNA hairpin loops, Dale et al. [36]performed optical melting studies for a series of hairpins. The study led to a set of different stability parameters for different loop sequences, such as GNRA and UUCG (where N is any nucleotides and R is a purine) tetraloops, hexaloops with UU first mismatches, and hairpin loop of iron responsive element, GAGUGC, all of which are significantly more stable than other hairpin loops of the same length.

The prediction of sequence-dependent loop free energy requires a model that goes beyond simple fitting with the experimentally measured empirical parameters. The formation of the intraloop base pairs and stacks would cause significant restriction of the loop conformational space and the loop entropy. It is practically impossible to exhaustively measure the loop free energy for all the different possible intraloop contacts for the different sequences and loop lengths. Therefore, evaluation of loop free energy with an ensemble of possible intraloop base pairing/stacking interactions cannot be achieved by experiment alone. We also need a computational model. The main purpose of the present study is to develop a statistical potential model that enables predictions of loop and junction three-dimensional structures from the sequence.

In general, there are two classes of physics-based models for RNA structure prediction: molecular dynamics (For review, see [37]) and Monte Carlo simulation methods and polymer statistical mechanics methods. Molecular dynamics simulations have provided much insights into the atomic details of intraloop interactions and their contributions to the loop stabilities [1], [38]–[40]. The polymer statistical mechanical models often employ low-resolution (coarse-grained) conformational models in order to capture the complete conformational ensemble. Along this line there have been different ways to construct the low-resolution RNA structures. For example, with a knowledge-based potential, Jonikas et al. [8] developed a structure filter model (N AST) where nucleotides are represented by the C_3^* atoms. In another coarse-grained model where nucleotide are represented by the P and C_3^* atoms, Keating and Pyle [41] developed a semi-automated approach to build RNA structures with a directed rotameric search strategy. Furthermore, in an attempt to develop a high-resolution RNA model (HiRE-RNA), Pasquali and Derreumaux [11] used six to seven beads for each nucleotide (one bead for the phosphate P, four beads for the sugar O_5^*, C_5^*, C_4^*, C_1 respectively, and one bead for a pyrimidine base and two beads for a purine base).

Our RNA folding model is based on a virtual bond-based RNA conformational model (called "Vfold" model; Fig. 5) [42]. The Vfold model uses two virtual bonds $(C_4^* - P - C_4^*)$ for each nucleotide and samples RNA conformations through self-avoiding walks in a diamond lattice. It provides an effective tool to sample RNA conformations and to evaluate the conformational entropy. The model has shown a high promise in predicting the 2D and 3D structures and the folding stabilities from the sequence [14], [42]–[48]. However, the Vfold model does not account for the sequence-dependent conformational propensity of the loop, namely, the model assumes that for any given loop, all the loop conformations generated in the model have the same energy. Such a simplification could cause inaccuracy in the prediction of loop stability and structure [36], [49], [50]. Physically, the sequence-dependence of loop stability arises from the local interactions, which affect the loop flexibility and hence conformational propensity [50], as well as the nonlocal interactions between the different nucleotides. In this study, we develop a method to extract a set of virtual bond-based (coarse grained) statistical potentials (scoring functions) from a set of non-redundant RNA structures such as the RNA09 database [51] and the Leontis database (http://rna.bgsu.edu/nrlist/). Specifically, we aim to derive a set of dinucleotide statistical potentials $u_{ij}(\theta, \eta)$ as a function of the (4×4) types of the dinucleotides base i and j and the backbone pseudo-torsion angles (θ, η) of the dinucleotide conformation. We use the RNA09 database [51], the Leontis database, the Capriotti's database [52] and the PDB database [53], [54], respectively, to test the extracted potential functions. We note that the dinucleotide in this work refers to the two continuous nucleotides within the same loop. The goal is for a given RNA loop/junction sequence, to identify the lowest-RMSD structures from the Vfold-generated complete conformational ensemble.

RESULTS

Pseudo-torsion angles $(\theta,\ \eta)$

The θ and η values of dinucleotides in the 152 RNA loops/junctions are plotted as a 2D scatter plot (Fig. 6), where each point represents the θ-η coordinates for a dinucleotide. In contrast to the analysis in the previous studies [55], [56], here we find a distinct category of dinucleotides conformation around (θ=150°, η=225°). This region was once considered as the helical region, because the dinucleotides with θ-η coordinates located in this group are most likely found within the RNA helix. However, as we only count the dinucleotides in RNA loops and junctions, the plot shows that the loop/junction residues can also have the tendency to have the helix-like conformation. This observation provides a rational in the next step for building the 3D all-atom structures by adding the helical residues back to the coarse-grained backbone model.

Quasi-chemical approximation-based potentials

The 152 RNA loops/junctions within our training dataset consist of a total number of N_{total}=572 nucleotides, with the numbers of each type of the nucleotide $(\rho_A,\rho_C,\rho_G,\rho_U)$=(217,105,121,129) and the corresponding mole fraction of each nucleotide $(\chi_A,\chi_C,\chi_G,\chi_U)=(0.38,0.18,0.21,0.23)$.

From the dataset of the coarse-grained "correct" structures (Table 1), for each pair of the nucleotides i=(1,2,3,4) and j=(1,2,3,4) and pseudo-torsion angles (θ, η), we calculate the numbers $\rho_{ij}^{obs}(\theta,\ \eta)$ of the dinucleotides conformations and the total observed number $N(\theta,\ \eta)=\sum_{i,j}\rho_{ij}^{obs}(\theta,\ \eta)$. According to Eq. 8, we then calculate the expected number $\rho_{ij}^{exp}(\theta,\ \eta)$ of each type of dinucleotides (*i,j*) with pseudo-torsion angles (θ, η). If no pseudo-torsion angles (θ, η) is observed for dinucleotide (*i,j*), we assign an unfavorable potential $u_{ij}(\theta,\ \eta)=+2.0k_BT$ kcal/mol. From Eq. 7, we compute the potentials as a 4×4×3×3 tensor.

Table 1: The number of dinucleotides with torsion angles (θ, η)

	η		
θ	g^{+}	t	g^{-}
g^{+}	43	27	21
t	46	99	21
g^{-}	19	101	43

doi:10.1371/journal.pone.0048460.t001

Iterative approach-derived potentials

The convergence speed of the iteration depends on the selection of convergence criteria. For instance, if the convergence threshold parameter in Eq. 13 is set to 10^{-3}, the iterative process would converge after around 3000 iterations. In contrast, if the convergence threshold parameter is set to 10^{-2}, the iterative process would converge after 340 iterations. On an Intel(R) Xeon(R) CPU 5150 @ 2.66 GHz on Dell EM64T cluster system, the 3000-cycle iteration process took about 30 minutes and a 300-step iteration took less than 6 minutes.

Comparison between the two sets of the derived potential parameters shows that the "true" potentials SP_{fold} are less uniform than the "extracted" potentials (Fig. 2), suggesting that the "true" potentials SP_{fold} are more sensitive to the sequences and conformations of the dinucleotide and thus have the ability to discriminate the "correct" conformation from an ensemble of conformations for a given RNA loop/junction sequence.

Tests on training datasets

We test the statistical potentials for the accuracy in loop/junction structure prediction for a large number of sequences. For a given sequence, we generate the full ensemble of the virtual bond loop/junction conformations using our Vfold model. Each conformation is then scored by the total statistical potential, which is evaluated as the sum of the statistical potentials (SP_{native} or SP_{fold}) for the dinucleotide pairs in the conformation. The conformation with the lowest value of the total statistical potential is identified as the predicted native structure. If the predicted native structure is the same as the coarse-grained "correct" structure (i.e., the virtual bond structure that is closest to the PDB structure, evaluated by RMSD), the prediction is successful for the sequence; otherwise, the prediction fails.

We first apply the two sets of statistical potentials to the RNA loops and junctions in the training dataset. Such a test for structure prediction is nontrivial because the SP_{native} and SP_{fold} are derived based on the frequency $\rho_{ij}(\theta, \eta)$ (see Eqs. 7 and 10) instead of the structure. As described above, 152 loops and junctions are constructed from the 262 RNA structures in the RNA09 dataset, including 72 RNA hairpin loops, 25 internal/bulge loops, 14 pseudoknot loops, 35 multibranched loops and 6 junctions (the free segments other than the above 4 types) (Table 3A). Here we note that the two loops in an internal loops are counted separately in our calculation, as our model does not specify specific types of loops or junctions. This rule is also applied to the pseudoknot loops and multibranched loops.

We found that SP_{native} succeeded in finding 130 coarse-grained "correct" (lowest potential) structures out of 152 loops and junctions. In contrast, SP_{fold} can give successful predictions for all the 152 the coarse-grained "correct" conformations (Table 3A). SP_{fold} is more reliable than SP_{native} in structure prediction with success rate 100% vs 85.5%.

Test on the Capriotti's dataset

To rigorously test the reliability of the two sets of potential parameters, we need to perform the test on loops and junctions outside the training dataset. We will perform tests for several such test sets. We first choose a dataset collected by Capriotti, et al. [52], consisting of 85 structures with length $\geq$20 nucleotides and solved at resolution better than 3.5 Å. The 3DNA software determined a total of 72 loops/junctions with lengths ranging from 3 nt to 8 nt, excluding the 5′/3′-terminal dangling regions. The 72 loops/junctions can be classified into 37 hairpin loops, 14 internal/bulge loops, 5 pseudoknot loops, 1 junction and 15 multibranched loops (Table 3B).

Our results indicate that SP_{native} and SP_{fold} potentials can successfully find out 62 and 70 coarse-grained "correct" conformations, respectively, out of the 72 test cases (Table 3B). The result indicates that the "true" potential parameters SP_{fold} are more reliable than the directly extracted potentials SP_{native} (success rate 97% vs 86%). One of the two failed predictions for SP_{native} is for a pseudoknot loop, which has special loop structure due to the tertiary interactions (base triplets) with the helices. Another failed prediction is a junction located in a large RNA molecule and the junction structure is determined by not only its sequence content but also the surrounding structural environment.

Test on the PDB dataset

The January 2012 version of PDB database contains 2227 structures that contain at least one strand of RNA sequence. These structures range from hairpin-loop structures to RNA-protein complexes or RNA-DNA hybrids. We found 8452 loops and junctions in the 2227 PDB structures (*TEST-I*). All the loops and junctions (excluding the 3′/5′-terminal dangling regions) have lengths from 3 nt to 8 nt. Within these 2227 RNA molecules, 1609 have structures determined by X-ray crystallography, which contain 7459 RNA loops and junctions (*TEST-II*), and 934 of these 1609 RNAs have high-resolution structures (≤3.0 Å), which contain 1119 RNA loops and junctions (*TEST-III*).

Our test results show that the numbers of the correct predictions with the statistical potentials SP_{fold} and SP_{native} are 7364 (success rate=87%) v.s. 6992 (success rate=83%) for *TEST-I*, 6553 (success rate=88%) v.s. 6222 (success rate=83%) for *TEST-II*, and 1070 (success rate=95%) v.s. 1001 (success rate=89%) for *TEST-III*, respectively. Fig. 10 shows illustrations for two of the results (a hairpin loop from PDB structure 1IVS and an internal loop from PDB structure 1JJ2).

If we include the top five lowest-potential conformations, the numbers of successfully determined loops and junctions are increased to 8006 with SP_{fold} v.s. 7857 with SP_{native} for *TEST-I*, 7089 v.s. 6949 for *TEST-II* and 1102 v.s. 1065 for *TEST-III* (Table 2). Fig. 11 shows the minimal-RMSD structure, the 9-th potential structure computed with SP_{native}, and the predicted structure with the lowest-potential, for multibranched loop from PDB structure 3CCM.

Table 2: The numbers of sequences with successfully predicted loop/junction structures for (I) all the 8452 RNA loops/junctions in TEST-I (II) the 7459 RNA loops and junctions in TEST-II and (III) the 1119 RNA loops and junctions in TEST-III

	I		II		III	
	$\#SP_{fold}$	$\#SP_{native}$	$\#SP_{fold}$	$\#SP_{native}$	$\#SP_{fold}$	$\#SP_{native}$
Top-1	7364	6992	6555	6222	1070	1001
Top-2	7686	7411	6819	6563	1083	1022
Top-3	7796	7715	6909	6826	1101	1060
Top-4	7956	7798	7043	6895	1102	1061
Top-5	8006	7857	7089	6949	1102	1065
TOTAL-#	8452		7459		1119	

For each dataset, the numbers in columns are calculated with the SP_{fold} ("$\#SP_{fold}$", "true" potential parameters) and SP_{native} ("$\#SP_{native}$", "extracted" potential parameters), respectively. The potentials are obtained from the RNA09 database. The "Top-#" in the first column means that the "correct" structure is in the top-# lowest-potential conformations.

doi:10.1371/journal.pone.0048460.t002

Moreover, we test the predictions of the first 20 lowest-potential conformations for the loops and junctions in all the three databases *TEST-I*, *TEST-II* and *TEST-III* as well as the influence of the convergence threshold parameter on the accuracy of the structure prediction. Fig. 12 shows the numbers of successfully determined loops and junctions in the first 20 lowest-potential conformations for the three databases. The comparisons between the predicted results with SP_{fold} and SP_{native} supports our conclusions in the previously two benchmark tests that SP_{fold} is more reliable than SP_{native}. The loops/junctions that we failed to predict mostly involve tertiary interactions with helices or other cofactors such as protein and DNA.

The predicted results using the statistical potentials SP_{native} and SP_{native}, extracted from the Leontis dataset also support the above conclusions (Fig. 9 and Table 4).

Fig. 13 shows the sensitivity of the convergence threshold parameter ($\Delta u_{ij}^{n}(\theta, \eta)$=0.001 v.s. 0.01) to the RNA structural predictions. The comparisons of the numbers of the successfully determined loops and junctions in the three databases show that the convergence threshold parameters $\Delta u_{ij}^{n}(\theta, \eta)$ do not have strong influence on the accuracy of the structure predictions.

Furthermore, we categorize the sequences in the dataset *TEST-III* and the predicted results according to the types of RNA loops and junctions. The 1119 RNA loops and junctions can be grouped into 408 hairpin loops, 166 internal/bulge loops, 148 pseudoknot loops, 329 multibranched loops and 68 junctions. The correct predictions with SP_{fold} for each type of RNA loops and junctions are 395, 165, 147, 297 and 66, respectively (Table 3C). For junctions, multibranched loops, pseudoknots, internal/ bulge loops and five of the hairpin loops, the failed predictions are due to the interactions beyond the dinucleotide context, such as the loop-helix tertiary interactions and RNA-protein interactions. The possible reason for other six failed predictions for hairpin loops is that these hairpin loops are not closed by canonical base pairs. These loops are closed by non-canonical pairs, such as AA or AC, which may lead to different hairpin loop structures.

Table 3: The types and numbers of loops and junctions in (A) the RNA09 dataset, (B) the Capriotti›s dataset and (C) the *TEST-III* dataset, and the number of the correct predictions with the "true" potentials SP_{fold}

	RNA09		Capriotti's		*TEST-III*	
Types	Total #	Correct #	Total #	Correct #	Total #	Correct #
Hairpin	72	72	37	37	408	395
Internal/bulge	25	25	14	14	166	165
Pseudoknot	14	14	5	4	148	147
Multibranched	35	35	15	15	329	297
Junction	6	6	1	0	68	66
TOTAL	152	152	72	70	1119	1070

doi:10.1371/journal.pone.0048460.t003

For the *TEST-I* (the 8452 RNA structures in PDB) and *TEST-II* (the 7459 x-ray structures in *TEST-I*) databases, we randomly selected 3229 RNA loops/junctions from the database *TEST-I*and categorize them according to the types of RNA loops and junctions. The 3229 RNA loops and junctions contain 1406 hairpin loops, 255 internal/bulge loops, 184 pseudoknot loops, 1265 multibranched loops and 119 junctions. The correct (top-1) predictions with SP_{fold} for each type of RNA loops and junctions are 1274 (success rate=90.6%), 245 (success rate=96.1%),

178 (success rate=96.7%), 1162 (success rate=91.9%) and 110 (success rate=92.4%), respectively. The total number of the successful predictions is 2969 (success rate=91.9%).

Statistical potentials and the Leontis dataset

The March 17, 2012 version of the Leontis dataset contains 642 RNA sequences with structures of resolution better than 4.0 Å. The 3DNA software identified a total of 435 RNA loops and junctions with lengths ranging from 3 nt to 8 nt, excluding the 5′/3′-terminal dangling regions. Our calculations show that the difference between the diamond lattice-represented structures and the PDB structures varies for the different loop lengths and sequence contents with RMSD from 0.74 Å to 3.93 Å (Fig. 8), and the mean and standard deviation of RMSD values are 1.37 Å and 0.29 Å, respectively. We use such coarse-grained structures to calculate the observed dinucleotide frequencies, to extract SP_{native} and to search for the "true" potential functions SP_{fold}.

Table 4: The numbers of sequences of the successfully predicted loop/junction structures for (I) all 8452 RNA loops/junctions in *TEST-I* **(II) the 7459 RNA loops and junctions in *TEST-II* and (III) the 1119 RNA loops and junctions in *TEST-III***

	$\#_{SP_{fold}}$(A)	$\#_{SP_{fold}}$(B)	$\#_{SP_{native}}$(A)	$\#_{SP_{native}}$(B)
TEST-I				
Top-1	7364	7369	6992	7102
Top-2	7686	7686	7411	7500
Top-3	7796	7862	7715	7665
Top-4	7956	7969	7798	7844
Top-5	8006	8007	7857	7878
TOTAL-#	8452			
TEST-II				
Top-1	6555	6567	6222	6332
Top-2	6819	6861	6563	6657
Top-3	6909	6961	6826	6791
Top-4	7043	7055	6895	6947
Top-5	7089	7083	6949	6976
TOTAL-#	7459			
TEST-III				
Top-1	1070	1077	1001	1016
Top-2	1083	1084	1022	1053
Top-3	1101	1087	1060	1055
Top-4	1102	1092	1061	1058
Top-5	1102	1098	1065	1058
TOTAL-#	1119			

For each dataset, the numbers in columns "A" are calculated from SP_{fold} ("$\#_{SP_{fold}}$", "true" potential parameters) and SP_{native} ("$\#_{SP_{native}}$", "extracted" potential parameters), obtained from the RNA09 database, and the numbers in columns "B" are calculated from SP_{fold} ("$\#_{SP_{fold}}$") and SP_{native} ("$\#_{SP_{native}}$"), obtained from the Leontis' database respectively. The "Top-#" in the first column means that the "correct" structure is in the top # lowest-potential conformations.
doi:10.1371/journal.pone.0048460.t004

Figs. 3 and 4 show the comparison between the potentials SP_{native}/SP_{fold} extracted from the dataset RNA09 and from the Leontis dataset. We apply the "true" potential SP_{fold} extracted from the Leontis dataset to predict the loop/junction structures in three testing datasets: *TEST-I*,

TEST-II and *TEST-III*, constructed from the January 2012 version of PDB database. Our test results (Table 4) show a success rates of 87% (7369 out of 8452) for *TEST-I*, 88% (6567 out of 7459) for *TEST-II*, and 96% (1077 out of 1119) for *TEST-III*, respectively.

Moreover, the predictions of the top 20 lowest-potential conformations for the loops and junctions in all the three test datasets, *TEST-I*, *TEST-II* and *TEST-III*, are shown in Fig. 9, with comparisons with the predictions based on the SP_{fold} from the RNA09 dataset. The comparisons for the three test datasets (Table 4 and Fig. 9) show that the two "true" statistical potentials SP_{fold} extracted from the RNA09 dataset and the Leontis dataset lead to similar success rates in structure prediction and the predictions are not sensitive to the choice of the specific training set.

DISCUSSION

Motivated by the biological significance to predict sequence-dependent loop and junction structures, we have developed a knowledge-based scoring functions/potentials to predict the structures of RNA loops and junctions. We use a coarse-grained conformational model (the virtual bond model) to sample RNA loop/junction conformations. From the known RNA structures, we extract a set of sequence-dependent dinucleotide-based statistical potentials using two methods. In the first method, the statistical potentials (SP_{native}) are derived from the distributions of the dinucleotide conformations in the known (native) structures. In the second method, the statistical potentials (SP_{fold}) are derived based on the folding stability of the native structures (against all the other nonnative folds).

For a given sequence, the extracted statistical potentials enable ranking of the different conformations with the top ranked (the lowest-potential) structure as the predicted native structure. Extensive tests indicate that the statistical potentials can successfully predict the native structure for a large class of loop and junction sequences and that SP_{fold} consistently outperforms SP_{native} in structure prediction. Our test results also indicate that our results are not sensitive to the choice of the specific training dataset (Fig. 9 and Table 4).

The present approach has several advantages. First, the extraction of the statistical potential SP_{fold} is based on the sampling of the complete conformational ensemble, including the native and all the nonnative folds. Second, because we consider all the nonnative folds in the derivation of the statistical potential, our statistical potentials can be used to predict folding from the sequence. The build-up strategy for RNA loop structure in this study is a de novo approach. Compared with other 3D loop structure prediction models, such as ModeRNA [22] and RLooM [50], our loop/junction structure prediction method is based on the complete ensemble of the (coarse-grained) conformations and does not rely on the information of template structures or homologous RNA structures. The only input information for the prediction is the sequence. Therefore, the model can predict the low-resolution structure from the sequence if no known homologous conformations can be found in the PDB. With the predicted low-resolution scaffold, one may further predict the all-atom structures using the all-atom potential methods that are derived based on near-native structures [5], [15]. The present coarse-grained model offers a useful complement to the other all-atom based models.

We have performed extensive benchmark tests using the different databases. However, a direct comparison between our model other methods is not very straightforward. This is because our model is a coarse-grained model while other models mainly focus on the all-atom structures. Furthermore, our model aims to fold a low-resolution structure from the sequence without using any input information such as homologous templates, while other models mainly focus on the prediction of the native structures from near-native folds. Future development of our method, which may give all-atom structures from the low-resolution folds, would make direct comparison between our model and other models possible.

Applications of the present statistical potentials to the Vfold structure prediction model [14] may provide an effective strategy for better structure prediction. Despite the success, there are several limitations of our model. First, the current study is based on the coarse-grained Vfold model, which only provides a low-resolution approximation for the all-atom structure of RNA loops and junctions. Ultimately an all atom-based model is required to treat the detailed tertiary interactions. Future development of the model should address the issue to build all-atom RNA structures based on the low-resolution (Vfold-generated) representations. Second, the statistical potentials are derived for dinucleotide conformations. In realistic loop and junction structures, the long-range sequence effects within the loop, such as intra-loop interactions, can influence the loop structure. The present model cannot explicitly account for such long-range intraloop interactions, which occur frequently in loops and between the different loops. Following the same procedure outlined above, one may develop a trinucleotide or higher order many-body statistical potentials to address this issue. Third, in the current form of the model, we use the same set of statistical potentials for the different types of loops/junctions. More refined potentials according to RNA loop/junction types may lead to further improvement of the accuracy of model.

MATERIALS AND METHODS

Vfold model

We use the Vfold model to generate the full conformation ensemble of a given RNA loop or junction sequence. The Vfold model is a virtual bond-based RNA folding model [42], [45], [48], which is developed based on the two observations: the C-O torsion in the nucleotide backbone of RNA tend to adopt the trans (t) rotational isometric state (Fig. 5A) and the $P-O_5^*-C_5^*-C_4^*$ bonds and the $C_4^*-C_3^*-O_3^*-P$ bonds in the nucleotide backbone are approximately planar [57]. Therefore, the nucleotide backbone conformations can be reduced into two effective virtual bonds $P-C_4^*$ and C_4^*-P [58]–[60]. The length of each backbone virtual bond is about 3.9 Å. The virtual bonds show rotamer-like configurations gauche$^+(g^+)$, trans (t) and gauche$^-(g^-)$ (Fig. 5C) [55], [56], [61]. Such virtual bond conformations can be well represented by the bonds in a diamond lattice. Therefore, we can use self-avoiding walks on a diamond lattice to enumerate the conformations of RNA sequences. This approach promises proper treatment of the excluded volume effect between the different atoms and the complete sampling of the conformational ensemble.

The loop structures are enumerated on a diamond lattice through exhaustive self-avoiding walks. If the first atom (P) is fixed at position $\overrightarrow{X_0}$, then the coordinates $(\overrightarrow{X_N})$ of N-th atom (P or C_4^*) can be calculated with

$$\overrightarrow{X_N} = \overrightarrow{X_0} + \sum_{i=1}^{N} l_i \overrightarrow{b_i} \tag{1}$$

where, l_i=3.9A is the bond length of virtual bond b_i and $\overrightarrow{b_i}$ is the unit vector of virtual bond b_i. Also, the relation between the (i-1)-th virtual bond (b_{i-1}) and the i-th virtual bond (b_i) is:

$$\overrightarrow{b_i} = T(\beta_i, \theta_i) \cdot \overrightarrow{b_{i-1}} \tag{2}$$

where, β_i and θ_i are the related bond angles $\beta_i = 120^\circ$ and pseudo-torsion angle $\theta_i = g^+$ (60°), g^- (300°) or t (180°) (Fig. 5C). The matrix T is defined as

$$T(\beta_i, \theta_i) = \begin{bmatrix} 1 & 0 & 0 \\ 0 & \cos\beta_i & -\sin\beta_i \\ 0 & \sin\beta_i & \cos\beta_i \end{bmatrix} \times \begin{bmatrix} \cos\theta_i & -\sin\theta_i & 0 \\ \sin\theta_i & \cos\theta_i & 0 \\ 0 & 0 & 1 \end{bmatrix} \tag{3}$$

Therefore, if the unit vector of the first virtual bond $(P - C_4^*)$ is $\overrightarrow{b_1}$,), we can obtain the unit vector of the i-th virtual bond $(P - C_4^*$ or $C_4^* - P)$ with

$$\overrightarrow{b_i} = \prod_{j=2}^{i} T(\beta_j, \theta_j) \cdot \overrightarrow{b_1} \tag{4}$$

By considering excluded volume effect, i.e., two atoms cannot occupy the same site on a diamond lattice, we can generate the atomic coordinates using equations 1, 2, 3 and 4 and the conformation ensemble of RNA loops; see Ref. [42] for the detailed calculations. Here we give an example for illustration. In Fig. 5, following the $5' \rightarrow 3'$ direction, if the first atom P is fixed at $\overrightarrow{X_0}$ = col (0, 0, 3.9) (Å), the second atom C_4^* is fixed at $\vec{X}_1$=col (0, 0, 0) (Å), then, $\overrightarrow{b_1}$=col (0, 0, −1). According to Eq. 1, the coordinate of the third atom P can be computed:

$$\overrightarrow{X_2} = \overrightarrow{X_0} + l_1 \overrightarrow{b_1} + l_2 T(\pi - \beta_P, \eta) \cdot \overrightarrow{b_1}$$

where $l_1 = l_2 = 3.9$ Å are the bond lengths, $\beta_P(=120^\circ)$ is the bond angle and η is the torsion angle (we select 180° for illustration). According to Eq. 3, the matrix is computed as:

$$T(60^\circ, 180^\circ) = \begin{bmatrix} -1 & 0 & 0 \\ 0 & -1/2 & -\sqrt{3}/2 \\ 0 & -\sqrt{3}/2 & 1/2 \end{bmatrix}$$

Therefore, the coordinate of the third atom P is col (0, $\sqrt{3}$/2, −1/2) (3.9 Å). Following the procedure and considering the excluded volume between any two atoms, we can compute the coordinates for other atoms and generate the conformations for a given RNA loop.

Statistical potential

Assuming an equilibrium Boltzmann distribution for the structures in the PDB [53], [54] and NDB [62] database, one can extract the interaction potentials:

$$u(r) = -k_B T \ln \frac{\rho(r)}{\rho^*(r)} \tag{5}$$

Here, k_B is the Boltzmann constant, T is the absolute temperature, $\rho(r)$ is the observed density, and $\rho^*(r)$ is the density in a "reference" state where no interactions occur.

The above approach can give continuous distance-dependent pairwise potentials [15], [63]–[66]or discrete base pairs/stacks potentials [27], [67]. In this work, we extract the potentials for the different dinucleotide conformations described by the pseudo-torsion angles $\theta\ (P - C_4^* - P - C_4)$ and $\eta\ (C_4^* - P - C_4 - P)$ and the different dinucleotide sequences (Fig. 5). The potential $u_{ij}(\theta,\eta)$ is extracted based on the following formula:

$$u_{ij}(\theta, \eta) = -k_B T \ln \frac{\rho_{ij}(\theta, \eta)}{\rho_{ij}^*(\theta, \eta)} \tag{6}$$

where *i* and *j* denote the types of the nucleotides (bases) of the two consecutive nucleotides within the same loop and (θ, η) are the pseudo-torsion angles of the dinucleotide.

Reference state

The proper choice of the reference state is a key issue in deriving the statistical potentials from the known structures. As pointed out by Thomas and Dill [68], [69], an accurate ideal reference state is not achievable. Different approximations such as the quasi-chemical approximation [64] have been used to model the reference state. Furthermore, to circumvent the reference state problem, an iterative method for successive refinement of the statistical potentials was developed and was shown to give reliable scoring functions for protein folding [68], [69] and protein-ligand interactions [63]. Here, for comparison, we employ the above two approaches to extract two different sets of RNA knowledge-based potentials and apply the extracted potentials to RNA loop and junction structure prediction.

Quasi-chemical approximation-based approach.

In the quasi-chemical approximation [64] we use an "expected state" as the "reference state". Such an approximation leads to the Boltzmann relation (Eq. 6):

$$u_{ij}(\theta, \eta) = -k_B T \ln \frac{\rho_{ij}^{\text{obs}}(\theta, \eta)}{\rho_{ij}^{\text{exp}}(\theta, \eta)} \tag{7}$$

where $\rho_{ij}^{\text{obs}}(\theta, \eta)$ and $\rho_{ij}^{\text{exp}}(\theta, \eta)$ are the observed number and the expected number, respectively, of the dinucleotides (i, j) with pseudo-torsion angles (θ, η) in the entire training dataset. $\rho_{ij}^{\text{exp}}(\theta, \eta)$ is computed from the following formulas:

$$\rho_{ij}^{\text{exp}}(\theta, \eta) = N(\theta, \eta)\chi_i \chi_j \tag{8}$$

where N(θ,η) is the total observed number of dinucleotides with pseudo-torsion angles θ and η (calculated with a composition-independent scale, meaning each nucleotide is independent to other nucleotides in the state (Eq. 8)) [70], and χ_i and χ_j are the mole fraction of nucleotide *i* and *j* of the sequences in the entire training dataset, respectively [70].

The quasi-chemical approximation-based statistical potentials are the "extracted" energy-like parameters derived from the observed density in the database of native structures [68]. We denote such extracted statistical potentials as SP_{native}. The SP_{native} parameters are not the "true" potentials extracted from the criteria of identifying the "correct"/ native structure from an ensemble of the nonnative conformations. We denote the potentials that can account for the nonnative conformations as the "true" statistical potentials (SP_{fold}).

Convergent iterative approach.

To extract the "true" energy (SP_{fold}), Thomas and Dill developed an iterative approach based on the folding stability of the native structure [68], [69]. The method has the advantage of accounting for the effect of the distribution of the whole conformational ensemble, including both the native and the nonnative states, for a given sequence. The overall strategy of the iterative approach is to train a set of potential parameters iteratively until the collection of the native structures in the training dataset and the full conformational ensemble (native and nonnative) lead to the same frequencies of the different dinucleotide conformations [63], [68], [69].

In our calculation, the iterative process starts with a set of initial values for the potentials, $u_{ij}^{(0)}(\theta, \eta)$. The superscript (n) denotes the n-th iterative step. We use the quasi-chemical approximation-based statistical potentials SP_{native} as the initial input.

At each step, we compute the total potential/score for each conformation of loop/junction by summing over the potential of all the dinucleotides contained in the loop/junction:

$$E^{(n)} = \sum_{\text{dinucleotides}} u_{ij}^{(n)}(\theta, \eta) \tag{9}$$

Then the predicted frequencies (numbers) of the different dinucleotide conformations are computed from the Boltzmann average over all the conformations:

$$\rho_{ij}^{\text{pred}(n)}(\theta, \eta) = \sum_{s}^{S} \frac{\sum_{l}^{L} n_{ij}(\theta,\eta) \cdot e^{-E^{(n)}(s,l)/k_B T}}{\sum_{l}^{L} e^{-E^{(n)}(s,l)/k_B T}} \tag{10}$$

where $E^{(n)}(s, l)$ is the potential/score for the l-th conformation of the s-th RNA loop/junction sequence, computed with the potentials $u_{ij}^{(n)}(\theta, \eta)$ at the n-th step (Eq. 9), $n_{ij}(\theta, \eta)$ is the number of dinucleotide (i, j) with pseudo-torsion angles (θ, η) in the l-th conformation of the s-th RNA loop/junction, L is the number of conformations of the s-th RNA loop/junction in the training dataset, and S is the number of RNA loop/junction sequences in the training dataset.

In each step, we also calculate the difference between the observed and predicted dinucleotide frequencies:

$$\Delta u_{ij}^{(n)}(\theta, \eta) = -k_B T \ln \frac{\rho_{ij}^{\text{obs}}(\theta, \eta)}{\rho_{ij}^{\text{pred}(n)}(\theta, \eta)} \quad (11)$$

where $\rho_{ij}^{\text{pred}(n)}(\theta, \eta)$ is calculated by Eq. 10 and $\rho_{ij}^{\text{obs}}(\theta, \eta)$ is the dinucleotide frequency observed from the "correct" (native) structures in the training dataset.

In general, the initially guessed potential parameters are not equal to the "true" potentials and thus there are differences between the observed dinucleotide frequencies and the predicted dinucleotide frequencies $(\Delta u_{ij}^{(n)}(\theta, \eta) \gg 0)$. For each n-th step, we refine the potentials by accounting for the differences between the "observed" and the "predicted" frequencies of the dinucleotide conformations (see the parameter $\Delta u_{ij}^{(n)}(\theta, \eta)$ in Eq. 11):

$$u_{ij}^{(n)}(\theta, \eta) = u_{ij}^{(n-1)}(\theta, \eta) + \Delta u_{ij}^{(n-1)}(\theta, \eta) \quad (12)$$

We repeat the above iterative process until $|\Delta u_{ij}^{(n)}(\theta, \eta)|$ approaches 0 (smaller than a threshold number, e.g. 10^{-3}). The final set of potentials $u_{ij}^{(n)}(\theta, \eta)$ is the "true" potentials SP_{fold}.

Flowchart

The iterative method is summarized as follows (Fig. 1):

- Prepare the training dataset of the native structures. Download RNA structures from PDB database and extract the loops and junctions. For each RNA loop/junction in the training set, generate the full conformational ensemble of RNA backbone by self-avoiding exhaustive walks on a diamond lattice (Vfold model). The conformation ensemble is used in the iterative calculation.
- Calculate the observed dinucleotide frequencies $\rho_{ij}^{obs}(\theta, \eta)$ by summing the dinucleotide densities observed in the training dataset.
- Choose a set of initially guessed potentials. In this work, we start with the "extracted" quasi-chemical approximation-based potentials (SP_{native}).
- Calculate the potential/score for each generated conformation of each RNA loop/junction sequence by using Eq. 9 with iterative potentials $u_{ij}^{(n-1)}(\theta, \eta)$. Then calculate the weighted predicted frequencies $\rho_{ij}^{pred(n)}(\theta, \eta)$ of each dinucleotide (i, j) with pseudo-torsion angles (θ, η), by employing Eq. 10.
- Calculate the convergence parameter $\Delta u_{ij}^{(n-1)}(\theta, \eta)$ according to Eq. 11. If the following convergence condition is satisfied:
- $$|\Delta u_{ij}^{(n-1)}(\theta, \eta)| \leq C_0 \qquad (13)$$
- with a preset small value of the threshold parameter C_0, say, 10^{-3}, then skip to the final step and the "true" potentials SP_{fold} are obtained; otherwise, continue to the next step.
- Adjust the current potentials $u_{ij}^{(n-1)}(\theta, \eta)$ by adding the correction $u_{ij}^{(n-1)}(\theta, \eta)$ (Eq. 12), and obtain a new set of potentials $u_{ij}^{(n)}(\theta, \eta)$. Move to the next (n-th) iterative step and return to Step 4 for the next cycle.

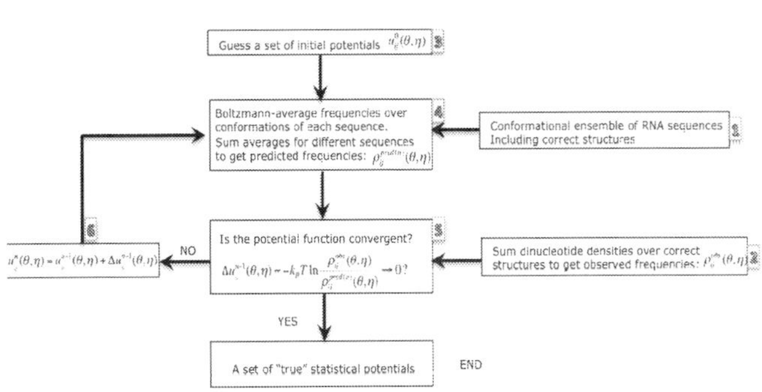

Figure 1: Flow chart of the iterative approach.

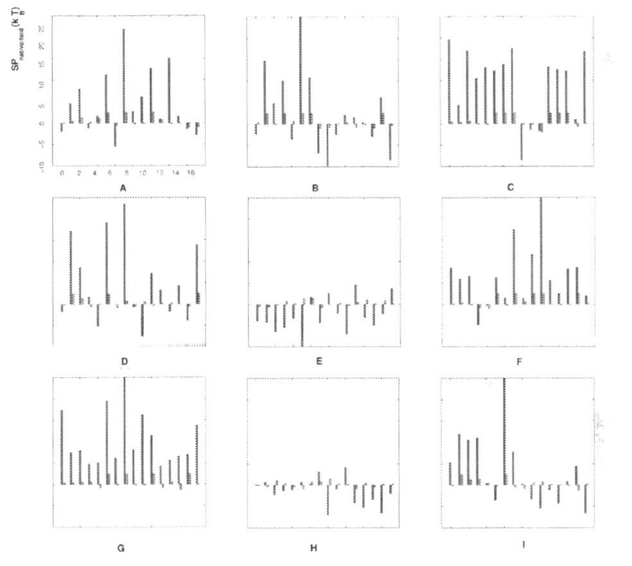

Figure 2: Comparison between the "extracted" statistical potentials SP_{native} **and the "true"** potentials SP_{fold}. = The figures from (A) to (I) stand for the torsion angles (θ, η =): (A) (g^+, g^+), (B) (g^+, t), (C) (g^+, g^-), (D) (t, g^+), (E) (t, t), (F) (t, g^-), (G) (g^-, g^+), (H) (g^-, t) and (I) (g^-, g^-) , respectively. In each figure, the red bars represent the statistical potentials SP_{fold}, the green bars represent the statistical potentials SP_{native}, both of which are obtained from the RNA09 dataset, and the x-axis stands for the dinucleotides with different nucleotides (i, j=) AA, AC, AG, AU, CA, CC, CG, CU, GA, GC, GG, GU, UA, UC, UG and UU from 1 to 16.

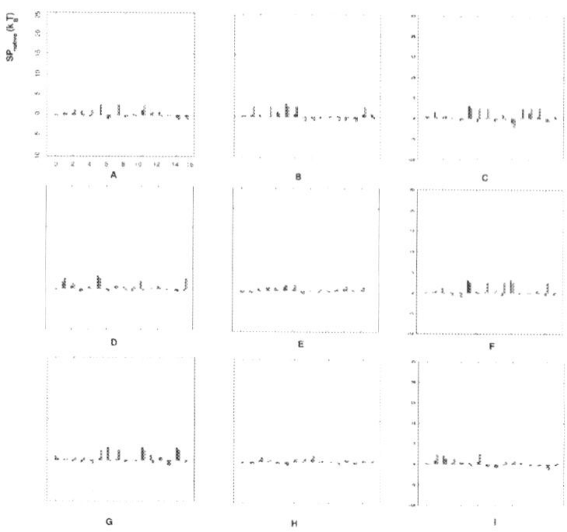

Figure 3: Comparison between the "extracted" statistical potentials SP_{native} **from the RNA09 dataset and the** Leontis dataset, respectively. The figures from (A) to (I) stand for the torsion angles (θ, η=): (A) (g^+, g^+), (B) (g^+, t), (C) (g^+, g^-), (D) (t, g^+), (E) (t, t), (F) (t, g^-), (G) (g^-, g^+), (H) (g^-, t) and (I) (g^-, g^-), respectively. In each figure, the red bars represent the statistical potentials SP_{native} extracted from the Leontis dataset, the green bars represent the ones from the RNA09 dataset, and the x-axis stands for the dinucleotides with different nucleotides (i, j=) AA, AC, AG, AU, CA, CC, CG, CU, GA, GC, GG, GU, UA, UC, UG and UU from 1 to 16.

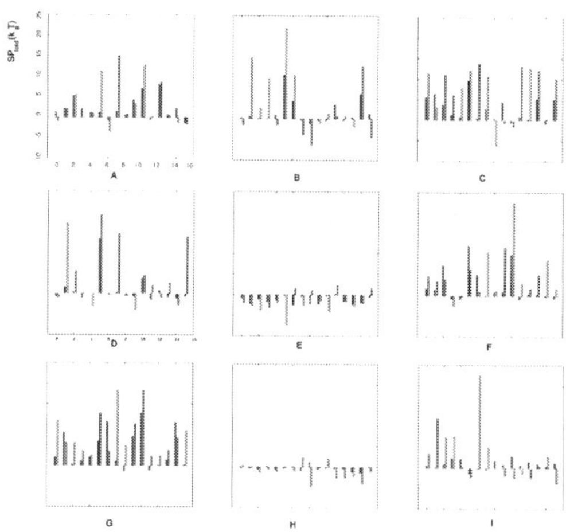

Figure 4: Comparison between the "true" statistical potentials SP_{fold} **obtained from the RNA09 dataset and the** Leontis dataset, respectively. The figures from (A) to (I) stand for the torsion angles (θ, η=): (A) (g^+, g^+), (B) (g^+, t), (C) (g^+, g^-), (D) (t, g^+), (E) (t, t), (F) (t, g^-), (G)

(g^-, g^+), (H) (g^-, t) and (I) (g^-, g^-), respectively. In each figure, the red bars represent the statistical potentials SP_{fold} extracted from the Leontis dataset, the green bars represent the ones from the RNA09 dataset, and the x-axis stands for the dinucleotides with different nucleotides (*i, j*=) AA, AC, AG, AU, CA, CC, CG, CU, GA, GC, GG, GU, UA, UC, UG and UU from 1 to 16.

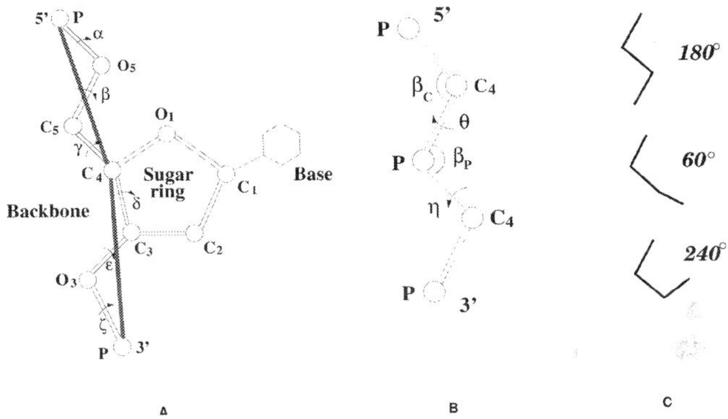

Figure 5: The pseudo-torsional angles. (A) The virtual bond scheme for RNA nucleotides. (B) The bond angles (β_C, β_P) and the pseudo-torsional angles (θ, η) of the virtual bonds. (C) The three preferred rotamer-like configurations of the virtual bonds: t, g^+ and g^-, with the torsional angles equal to 180°, 60° and 300°, respectively.

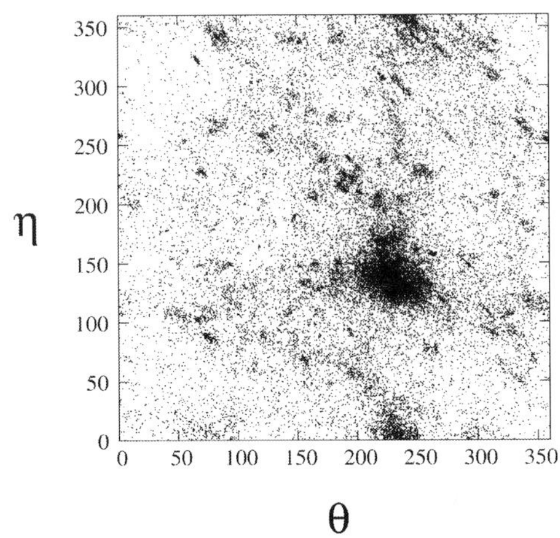

Figure 6: The distribution of the θ **and** η **values for the different types of** dinucleotides in RNA loops and junctions. The RNA loops and junctions are selected from the RNA09 dataset and the angles are calculated based on the PDB structures.

Preparation of the training dataset

We use the RNA 2009 database (RNA09; http://Kinemage.biochem.duke.edu/databases/rnadb.phb, an updated version of previous 2005 database (RNA05) [51]) to extract the SP_{native} potential and use the database as the training dataset to search for the "true" potential SP_{fold}. The RNA09 dataset consists of 262 RNA sequences with the experimentally determined structures of resolution ≤3.0 Å.

RNA loops and junctions are constructed by all the unpaired nucleotides in RNA structure. We detect the loops and junctions by removing the nucleotides involved in the base pairs. Furthermore, we remove all the loop structures with modified bases or non-RNA atoms and molecules. Advances in the knowledge of the structures of both the canonical (Watson-Crick and wobbles) base pairs and non-canonical base pairs (normally classified as tertiary interactions) [71], [72] enable the development of several automated tools to detect the base pairs in RNA structures. In this work, we use the 3DNA software package (http://3dna.rutgers.edu/home) [73] to search for all the possible base pairs from the RNA structures.

In addition, the time consumption to enumerate the loop conformations on a diamond lattice increases exponentially with the length of the loop [45]. Therefore, we only choose loops/junctions with length ≤8 nt in the dataset due to the long computer time for the exhaustive enumeration of the loop conformations for longer loops [45]. In this study, we use the algorithm reported in Ref. [74] to search for the best fits on diamond lattice for the RNA loops/junctions.

Larger loops often involve tertiary interactions with other subunits of RNA structures, which are not accounted for in this study. As a result, we found 152 loops/junctions with the length ranging from 3 nt to 8 nt in the RNA09 dataset. These loop and junction structures are used as the training dataset in our iterative approach.

To further test the influence of the use of the different training dataset on the statistical potentials and the predictive power of the statistical potentials, we also use another non-redundant high-resolution RNA structure dataset, collected by Leontis› lab (http://rna.bgsu.edu/nrlist/).

Conformational ensembles of RNA loops/junctions

We use the Vfold model to generate the full conformation ensemble of a given RNA loop or junction sequence. In the Vfold-generated loop/junction conformational ensemble, because each of the two pseudo-torsion angles (θ and η) of a dinucleotide occupies three torsional states in a diamond lattice, there exist 4×4×3×3 parameters for the potentials for the four types of each nucleotide (A, C, G and U) and the three possible rotamer-like states for each torsional angle (and hence 3×3 types of the pseudo-torsion angle pairs θ and η).

For each PDB structure of the loop/junction, we find the minimum-RMSD fit of the virtual bond conformation on the diamond lattice. We call such a coarse-grained (correct) structure as the "coarse-grained correct structure"; see Step 1 in Fig. 1. Our calculations show that the differences between the diamond lattice-represented structures and the PDB structures varies for the different loop lengths and sequence contents with RMSD in the range from 0.67 Å to 2.07 Å (Fig. 7), the mean and standard deviation of RMSD values are 1.35 A and 0.30 Å, respectively.. We will use the coarse-grained correct structure to calculate the observed dinucleotide frequencies, from which SP_{native} and SP_{fold} are extracted.

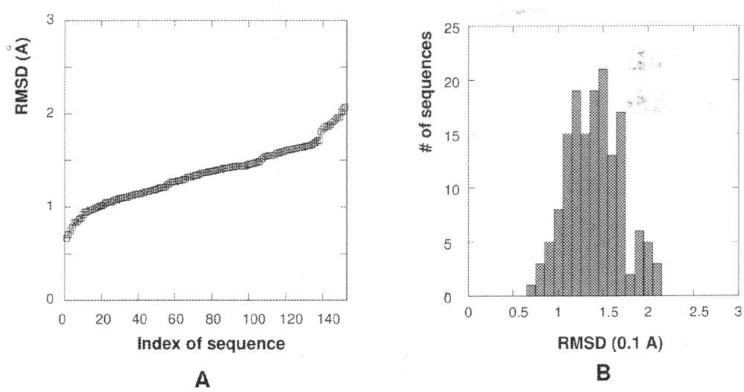

Figure 7: (A) The RMSD between the PDB structures and the diamond lattice-represented structures for RNA loops and junctions in the RNA09 dataset. The x-axis represents the index of RNA loops/junctions in the RNA09 dataset. (B) The number of RNA loops/junctions within each RMSD-value bin (0.1 Å). The mean and standard deviation of RMSD values for the 152 RNA loops/junctions are 1.35 Å and 0.30 Å, respectively.

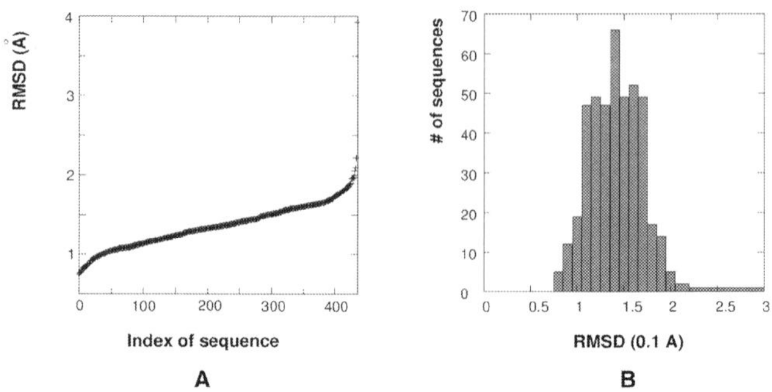

Figure 8: (A) The structural differences (RMSD values) between PDB structures and diamond lattice-represented structures for the RNA loops/junctions within the Leontis dataset.

The x-axis represents the index of RNA loops/junctions in the Leontis dataset. (B) The number of RNA loops/junctions within each RMSD-value bin (0.1 Å). The mean and standard deviation of RMSD values for the 435 RNA loops/junctions are 1.37 Å and 0.29 Å, respectively.

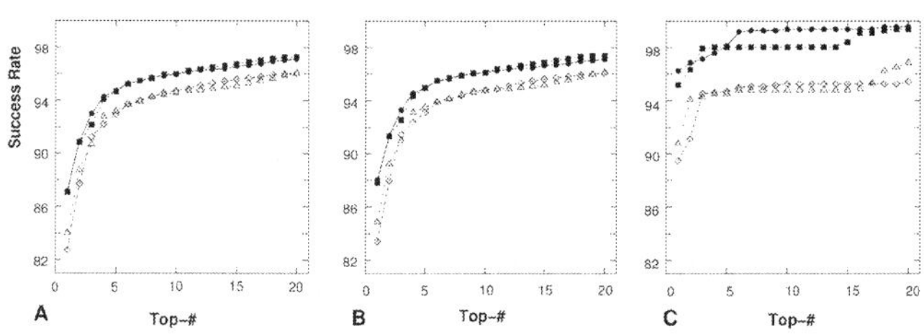

Figure 9: The success rate for the prediction of the coarse-grained correct loop/junction structures for (A) all the 8452 RNA loops and junctions in *TEST-I*, (B) the 7459 RNA loops and junctions in *TEST-II* **and (C) the 1119 RNA loops and junctions in** *TEST-III*. The "Top-#" in x-axis means that the "correct" structure is in the top # lowest-potential conformations. In each figure, "● " and "▪" represent the success rate with SP_{fold}extracted from the Leontis' database and the RNA09 dataset, respectively; while, "Δ" and "◊" represent the success rate with SP_{native} extracted from the Leontis› database and the RNA09 dataset, respectively.

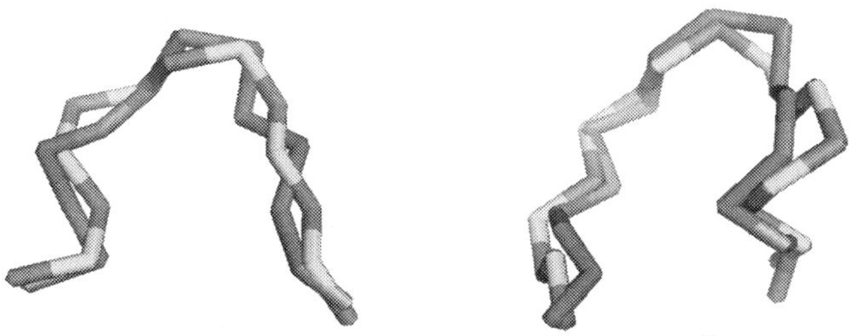

A B

Figure 10: Comparison between the virtual bond-based PDB structure and correctly predicted structure with statistical potentials SP_{fold}. The loops are (A) a hairpin loop from PDB structure 1IVS and (B) an internal loop from PDB structure 1JJ2, respectively. In both figures, the structures shown in brown (P) and yellow (C_4^*) stand for the PDB structure, and the ones shown in brown (P) and purple (C_4^*) represent the correctly predicted structure, which have the lowest potential and the minimal RMSD. The RMSD values are (A) 1.52 Å and (B) 1.96 Å, respectively. The atomic structures are illustrated with Pymol software (http://www.pymol.org/).

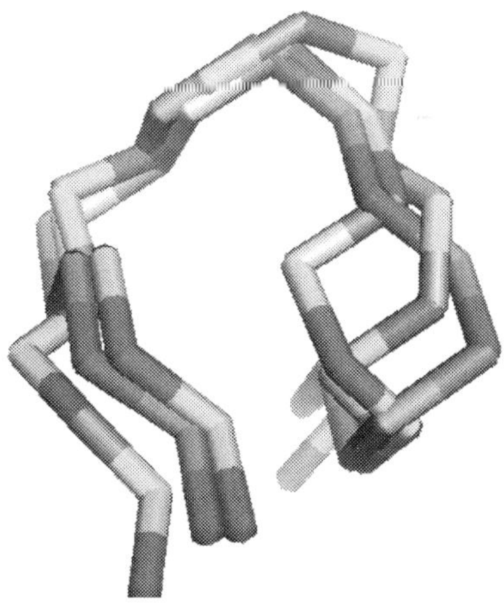

Figure 11: Comparison between the virtual bond-based PDB structure and top-ranked predicted structures with statistical potentials SP_{fold}. The loop is a multibranched loop from PDB structure 3CCM. In the figure, the structure shown in brown (P) and yellow (C_4^*) stands for the PDB structure, the one shown in brown (P) and pink (C_4^*) represents the minimal-RMSD structure (RMSD=2.42 Å), and the one shown in brown (P)

and lightgray (C_4^*) represents the predicted structure with lowest potentials (RMSD=2.87 Å), computed with statistical potentials SP_{fold}. The minimal-RMSD structure has the 9-th lowest potential. The atomic structures are illustrated with Pymol software (http://www.pymol.org/).

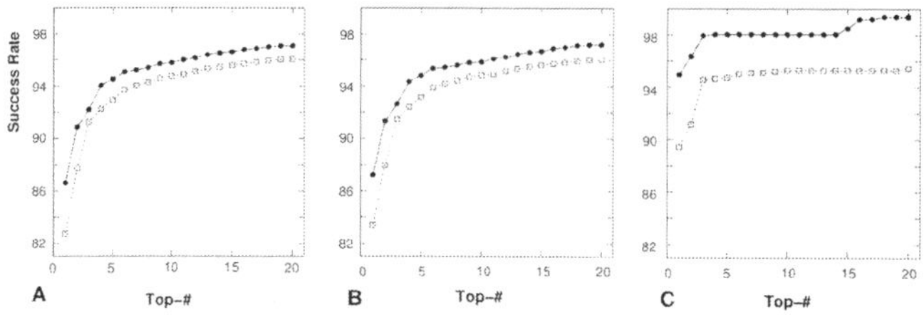

Figure 12: The success rate of coarse-grained correct loop/junction structure predictions for (A) all the 8452 RNA loops/junctions in *TEST-I*, (B) the 7459 RNA loops and junctions in *TEST-II*, and (C) the 1119 RNA loops and junctions in *TEST-III*. The "Top-#" in x-axis means that the "correct" structure is in the top # lowest-potential conformations. In each figure, ● and □ represent the success rate with SP_{fold} and SP_{native}, respectively.

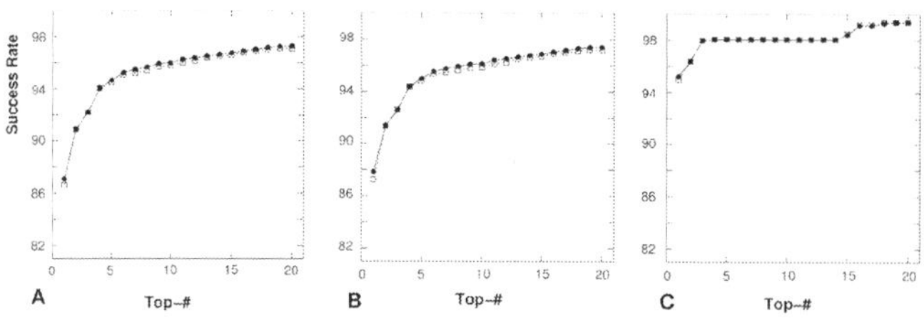

Figure 13: The sensitivity of the structure prediction to the change of the convergence threshold parameter. : $\Delta u_{ij}^n(\theta, \eta)$ =0.001 v.s. 0.01. The tests are based on the predictions of the RNA loops and junctions in all the three databases: (A) *TEST-I*, (B) *TEST-II* and (C) *TEST-III*. In each figure, ● and □ represent the success rate of the SP_{fold}-based structure prediction with $\Delta u_{ij}^n(\theta, \eta)$ =0.001 and 0.01, respectively.

ACKNOWLEDGMENTS

Most of the numerical calculations involved in this research were performed on the HPC resources at the University of Missouri Bioinformatics Consortium (UMBC).

AUTHOR CONTRIBUTIONS

Conceived and designed the experiments: LL SJC. Performed the experiments: LL. Analyzed the data: LL SJC. Wrote the paper: LL SCJ.

REFERENCES

1. Ding F, Sharma S, Chalasani P, Demidov V, Broude N, et al. (2008) Ab initio RNA folding by discrete molecular dynamics: from structure prediction to folding mechanisms. RNA 14: 1164–1173.
2. Zhang J, Dundas J, Lin M, Chen R, Wang W, et al. (2009) Prediction of geometrically feasible three-dimensional structures of pseudoknotted RNA through free energy estimation. RNA 15: 2248–2263.
3. Gherghe C, Leonard C, Ding F, Dokholyan N, Weeks K (2009) Native-like RNA tertiary structures using a sequence-encoded cleavage agent and refinement by discrete molecular dynamics. J Am Chem Soc 131: 2541–2546.
4. Hajdin C, Ding F, Dokholyan N, Weeks K (2010) On the significance of an RNA tertiary structure prediction. RNA 16: 1340–1349.
5. Das R, Baker D (2007) Automated de novo prediction of native-like RNA tertiary struc22 tures. Proc Natl Acad Sci U S A 104: 14664.
6. Shapiro B, Yingling Y, Kasprzak W, Bindewald E (2007) Bridging the gap in RNA structure prediction. Curr Opin Struct Biol 17: 157–165.
7. Parisien M, Major F (2008) The MC-fold and MC-Sym pipeline infers RNA structure from sequence data. Nature 452: 51–55.
8. Jonikas M, Radmer R, Laederach A, Das R, Pearlman S, et al. (2009) Coarse-grained modeling of large RNA molecules with knowledge-based potentials and structural filters. RNA 15: 189–199.
9. Das R, Karanicolas J, Baker D (2010) Atomic accuracy in predicting and designing noncanonical RNA structure. Nat Methods 7: 291–294.
10. Yang S, Parisien M, Major F, Roux B (2010) RNA structure determination using SAXS data. J Phys Chem B 114: 10039–10048.
11. Pasquali S, Derreumaux P (2010) HiRE-RNA: a high resolution

coarse-grained energy model for RNA. J Phys Chem B 114: 11957–11966.

12. Xia Z, Gardner D, Gutell R, Ren P (2010) Coarse-grained model for simulation of RNA three-dimensional structures. J Phys Chem B 114: 13497–13506.
13. Flores S, Altman R (2010) Turning limited experimental information into 3D models of RNA. RNA 16: 1769.
14. Cao S, Chen SJ (2011) Physics-based de novo prediction of RNA 3D structures. J Phys Chem B 115: 4216–4226.
15. Bernauer J, Huang X, Sim A, Levitt M (2011) Fully differentiable coarse-grained and all-atom knowledge-based potentials for RNA structure evaluation. RNA 17: 1066–1075.
16. Sripakdeevong P, Kladwang W, Das R (2011) An enumerative stepwise ansatz enables atomic-accuracy RNA loop modeling. Proc Natl Acad Sci U S A 108: 20573–20578.
17. Parisien M, Major F (2012) Determining RNA three-dimensional structures using low resolution data. J Struct Biol in press.
18. Bida J, Maher L III (2012) Improved prediction of RNA tertiary structure with insights into native state dynamics. RNA 18: 385–393.
19. Capriotti E, Marti-Renom M (2008) Computational RNA structure prediction. Current Bioinformatics 3: 32–45.
20. Laing C, Schlick T (2010) Computational approaches to 3D modeling of RNA. J Phys: Condens Matter 22: 283101.
21. Laing C, Schlick T (2011) Computational approaches to RNA structure prediction, analysis, and design. Curr Opin Struct Biol 21: 306–318.
22. Rother K, Rother M, Boniecki M, Puton T, Bujnicki J (2011) RNA and protein 3D structure modeling: similarities and differences. Journal of molecular modeling 1–12.
23. Cruz J, Blanchet M, Boniecki M, Bujnicki J, Chen S, et al. (2012) RNA-Puzzles: A CASP-like evaluation of RNA three-dimensional structure prediction. RNA 8: 1–16.
24. Leontis N, Westhof E (2012) RNA 3D Structure Analysis and Prediction. Springer-Verlag

25. Dima R, Hyeon C, Thirumalai D (2005) Extracting stacking interaction parameters for RNA from the data set of native structures. J Mol Biol 347: 53–69.
26. Mathews D, Sabina J, Zuker M, Turner D (1999) Expanded sequence dependence of thermodynamic parameters improves prediction of RNA secondary structure1. J Mol Biol 288: 911–940.
27. Wu J, Gardner D, Ozer S, Gutell R, Ren P (2009) Correlation of RNA secondary structure statistics with thermodynamic stability and applications to folding. J Mol Biol 391: 769–783.
28. Nussinov R, Jacobson A (1980) Fast algorithm for predicting the secondary structure of single-stranded RNA. Proc Natl Acad Sci U S A 77: 6309.
29. Zuker M (1989) On finding all suboptimal foldings of an RNA molecule. Science 244: 48–52.
30. Hofacker I, Fontana W, Stadler P, Bonhoeffer L, Tacker M, et al. (1994) Fast folding and comparison of RNA secondary structures. Monatshefte für Chemie/Chemical Monthly 125: 167–188.
31. Hofacker I (2003) Vienna RNA secondary structure server. Nucleic Acids Res 31: 3429–3431.
32. Mathews D, Turner D (2006) Prediction of RNA secondary structure by free energy minimization. Curr Opin Struct Biol 16: 270–278.
33. Serra M, Barnes T, Betschart K, Gutierrez M, Sprouse K, et al. (1997) Improved param eters for the prediction of RNA hairpin stability. Biochemistry 36: 4844–4851.
34. Giese M, Betschart K, Dale T, Riley C, Rowan C, et al. (1998) Stability of RNA hairpins closed by wobble base pairs. Biochemistry 37: 1094–1100.
35. Mathews D, Disney M, Childs J, Schroeder S, Zuker M, et al. (2004) Incorporating chemical modification constraints into a dynamic programming algorithm for prediction of RNA secondary structure. Proc Natl Acad Sci U S A 101: 7287.
36. Dale T, Smith R, Serra M (2000) A test of the model to predict unusually stable RNA hairpin loop stability. RNA 6: 608–615.
37. Schlick T, Collepardo-Guevara R, Halvorsen L, Jung S, Xiao X (2011) Biomolecular mod eling and simulation: a field coming of

age. Q Rev Biophys 44: 191.

38. Hyeon C, Thirumalai D (2007) Mechanical unfolding of RNA: from hairpins to structures with internal multiloops. Biophys J 92: 731–743.
39. Villa A, Stock G (2006) What NMR relaxation can tell us about the internal motion of an RNA hairpin: a molecular dynamics simulation study. J Chem Theory Comput 2: 1228–1236.
40. Sorin E, Rhee Y, Nakatani B, Pande V (2003) Insights into nucleic acid conformational dynamics from massively parallel stochastic simulations. Biophys J 85: 790–803.
41. Keating K, Pyle A (2010) Semiautomated model building for RNA crystallography using a directed rotameric approach. Proc Natl Acad Sci U S A 107: 8177.
42. Cao S, Chen SJ (2005) Predicting RNA folding thermodynamics with a reduced chain representation model. RNA 11: 1884–1897.
43. Cao S, Chen SJ (2006) Predicting RNA pseudoknot folding thermodynamics. Nucleic Acids Res 34: 2634.
44. Cao S, Chen SJ (2009) Predicting structures and stabilities for h-type pseudoknots with interhelix loops. RNA 15: 696–706.
45. Liu L, Chen SJ (2010) Computing the conformational entropy for RNA folds. J Chem Phys 132: 235104.
46. Cao S, Giedroc D, Chen SJ (2010) Predicting loop–helix tertiary structural contacts in RNA pseudoknots. RNA 16: 538–552.
47. Cao S, Chen SJ (2012) Predicting kissing interactions in microRNA–target complex and assessment of microrna activity. Nucleic Acids Res 40: 4681–4690.
48. Chen SJ (2008) RNA folding: conformational statistics, folding kinetics, and ion electro statics. Annu Rev Biophys 37: 197.
49. Vecenie C, Morrow C, Zyra A, Serra M (2006) Sequence dependence of the stability of RNA hairpin molecules with six nucleotide loops. Biochemistry 45: 1400–1407.
50. Schudoma C, May P, Nikiforova V, Walther D (2010) Sequence–structure relationships in RNA loops: establishing the basis for loop homology modeling. Nucleic Acids Res 38: 970–980.
51. Murray L, Arendall W, Richardson D, Richardson J (2003) RNA

backbone is rotameric. Proc Natl Acad Sci U S A 100: 13904.

52. Capriotti E, Norambuena T, Marti-Renom M, Melo F (2011) All-atom knowledge-based potential for RNA structure prediction and assessment. Bioinformatics 27: 1086.
53. Bernstein F, Koetzle T, Williams G, Meyer E, Brice M, et al. (1977) The protein data bank: A computer-based archival file for macromolecular structures. J Mol Biol 112: 535–542.
54. Berman H, Olson W, Beveridge D, Westbrook J, Gelbin A, et al. (1992) The nucleic acid database. A comprehensive relational database of three-dimensional structures of nucleic acids. Biophys J 63: 751.
55. Duarte C, Pyle A (1998) Stepping through an RNA structure: a novel approach to conformational analysis1. J Mol Biol 284: 1465–1478.
56. Wadley L, Keating K, Duarte C, Pyle A (2007) Evaluating and learning from RNA pseudotorsional space: quantitative validation of a reduced representation for RNA structure. J Mol Biol 372: 942–957.
57. Bloomfield V, Crothers D, Tinoco I (2000) Nucleic acids: structures, properties, and functions. Univ Science Books
58. Olson W, Flory P (1972) Spatial configurations of polynucleotide chains. I. steric interactions in polyribonucleotides: a virtual bond model. Biopolymers 11: 1–23.
59. Olson WK (1975) Configuration statistical of polynucleotide chains: A single virtual bond treatment. Macromolecules 8: 272–275.
60. Olson WK (1980) Configurational statistics of polynucleotide chains: An updated virtual bond model to treat effects of base stacking. Macromolecules 13: 721–728.
61. Flory P, et al. (1969) Statistical mechanics of chain molecules. Biopolymers 8: 699–700.
62. Berman H, Westbrook J, Feng Z, Gilliland G, Bhat T, et al. (2000) The protein data bank. Nucleic Acids Res 28: 235–242.
63. Huang S, Zou X (2006) An iterative knowledge-based scoring function to predict protein–ligand interactions: I. derivation of interaction potentials. J Comput Chem 27: 1866–1875.
64. Lu H, Skolnick J (2001) A distance-dependent atomic knowledge-

based potential for improved protein structure selection. Proteins: Struct, Funct, Bioinf 44: 223–232.

65. Zhou H, Zhou Y (2002) Distance-scaled, finite ideal-gas reference state improves structure-derived potentials of mean force for structure selection and stability prediction. Protein Sci 11: 2714–2726.
66. Zhao H, Yang Y, Zhou Y (2011) Structure-based prediction of RNA-binding domains and RNA-binding sites and application to structural genomics targets. Nucleic Acids Res 39: 3017–3025.
67. Sharma S, Ding F, Dokholyan N (2008) iFoldRNA: three-dimensional RNA structure prediction and folding. Bioinformatics 24: 1951–1952.
68. Thomas P, Dill K (1996) Statistical potentials extracted from protein structures: how accurate are they? J Mol Biol 257: 457–469.
69. Thomas P, Dill K (1996) An iterative method for extracting energy-like quantities from protein structures. Proc Natl Acad Sci U S A 93: 11628–11633.
70. Skolnick J, Kolinski A, Ortiz A (2000) Derivation of protein-specific pair potentials based on weak sequence fragment similarity. Proteins: Struct, Funct, Bioinf 38: 3–16.
71. Leontis N, Westhof E (2001) Geometric nomenclature and classification of RNA base pairs. RNA 7: 499–512.
72. Leontis N, Stombaugh J, Westhof E (2002) Motif prediction in ribosomal RNAs lessons and prospects for automated motif prediction in homologous RNA molecules. Biochimie 84: 961–973.
73. Lu X, Olson W (2003) 3DNA: a software package for the analysis, rebuilding and visualization of three-dimensional nucleic acid structures. Nucleic Acids Res 31: 5108.
74. Ferro D, Hermans J (1977) A different best rigid-body molecular fit routine. Acta Crystallographica Section A: Crystal Physics, Diffraction, Theoretical and General Crystallography 33: 345–347.

PART IV COARSE-GRAINED MODELS FOR LIPIDS

CHAPTER 11

THE EFFECT OF TETHERS ON ARTIFICIAL CELL MEMBRANES: A COARSE-GRAINED MOLECULAR DYNAMICS STUDY

William Hoiles[1], Rini Gupta[1], Bruce Cornell[2], Charles Cranfield[3], Vikram Krishnamurthy[4]

[1] Department of Electrical and Computer Engineering, The University of British Columbia, Vancouver, British Columbia, Canada

[2] Director of Science and Technology, Surgical Diagnostics Pty Ltd., Unit 6 30-32 Barcoo Street, Roseville, New South Wales, 2069, Australia

[3] School of Life Sciences, University of Technology Sydney, Broadway, New South Wales, Australia

[4] Electrical and Computer Engineering, Cornell University, New York, New York, United States of America

ABSTRACT

Tethered bilayer lipid membranes (tBLMs) provide a stable platform for modeling the dynamics and order of biological membranes where the tethers mimic the cytoskeletal supports present in biological cell membranes. In this paper coarse-grained molecular dynamics (CGMD)

is applied to study the effects of tethers on lipid membrane properties. Using results from the CGMD model and the overdamped Fokker-Planck equation, we show that the diffusion tensor and particle density of water in the tBLM is spatially dependent. Further, it is shown that the membrane thickness, lipid diffusion, defect density, free energy of lipid flip-flop, and membrane dielectric permittivity are all dependent on the tether density. The numerically computed results from the CGMD model are in agreement with the experimentally measured results from tBLMs containing different tether densities and lipids derived from Archaebacteria. Additionally, using experimental measurements from Escherichia coli bacteria and Saccharomyces Cerevisiae yeast tethered membranes, we illustrate how previous molecular dynamics results can be combined with the proposed model to estimate the dielectric permittivity and defect density of these membranes as a function of tether density.

INTRODUCTION

Biological membranes are essential for sensing and signalling in biology and act to separate the intracellular and extracellular domain of the cell. The complexity associated with studying biological membranes has driven studies of well defined synthetic bilayer lipid membranes. These include black lipid membranes [1, 2], vesicles [3], and supported bilayers [4, 5]. A new class of experimental work has focused on the use of tethered bilayer lipid membranes (tBLMs) which include tethers that mimic the physiological response of the cytoskeletal networks found in all biological cell membranes [6, 7]. A classic example is the Spectrin-Ankyrin–Band 3 network in Red Blood Cell (RBC) membranes. The Spectrin cytoskeletin is tethered to the RBC membrane via the linker protein Ankyrin, and subsequently to the Band 3 protein that traverses the RBC membrane. RBCs survive in the human body for over 120 days and undergo multiple changes in geometry as the cells pass through the heart and major blood vesicles and circulate through the microcapilliaries requiring thousands of alterations in aspect ratio for each passage around the body. Useful features of the tBLMs for experimentalists are that they are stable over weeks to months, are solvent free and are sufficiently distant from their solid supporting substrate due to the tether groups that they are not distorted by adsorption to the substrate. Their geometry permits convenient electrical access to both side of the membrane and measure by electrical impedance spectroscopy of the bilayer conductance and capacitance. In addition the stabilizing tethers mimic the behavior of the cytoskeletal supports providing one of the best models for studying the dynamics and order in biological membranes. The key question we focus on in this paper is how does the gold bioelectronic interface and the tether density impact membrane dynamics and biomechanics.

Models of tethered membranes have been reported using coarse-grained molecular dynamics (CGMD) [8, 9]. CGMD models provide reasonable accuracy given the current computational efficiency available on supercomputing clusters. In [8] the authors study properties associated with DOPC membranes tethered to a gold surface via polyethylene glycol chains. It was shown that diffusivity of lipids in the distal layer (adjacent to the electrolyte bath) are larger than the proximal layer (adjacent to the bioelectronic interface). This result is not unexpected as the tethered lipids impede the movement of lipids in the proximal layer. The results in [8] suggest that to minimize the number of membrane defects, a tether with 5-10 polyethylene glycol moieties and tether density below 22% should be used. For longer tethers and higher tether density the membrane undulates, forms significant defects, and dissociates from the surface. However, experimental measurements from tBLMs containing up to 11 polyethylene glycol moieties do not contain significant defects and have a lifetime of several weeks [10]. Using the CGMD model and experimental measurements from tethered bilayer lipid membranes, we show that for an increase in tether density the membrane stabilizes–that is, there is a reduction in the number of membrane defects as the tether density increases from 1% to 100%. In [9] the formation dynamics of a DPPC membrane tethered to gold via polyethylene glycol chains is presented. The lipid concentration selected in the formation process is critical to minimize the formation of defects in the membrane (if the concentration is too low), or the formation of micelles on the membrane surface (if the concentration is too high). Though [8, 9] provide valuable insight into tethered membrane dynamics, several questions remain. How does the tethering architecture (i.e. bioelectronic interface) impact the tethered membrane dynamics? What effect do different lipids have on the tethered membrane stability? Additionally what effect does the tether density (i.e. cytoskeletal support) have on the stability of membranes?

In this paper a CGMD model of a tethered archaebacterial membrane is generated based on the MARTINI force field [11, 12]. The CGMD model is used to gain key insights into the diffusion of water and lipids, membrane thickness, defect density, and the free energy of lipid flip-flop as a function of tether density. The numerically computed results from the CGMD model are in agreement with the results from experimental measurements from tethered bilayer lipid membranes. Using the results from experimental measurements we show that a combination of diffusion-limited charge transfer, and ionic adsorption is present at the bioelectronic interface. These double-layer charging effects can be modeled using fractional order operators [13]. It is shown that the dielectric permittivity of tethered membranes is dependent on the tether density. To investigate the effect tethers and lipid architecture have on membrane stability, we construct tethered bilayer lipid membranes composed of archaebacterial lipids, and lipids from E. coli and Saccharomyces cerevisiae (S. cerevisiae) membranes. We show that archaebacterial lipids are the most resistant to electroporation, with E. coli having a higher resistance compared to lipids from S. cerevisiae.

METHODS

Coarse-Grained Molecular Dynamics Model of Tethered Bilayer Lipid Membrane

The coarse-grained molecular dynamics model of the tethered bilayer lipid membrane is constructed using the MARTINI force field [11, 12]. Parameters modeled include the spatially dependent diffusion tensor **D**, the thickness of the membrane h_m, the surface tension σ, and the line tension γ, defect density, and free energy of lipid flip-flop which are all dependent on the tethering density.

To construct the CGMD model the molecular components of the tethered bilayer lipid membrane are mapped into the MARTINI [11, 12] force field. The molecular components include the zwittrionic C20 diphytanyl-ether-glycero-phosphatidylcholine lipid (DphPC), C20 diphytanyl-diglyceride ether lipid (GDPE), benzyl disulphide connected to an eight-oxygen-ethylene-glycol group terminated by a C20 hydrophobic phytanyl chain (tether), benzyl disulphide connected to a four-oxygen-ethylene-glycol group terminated by an OH (spacer), and the gold surface. The mobile lipid phase is physically tethered to the chemically attached layer is composed of a 30% GDPE to 70% DphPC lipid membrane. To explore the effects of tether density on the dynamics and stability of the tethered membrane, 0% and 25% tether density (i.e. a 25% area tether density reflects one spacer molecular for every three tethering molecules) CGMD models of the tethered membrane are constructed. The CGMD model of the 25% tethered membrane is provided in Fig 1. A detailed description of the 0% and 25% CGMD models is provided below. Note that unless otherwise specified, the bead types and interactions are provided in [11, 12].

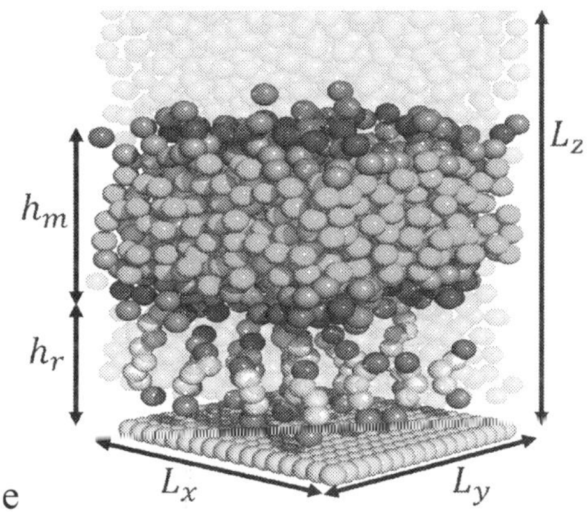

e

Figure 1: Schematic of the CGMD model of the tethered bilayer lipid membrane. The green beads represent the phytanyl tail of the GDPE and DphPC lipids, NC_3 bead is displayed in blue, the PO_4 bead in orange, OH bead in red, the COC bead as pink, polyethylene glycol beads as white, benzyl disulphide bead as brown, gold beads as yellow, and the water beads as a translucent blue. L_z is the height of the simulation cell, L_y and L_x are the length and width of the simulation cell, h_r is the tethering reservoir height, and h_m is the membrane thickness.

Lipids: The phosphatidylcholine headgroup of the DphPC lipid is represented by two beads: the positive choline by the Q_0 bead, and the negative phosphate by the Q_a bead. The ether-glycol is represented by a SN_a bead, and each of the phytanyl tails by four C_1 beads. The phytanyl and ether glycerol moieties of GDPE are represented by the same mapping as the DphPC, however the hydroxyl headgroup of GDPE is represented by a P_4 bead. In total the DphPC lipid is composed of 12 coarse-grain beads, and the GDPE lipid by 11 coarse-grain beads.

Tethers and Spacers: The eight ethylene glycol molecules of the tethers are represented by 8 PEG beads. The interaction of the PEG beads is provided in [9, 14]. The benzyl disulphide group is represented by a C_5 bead which has the highest polar affinity of the MARTINI coarse-grain beads [11]. The phytanyl tail of the tethers is represented by four C_1 beads. The spacers are mapped using an identical method as the tethers however the hydroxyl group is represented by the P_4 bead.

Gold Surface: The gold surface is composed of a square lattice with custom P_f beads spaced 0.3 nm apart. The interaction between P_f and P_4 is 1/3 the value between P_4 and P_4, and the interaction between P_f and other bead types is ~12% of the MARTINI value between P_4 and respective bead types. The following interactions are excluded: interaction between P_f beads, and between the C_5 beads of the tethers and spacers, and P_4 and Q_0 beads of the lipids. The interactions and spacing of the P_f beads are selected to reduce the effects of excess adsorption of beads to the surface– that is, the surface mimics a hydrophobic gold surface. Note that from the ab initio molecular dynamics results in [15], there is a wetting layer of water at the gold surface, however beyond 3.2 Å one of the hydrogen molecules in water points towards the gold surface which is typical for many hydrophobic surfaces. Given that in the CGMD model each water bead represents four water molecules, the water bead interaction with the gold bead surface is selected to be hydrophobic.

Diffusion Tensor from Coarse-Grained Molecular Dynamics Model

The simulation results from the CGMD model are the trajectories of coarse-grained beads as a function of time. In this section a method is provided to compute the spatially dependent diffusion tensor from the CGMD trajectories.

To estimate the diffusion tensor **D** of species from CGMD trajectories requires a probabilistic model. Let Ψ denote the conditional probability density function of observing a bead at position x at time t that was initially at position x_0 at time t_0. Assuming the bead's dynamics satisfy a

Weiner process, then the time evolution of Ψ is given by the overdamped Fokker-Planck equation [16]:

$$\frac{\partial \Psi}{\partial t} = \nabla \cdot \mathbf{D} \cdot [\nabla + \beta(\nabla F_w)]\Psi. \quad (1)$$

In Eq (1), **D** is the diffusion tensor, F the free energy, and β the inverse of the thermal free energy (i.e. $\beta = 1/k_B T$). Consider the CGMD model of the tethered membrane in Fig 1. The boundary conditions of Eq (1) in z are composed of no-flux boundary conditions at the gold bioelectronic interface, denoted by z_b, and the surface of the tethered membrane, denoted by z_t, with the others defined as infinite boundary conditions. From the coordinate axis of the CGMD model, refer to Fig 1, the diffusion tensor is diagonal with $D_{xx} = D_{yy} \neq D_{zz}$. The translational invariance parallel to the confining surface ensures that F_w in the parallel direction is constant (i.e. F_w is invariant in x and y). The free energy of water along the z-axis is related to the equilibrium density of water by $F_w(z) = -k_B T \log(\rho(z)/\rho_o)$ with ρ_o the bulk density. As a result of the no-flux boundary conditions in z, the diffusion coefficients in **D**, defined below Eq (1), will be a function of z. If we discretize the z-dimension sufficiently into layers, then in each layer the diffusion coefficients will not vary substantially. This allows Ψ Eq (1) to be decoupled such that $\Psi = \Psi_x \Psi_y \Psi_z$ with $\Psi_x(x, t|x_o, t_o)$, $\Psi_y(y, t|y_o, t_o)$, $\Psi_z(z, t|z_o, t_o)$ denoting the time evolution of the particle in each respective dimension.

The partial differential equation modeling the time evolution of $\Psi_x(x, t|x_o, t_o) = \Psi_y(y, t|y_o, t_o)$ is given by Fick's second law of diffusion. For an infinite boundary condition, the analytical solution of Ψ_x and Ψ_y is given in terms of the Green's function. This allows $D_{xx} = D_{yy}$ to be estimated by the evaluation of the second central moment (i.e. for D_{xx} the relation $\langle(x - x_o)^2\rangle = 2D_{xx}(t - t_o)$ holds with the ensemble average taken over time). The time evolution of Ψ_z is given by:

$$\frac{\partial \Psi_z}{\partial t} = \frac{\partial}{\partial z}\left[D_{zz}e^{-\beta F(z)}\frac{\partial}{\partial z}(\Psi_z e^{\beta F(z)})\right]. \quad (2)$$

No analytical solution exists for Eq (2) given the no-flux boundary conditions at z_b and z_t. If we consider the round-trip time, denoted by τ_{rt}, then D_{zz} can be evaluated using [17]:

$$D_{zz}(z) = \frac{e^{\beta F(z)}}{\partial \tau_{rt}(z)/\partial z} \int_{z_b}^{z_t} e^{-\beta F(z')} dz'. \tag{3}$$

In Eq (3), τ_{rt} denotes the time needed to start at z, reach a position z_*, start from z_* again, and reach z (i.e. the round-trip time). The estimation of τ_{rt} can be obtained directly from the CGMD trajectories allowing Eq (3) to be used to estimate $D_{zz}(z)$.

Coarse-Grained Molecular Dynamics Simulation Details

The coarse-grained molecular dynamics simulations were performed using GROMACS [18] version 4.6.2 (double precision) with the MARTINI force field [11, 12]. Unless otherwise stated, the CGMD simulation parameters are provided in [8, 9]. The interaction of the CGMD beads are defined by the Lennard-Jones (LJ) potential, and harmonic potentials are utilized for bond and angle interactions. A shift function is added to the Coulombic force to smoothly and continuously decay to zero from 0 nm to 1.2 nm. The LJ interactions were treated likewise except that the shift function was turned on between 0.9 nm and 1.2 nm. The grid-type neighbour searching algorithm is utilized for the simulation–that is, atoms in the neighbouring grid were updated every ten time steps. The equations of motion are integrated using the leapfrog algorithm with a timestep of 20 fs. The temperature is held constant at 320 K using a velocity rescaling algorithm [19] with a time constant of 0.5 ps. The lipids, tethers, spacers and water molecules are coupled separately for temperature control. Production runs are performed in the NVT ensemble using periodic boundary conditions in three-dimensions with a total simulation time horizon of 1.5 μs. Visualization of the results are reported using VMD and PyMOL. The numerical method used to evaluate D_{zz} Eq (3) is provided in [20].

Prior to the final NVT production we first equilibrate the CGMD system sufficiently to model the experimental conditions. Currently there are no methods to perform an NPT ensemble simulation for CGMD systems that contain a substrate as the interaction site density of the surface must be kept constant–equivalently, it is desirable to keep the interaction strength of the substrate constant [8, 21–23]. Therefore, either the NVT or NAP_zT ensembles are used to perform simulations with a substrate surfaces present [8, 21–23]. To minimize the induced pressure effects caused by the constant area constraint, the following steps are performed prior to the NVT production runs. First, the membrane and solvent, with no substrate, are equilibrated for 50 ns using the steepest descent method in GROMACS. Then a production run in the NPT ensemble is performed for 1 μs to ensure that the membrane and solvent reach a sufficiently equilibrated pressure of 1 bar. Second, the substrate is added to the top and bottom of the simulation cell. The gold substrate area and particle density are chosen as a best compromise between the equilibrium area per head-group of a free-standing DphPC bilayer (69 Å) and the commensurability with the gold substrate surface. The initial dimensions of the simulation cell are $L_x = L_y = 10.8$ nm, and $L_z = 15.2$ nm. Another 1 μs production run is performed using the NAP_zT ensemble ensuring the substrate area is constant while allowing the normal pressure to fluctuate around the reference value of 1 bar. The pressure is coupled semi-isotropically in the NAP_zT with P_z coupled to 1 bar by the Berendsen barostat using $\tau = 5.0$ ps and a compressibility of 3×10^{-5} bar^{-1}. After the 1 μs NAP_zT ensemble simulation the final simulation cell size is $L_x = L_y = 10.8$ nm, and $L_z = 16.0$ nm. An NVT ensemble with these simulation cell dimensions is used for all subsequent production runs.

Is the simulation cell size for the NVT ensemble sufficiently large to accurately estimate the lateral diffusion coefficient of the lipids? The simulation cell size should be sufficiently large to reduce the effects caused by highly correlated transitions of lipids. This results as the movement of one lipid perturbs its neighbours and propagates across the periodic boundaries resulting in highly correlated lipid movements. Generally this effect is not present for system that contain over 100 lipid molecules [24–26]. However, currently there are no theoretical guarantees of simulation cell sizes that can eliminate this correlation effect. For example, it may be the case that lipids undergo collective diffusion on length scales ranging from a few to several tens of nanometers [27–31]. To minimize the probability of such an effect while still ensuring computational tractability we use a simulation cell size of $L_x = L_y = 10.8$ nm, and $L_z = 16.0$ nm which contains 320 lipid molecules. Assuming that the Saffman-Delbrück hydrodynamic model is valid (e.g. assuming no tethers, mean-field approximations hold, and the elementary model of a bilayer [32] hold), the estimated lateral diffusion coefficient with this simulation cell size should be approximately equal to the actual diffusion coefficient with a maximum error of 10% to 20% [28].

Tethered Bilayer Lipid Membrane Preparation

To validate the results from the CGMD model and to better understand the effect of tethers and the solid gold bioelectronic interface have on the membrane dynamics and order, we utilize experimental measurements from tethered bilayer lipid membranes. The fabrication procedure of the tethered membrane id described below.

The tethered membrane is supported by a 25 × 75 × 1 mm polycarbonate slide. Six 100 nm thick sputtered gold electrodes, each with dimensions 0.7 × 3 mm, are patterned on the polycarbonate slide. Each electrode is contained within an isolated flow cell with a common opposing, large area gold return electrode. The formation of the tethered membrane is performed in two stages using the solvent-exchange technique presented in [7]. The first stage of formation involves anchoring of the tethers and spacers to the gold surface. The tethers provide structural integrity to the membrane and mimic the cytoskeletal supports of natural cell membranes. The spacers laterally separate the tethers allowing patches of mobile lipids to diffuse in the membrane. The spacer is composed of a benzyl disulphide connected to a four-oxygen-ethylene-glycol group terminated by an OH; the tethers are composed of a benzyl disulphide connected to an eight-oxygen-ethylene-glycol group terminated by a C20 hydrophobic phytanyl chain. To form the anchoring layer, an ethanolic solution containing 370 μM of engineered ratios of benzyl disulphide components is prepared. This solution is exposed to the gold surface for 1 to 2 hours, then the surface is flushed with ethanol and air dried for approximately 2 min. Note that in the special case of 100% tethering, the engineered tethered membrane is composed of a tethered monolayer with no spacer molecules. The second stage is the formation of the tethered membrane. The mobile lipids are composed of synthetic mixtures of zwittrionic C20 diphytanyl-either-glycero-phosphatidylcholine lipid (DphPC), and C20 diphytanyl-diglyceride either (GDPE). The solution containing the mixture of mobile lipids is added to the gold bonded components from the first stage. Several lipid solvents can be used

[7, 33], however in most cases the lipids selected to form the bilayer are soluble in ethanol. As an example, 8 μL of 3 mM mixture of 70% DphPC and 30% GDPE in ethanol is added to the flow cell chamber containing the gold bonded layer. The solution is incubated for 2 min at 20°C in which the tethered membrane forms. Following a 1.5 to 2 min incubation, 300 μL of phosphate buffered saline is flushed through each flow cell chamber. The tethered membrane is equilibrated for 30 min prior to performing any experimental measurements. The formation of the E. coli, and S. caerevisae tethered membranes follows a similar procedure.

All experimental measurements were conducted at 20°C in a phosphate buffered solution with a pH of 7.2, and a 0.15 M saline solution composed of Na^+, K^+, and Cl^-. At this temperature the tethered membrane is in the liquid phase. A pH of 7.2 was selected to match that typically found in the cellular crystalline fluid phase. The quality of the tethered membrane is measured continuously using an SDx tethered membranes tethaPod™ swept frequency impedance reader operating at frequencies of 1000, 500, 200, 100, 40, 20, 10, 5, 2, 1, 0.5, 0.1 Hz and an excitation potential of 25 mV (SDx Tethered Membranes, Roseville, 2069, NSW, Sydney, Australia).

RESULTS

Order and Dynamics of Tethered Bilayer Lipid Membranes

The structure and order of the tethered bilayer lipid membranes (tBLMs) is modelled using simulation results from the CGMD model (Fig 1). Computed properties include the spatially dependent density and diffusion tensor **D**, membrane thickness h_m, surface tension σ, and line tension γ are computed.

The density of water for the 0% and 25% tethered membranes is provided in Fig 2a. As expected, the density of water is attenuated at the gold interface and membrane interface for both the 0% and 25% tethered membranes. However, by comparing the density of the 0% and 25% tethered membranes, it is apparent that the density of water is dependent on the tether density. Given the water density is position-dependent, this strongly suggests the self-diffusion tensor of the water will also be position dependent. Using Eq (3), the position dependent perpendicular diffusion coefficient $D_\perp(z)$ is provided in Fig 2b. From Eq (3), the diffusion coefficient $D_\perp(z)$ is dependent on the round-trip time $\tau_{rt}(z)$ and the density profile $\rho(z)$. Intuitively, the slope of the round-trip time $\tau_{rt}(z)$ is inversely proportional to $D_\perp(z)$. In Fig 2b the initial peak at 1.05 nm and second peak at 1.52 nm of the 25% diffusion profile match the location of the density peaks at 1.05 nm and 1.52 nm in Fig 2a. The initial peak at 1.25 nm for the 0% diffusion profile results as the round-trip time of a water bead in this region is approximately constant. Additionally, for the 0% tethered membrane, the density profile remains approximately constant for $z = 2.0$ nm to $z = 2.5$, however the round-trip time does not remain constant. The combined effect of the constant density but varying round-trip time causes $D_\perp(z)$ to be non-constant in this region. In the the water density and spatially dependent diffusion coefficient for the gold interface with no membrane present is provided. From the, the diffusion coefficient $D_\perp(z)$ approximately reaches the bulk diffusion coefficient for $z \geq 2.5$ nm from the gold interface with no membrane present.

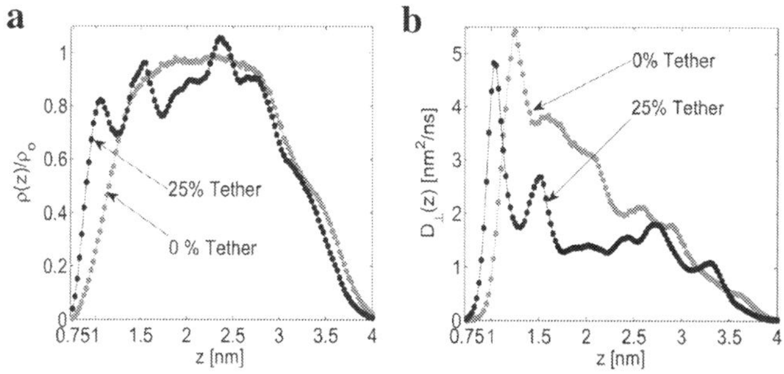

Figure 2: Normalized water density and perpendicular diffusion $D_\perp(z)$. The normalized density $\rho(z)/\rho_o$ (a) and perpendicular diffusion coefficient $D_\perp(z)$ (b) are computed from the CGMD bead trajectories.

The computed lateral diffusion of the DphPC lipids, GDPE lipids, and water for the 0% and 25% tethered membrane are provided in Table 1.

Table 1: Lateral Diffusion Coefficient $D_{||}$ **($nm^2/\mu s$)**

Tethering density		0%	25%
Proximal Layer	DphPC	290±23	87±10
	GDPE	256±36	116±16
Distal Layer	DphPC	289±24	115±12
	GDPE	256±38	128±13
Bulk Water		2105±288	2010±233
Tethering Reservoir Water		2730±258	1236±110

doi:10.1371/journal.pone.0162790.t001

The computed lateral diffusion of the DphPC and GDPE lipids in the proximal (i.e. adjacent to the tethering reservoir) layer and distal (i.e. adjacent to the bulk water) are nearly identical. The diffusion of DphPC is related to that of GDPE by a multiplicative factor of 1.13. The effects of the tethers cause the lateral diffusion coefficient of DphPC to decrease by a factor of approximately 3.3 in the proximal layer, and 2.5 in the distal layer. Similarly for GDPE the decrease is 2.2 in the proximal layer, and 2.2 in the distal layer. This result is in agreement with the experimental results reported in [34, 35] for different lipids and tethering densities. The lateral diffusion of water in the bulk region is related to the tethering reservoir lateral diffusion by a factor of approximately 1.6 for the 25% tethered membrane. It is interesting that the lateral diffusion of water in the tethering reservoir is 2.7 nm^2/ns which is close to but faster than that in the bulk of 2.1 nm^2/ns for the 0% tethered membrane. The interplay between hydrogen bond breaking and cooperative rearrangement of regions of approximately 1 nm in size cause the lateral diffusion to significantly increase in nanoconfined water regions [36]. Since explicit hydrogen bonds are not included in the CGMD model [12], we attribute the increased diffusion coefficient to the cooperative rearrangement of water molecules. A unique result of this study is that no anomalous diffusion was detected for water and lipid headgroups near the outer surface of bilayer lipid membrane. This is in contrast to other dynamical studies [37–40] on different membranes which report an anomalous water and lipid diffusivity in the vicinity of the membrane surface.

To compute the membrane thickness h_m (Fig 1) the particle density of the lipid headgroups is used. The thickness of the 25% and 0% tethered membrane is h_m = 3.53 nm and h_m = 3.48 respectively. The thickness of the phytanyl tails (i.e. hydrocarbon tails) was also computed for the 25% and 0% tethered membranes and is 2.15 nm and 2.11 nm respectively. The reservoir thickness h_r (Fig 1) of the 25% and 0% tethered membrane is h_r = 3.30 nm respectively. These numerically computed values are consistent with the experimentally measured thickness for DphPC based tethered membranes [41].

To compute the line tension γ, and surface tension σ the method provided in [42] was utilized. The resulting values are: $\sigma = 15$ mN/m, and $\gamma = 12$ pN. These values are in agreement with the experimental results [43] and simulation results [44–46] reported in the literature for similar DphPC based membranes. Note that surface tension σ is a key mechanical property in regulating the motility and the reshaping of bilayer membranes. A bilayer membrane with no tethers simulated in a NPT ensemble will have zero surface tension $\sigma = 0$ mN/m as there is no induced stress on the membrane [47, 48]. However, if an external stimuli is applied then a non-zero surface tension will result [49–51]. The induced surface tension in a tethered membrane results from the pressure differences across the membrane surface, and the adhesion of the membrane to the tethered supports. Note that the induced surface tension from the tethers is analogous to the induced surface tension in cellular membranes from the cytoskeletal supports [52–54]. The computed surface tension of the tethered DphPC membrane is $\sigma = 15$ mN/m which indicates that the tethers introduce a non-zero stress on the DphPC membrane. As $\sigma > 0$ mN/m the tethers slightly increase the area-per-lipid from their equilibrium value–that is, the tethers cause a slight increase in DphPC membrane area compared to the untethered DphPC membrane area.

Lipid Energetics and Defect Density

The potential of mean force (PMF) of lipids in a lipid bilayer is a key thermodynamic property that can be used to estimate the defect density and the free energy of lipid flip-flop. Lipid flip-flop occurs when a lipid molecule flips from one side of the lipid bilayer to the other. Here the umbrella sampling method [55] is used to compute the PMF for moving a single DphPC lipid along the normal to the membrane surface for both the 0% and 25% tethered membranes. The umbrella potential acts on the center of mass of the phosphate group of the DphPC lipid with a harmonic restraint with a force constant of 500 kJ/mol/nm^2 in the direction normal to the membrane surface. The spacing between biasing potentials was selected as 0.15 nm and 0.1 nm for the 0% and 25% tethered membranes. This resulted in 37 and 48 parallel simulations for the 0% and 25% tethered membranes respectively. The starting structures corresponding to each umbrella window were created by pulling a lipid to their respective position using an umbrella potential with a force constant of 500 kJ/mol/nm^2 in a 1 ns simulation. Each umbrella window was then equilibrated for 50 ns with a 1000 kJ/mol/nm^2 force constant, followed by a 1 μs simulation. The PMF profile was constructed from the biased distributions of the center of mass of the lipids using the weighted histogram analysis method [56] with a relative tolerance of 10^{-4}.

The computed PMF for both tethering densities is provided in Fig 3. As expected, the minimum of the PMFs at -1.82 nm and 1.82 nm corresponds to the equilibrium position of the DphPC lipid. There is an increase in the PMF from these equilibrium positions as a result of the hydrophobic tail coming into contact with the water, and the hydrophilic headgroups coming into contact with the hydrophobic phytanyl tails of the adjacent lipids. The difference in energy between the equilibrium position and when fully solvated in the water solution corresponds to the free energy of lipid desorption. The free energy of desorption for the 0% and 25% membrane is 85.93 kJ/mol and 91 kJ/mol respectively. As expected, tethers cause the associated energy of lipid desorption to increase as compared with the untethered membrane. As the lipid is moved to the center of the membrane, the steep slope in the PMF is caused by the lipid headgroup interacting with water and the hydrophobic interior of the membrane interior. This characteristic of the PMF has been observed when other charged/polar molecules are transferred into the hydrophobic interior of the membrane and is associated with the formation of an aqueous pore (i.e. membrane defect) [57–60].

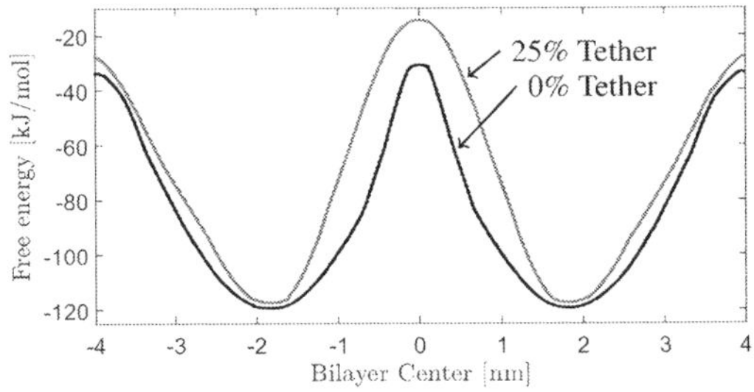

Figure 3: PMF of DphPC lipid. The PMFs for a 0% and 25% tether density DphPC lipid membrane.

Using the PMF (Fig 3), the free energy required for complete flip-flop of a single lipid is equal to the energy required to move a lipid from it's equilibrium position to the center of the bilayer, then to the other leaflet's equilibrium position. Therefore, from the maxima in the PMF's between the equilibrium position and the bilayer center, the free energy barrier for flip-flop of the DphPC lipid increases from 89 kJ/mol for the free bilayer to 103.17 kJ/mol for the 25% tethered membrane. Notice that the lipid flip-flop energy of the 0% membrane is higher then for DOPC (87 kJ/mol) [60], DPPC (78 kJ/mol), DMPC (45 kJ/mol), and DLPC (17 kJ/mol) [57] as expected because archaebacterial lipids have a higher stability compared to these lipids. Additionally comparing the lipid flip-flop energy between the 0% and 25% tethered membrane, these results show that inclusion of tethers in the DphPC membrane prevents the formation of defects and increases the free energy for the translocation of charged headgroups across the membrane.

How is the energetics of lipid flip-flop associated with the membrane conductance? Assuming that ions only cross the membrane interface via membrane defects, then as the membrane defect density increases it is expected that the associated membrane conductance will also increase. Using the methods presented in [57, 61], the defect density $\rho 0 = e^{-\beta \Delta G_p}/A_L$, with ΔG_p the associated energy required to form a defect, can be estimated. As illustrated in [57, 61–64], and in agreement with our CGMD simulation results, when the lipid headgroup is positioned in the center of the membrane a membrane defect forms. Therefore, we associated the energy required for lipid flip-flop with the energy required to form a membrane defect. Additionally, as the membrane defect density increases the associated conductance of the membrane increases as more pathways for ion transport through the membrane are present [57, 61]. Given that as the tether density increases the associated defect density decreases, it is expected that the membrane conductance decreases for increasing tether density. This is in agreement with our experimental results which show that as the tether density increases the associated membrane conductance decreases.

Measured Dynamics and Order of Tethered Bilayer Lipid Membranes

The surface of the gold electrode anchors the self-assembled monolayer of both the spacer and tether molecules. To investigate the effect of this surface on diffusion processes within the tBLM, an experimental test element was constructed employing only spacer molecules attached to the gold surface, denoted as the spacer surface. The phase behavior of ions charging and discharging the spacer surface for electrolyte concentrations of 2 M, 1 M, 500 mM, 200 mM NaCl is experimentally measured and modeled numerically. Results are given in Fig 4a. In the absence of a diffusion limited process $\angle Z(f) = 90°$ for $f < 1$ Hz where the capacitive effects of the bioelectronic interface are dominant. As seen from Fig 4a, for $f < 1$ Hz $\angle Z(f) = 76°$, therefore for all concentrations measured there is a diffusion-limited process present with a fractional order parameter $p = 0.86$. To illustrate the accuracy of the fractional order macroscopic model, Fig 4b provides the experimentally measured and numerically modeled current response of the spacer only electrode for a drive potential V_s defined by a 100 V/s rise for 5 ms, and a -100 V/s fall for 5 ms. As seen, the fractional-order differential equation fractional order operator $p = 0.83$ provides an accurate prediction of the current response. The fractional-order behavior at the electrode surface may be caused by: diffusion-limited charge transfer and adsorption on the electrode [38]. Though the source of the behavior is unknown, the dynamics of the interface can be modeled using the fractional order differential equation as illustrated in Fig 4a and 4b.

The tether density would be anticipated to impact membrane stability. As suggested from the CGMD results, the higher the tethering density the more resistant the membrane is to membrane defects. To validate this result we have fabricated tethered membranes with different tethering densities from 1%, 10%, and 100% and applied different transmembrane potentials to observe the dynamics of the defect density. Fig 4c to 4f show the results for 1%, 10%, and 100% tethering densities and various drive potentials. As expected, negligible defects are present for drive potentials below 250 mV as seen in Fig 4c and 4d. For larger drive potentials the formation of defects becomes pronounced, as seen in Fig 4c and 4d. It was found that the defect density in the 10% tethered membrane was always less than in the 1% tethered membrane, in agreement with the CGMD results. In the limiting case of a tethered membrane with 100% tether density, it was anticipated they would be the most resistant to membrane defects. Fig 4e shows that the 100% tethered membrane can withstand drive potentials of up to 800 mV. To determine whether the membrane was irreversibly damaged, the current response was compared to that of the numerical model for potential steps of 50 mV, 100 mV, 150 mV, and 200 mV. Fig 4f shows excellent agreement between the experimentally measured current and numerically computed current suggesting all current pathways formed by the voltage gradient reseal.

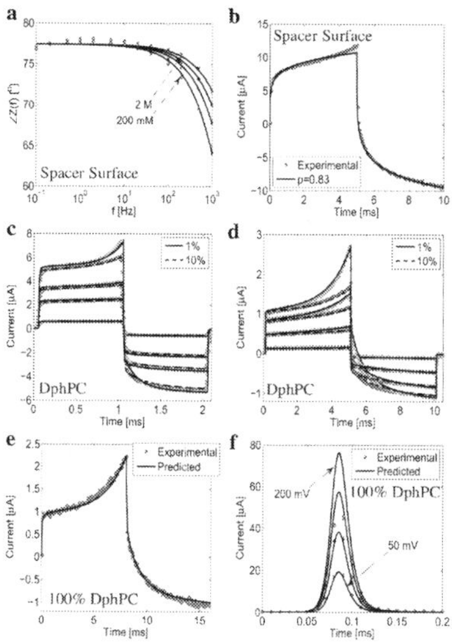

Figure 4: Experimentally measured and numerically computed current response of the tethered membrane. (a) and (b) provide the measured and computed phase and current response of the spacer surface. Note $C_m \rightarrow \infty$ and $G_m = 0$ as the membrane is not present. (c) to (f) provide the experimentally measured and numerically computed current response I(t) for 1%, 10%, and 100% tethered DphPC membranes. In (c) the drive potential $V_s(t)$ is defined by a 1 ms linearly increasing potential with a slope of 10 V/s, 40 V/s, 70 V/s, and 90 V/s followed by a 1 ms linearly decreasing potential with identical slope. In (d) $V_s(t)$ is a 5 ms linearly increasing potential with slope of 50 V/s, 200 V/s, 300 V/s, and 450 V/s followed by a 5 ms linearly decreasing potential with identical slope. In (e) $V_s(t)$ is defined by a 8 ms linearly increasing potential with a slope of 100 V/s followed by a 8 ms decreasing potential with identical slope. In (f) $V_s(t)$ is a step of 50 mV, 100 mV, 150 mV, and 200 mV.

The above analysis shows that there exists a diffusion-limited process at the bioelectronic gold interface of tethered membranes, and that as the tethering density increases the number of membrane defects decreases.

Effect of Variations in Membrane Composition and Tether Density

In this section the results from the CGMD model and experimental measurements are utilized to gain insight into important biological parameters as a function of tether density and lipid composition. The results presented are in agreement with the molecular dynamics results reported in [65, 66]. We consider tethered membranes composed of DphPC, S. cerevisae, and E. coli lipids with a 1% and 10% density are constructed. Fig 5 provides the numerically computed and experimentally measured current response. As seen, excellent agreement is obtained between the computed and measured current response. This allows the fractional order macroscopic model, and the CGMD model to be used to estimate important biological parameters.

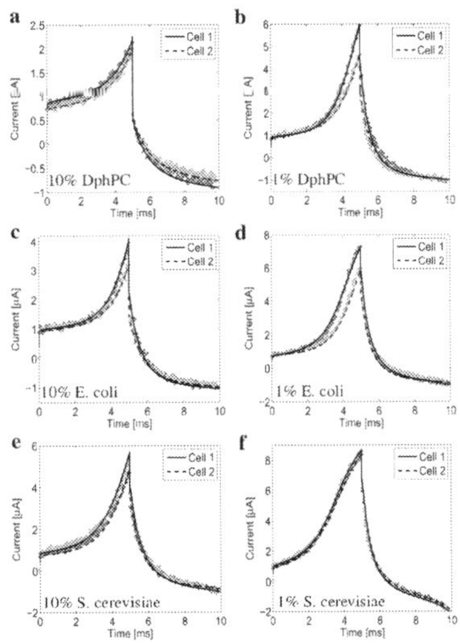

Figure 5: The measured and numerically computed current response of the 1% and 10% tether density DphPC, S. cerevisiae, and E. coli membranes. The excitation potential V_s is defined by a linear ramp of 100 V/s for 5 ms followed by a -100 V/s for 5 ms. Cell 1 and Cell 2 denote tethered membranes prepared using an identical procedure.

Consider the relative dielectric permittivity ε_m of the tethered DphPC, E. Coli, and S. cerevisiae membranes. For homogeneous hydrocarbon tails the permittivity is approximately 2, however the lipid headgroups cause the effective permittivity ε_m to be larger. For the 10% tethered DphPC membrane $h_m = 3.5$ nm, $A_m = 2.1$ mm^2, and C_m is the range of 12.5 nF to 15.5 nF. Therefore the relative permittivity is $\varepsilon_m = 2.35 - 2.92$. For the 1% tethered DphPC C_m is in the range of 16 nF to 17.5 nF, and assuming the thickness of identical to that of the 0% tethered DphPC $h_m = 3.4$ nm, the relative permittivity is ε_m is in the range 2.92 to 3.20. Note that the relative permittivity of the membrane ε_m must be between the values of 2 for pure hydrocarbon and 80 for an electrolyte at physiological concentrations. Therefore as the density of water increases between the distal and proximal layer, or the hydrocarbon thickness decreases it is expected that the relative permittivity will increase. Using the CGMD results in the Structure and Biomechanics section the decrease in h_m between the tethered and untethered DphPC membrane is a result of changes in the thickness of the hydrocarbon region. Therefore the decrease in ε_m between the 10% and 1% tethered DphPC membranes results from changes in the hydrocarbon thickness. This shows that the dielectric permittivity is dependent on the tether density. To estimate the permittivity of the tethered E. coli and S. cerevisae membranes we chose $h_m = 3.29$ nm for the E. coli, and $h_m = 4.30$ nm for the S. cerevisae as justified by the molecular dynamics results [66, 67]. The associated permittivity of E. coli is $\varepsilon_m = 2.5 - 3.0$ and S. cerevisae $\varepsilon_m = 3.2 - 4.2$. These are in excellent agreement with the experimentally measured results of E. coli and S. cerevisae cell membranes [68].

Comparing the current response of the 1% and 10% tethered DphPC, E. coli, and S. cerevisae membranes we see that the resistance to membrane defects from highest to lowest is: DphPC, E. coli, and S. cerevisae. As expected the resistance to membrane defects increases as the tether density increases. The difference in the resistance to membrane defects of DphPC compared to that of E. coli and S. cerevisae is a result of the phytanyl chain packing properties and the either linking the phytanyl to the lipid headgroup. The DphPC and GDEP ether-bound lipids found in archaebacterial membranes are known to have a lower lateral diffusion coefficient compared to common phospholipids found in prokaryotes and eukaryotic membranes [69]. It is suggested in [70] that the stability of the

DphPC membrane is closely related to the slow conformational motion of the phytanyl chains. Therefore we conclude that the difference in the resistance to membrane defects between the DphPC, and E. coli and S. cerevisae membranes is a result of the unique dynamics of the ether-phytanyl group which may result from a hydrogen bonding difference. In addition, using the CGMD model the conformational motion of the phytanyl tail in the DphPC decreases as the tether density increases which results in the larger tether density membrane having a higher resistance to membrane defects. A surprising observation is that the E. coli membrane is more resistant to membrane defects compared to that of S. cerevisae even though the thickness of S. cerevisae is larger then that of E. coli. A possible mechanism for this difference is that defects in the E. coli membrane are primarily formed by the flip-flop of specific groups of phospholipids [66].

DISCUSSION

Using the CGMD model and experimental measurements from tethered bilayer lipid membranes, we summarize several key findings on how tethers and the bioelectronic gold interface impact the tethered membrane dynamics. First, in contrast to previous molecular dynamics results [71], no anomalous diffusion was detected at the surface of the membrane or in the tethering reservoir between the gold and membrane surfaces. This allows the transport of ions away from the electrode surface to be modeled using continuum theories that do not include fractional order operators [13]. For example this validates the use of the Generalized Poisson-Nernst-Planck model for computing the aqueous pore conductance in a tethered membrane as done in [6]. Second, the diffusion tensor of water and particle density are both spatially dependent. Third, the membrane thickness, lipid diffusion adjacent to the tethering reservoir, defect density, the free energy barrier for lipid flip-flop, and effective dielectric permittivity of the membrane are dependent on the tether density. This suggests that the membrane dynamics are strongly dependent on the density of cytoskeletal supports in the membrane. Fourth, experimental measurements show that a combination of diffusion-limited charge transfer and ionic adsorption are present at the surface of the gold electrodes. These double-layer charging effects can be modeled using fractional order operators [13]. Fifth, the numerical results from the CGMD model are in agreement with the results from experimental measurements from tethered bilayer lipid membranes composed of DphPC, E. coli, and S. cerevisae lipids. This supports the use of molecular dynamics to go from structure to function. All these results provide valuable insights into the dynamic properties of membranes, and into the diffusive transport of lipids and ions in natural membranes that contain cytoskeletal supports. They also yield design rules for future tethered membrane based biosensors.

ACKNOWLEDGMENTS

All coarse-grained molecular dynamics simulations were performed using WestGrid (www.westgrid.ca) and Compute Canada Calcul Canada (www.computecanada.ca) supercomputing resources.

AUTHOR CONTRIBUTIONS

- **Conceptualization:** WH VK.
- **Data curation:** BC CC.
- **Formal analysis:** WH VK.
- **Investigation:** WH VK.
- **Methodology:** WH VK.
- **Project administration:** VK.
- **Software:** WH RG.
- **Supervision:** VK.
- **Validation:** WH VK RG.
- **Visualization:** WH RG.
- **Writing – original draft:** WH RG VK.
- **Writing – review & editing:** WH VK.

REFERENCES

1. Malmstadt N, Jeon T, Schmidt J. Long-Lived Planar Lipid Bilayer Membranes Anchored to an In Situ Polymerized Hydrogel. Advanced Materials. 2008;20(1):84–89.
2. Beerlink A, Mell M, Tolkiehn M, Salditt T. Hard x-ray phase

contrast imaging of black lipid membranes. Applied Physics Letters. 2009;95(20):203703.

3. Dietrich C, Bagatolli L, Volovyk Z, Thompson N, Levi M, Jacobson K, et al. Lipid rafts reconstituted in model membranes. Biophysical journal. 2001;80(3):1417–1428. pmid:11222302
4. Lin W, Blanchette C, Ratto T, Longo M. Lipid asymmetry in DLPC/DSPC-supported lipid bilayers: a combined AFM and fluorescence microscopy study. Biophysical journal. 2006;90(1):228–237. pmid:16214871
5. Castellana E, Cremer P. Solid supported lipid bilayers: From biophysical studies to sensor design. Surface Science Reports. 2006;61(10):429–444.
6. Hoiles W, Krishnamurthy V, Cranfield C, Cornell B. An Engineered Membrane to Measure Electroporation: Effect of Tethers and Bioelectronic Interface. Biophysical journal. 2014;107(6):1339–1351. pmid:25229142
7. .Cornell B, Braach-Maksvytis V, King L, Osman P, Raguse B, Wieczorek L, et al. A biosensor that uses ion-channel switches. Nature. 1997;387(1):580–583. pmid:9177344
8. Liu C, Faller R. Conformational, Dynamical. and Tensional Study of Tethered Bilayer Lipid Membranes in Coarse-Grained Molecular Simulations. Langmuir. 2012;28:15907–15915. pmid:23092246
9. Wang S, Larson RG. Coarse-grained molecular dynamics simulation of tethered lipid assemblies. Soft Matter. 2013;9:480–486.
10. Raguse B, Braach-Maksvytis V, Cornell B, King L, Osman P, Pace R, et al. Tethered lipid bilayer membranes: Formation and ionic reservoir characterization. Langmuir. 1998;14(3):648–659.
11. Marrink S, Risselada H, Yefimov D, Tieleman D, Vries A. The MARTINI force field: Coarse Grained Model for Biomolecular Simulations. J PhysChemB. 2007;111:7812–7824. pmid:17569554
12. Marrink S, Tieleman P. Perspective on the MARTINI model. Chem Soc Rev. 2013;42:6801–6822. pmid:23708257
13. Bressloff P, Newby J. Stochastic models of intracellular transport. Review Modern Physics. 2013;85:135–196.

14. Lee H, Vries A, Marrink S, Pastor R. A Coarse-Grained Model for Polyethylene Oxide and Polyethylene Glycol: Conformation and Hydrodynamics. The Journal of Physical Chemistry B. 2009;113(40):13186–13194. pmid:19754083
15. Cicero G, Calzolari A, Corni S, Catellani A. Anomalous wetting layer at the Au (111) surface. The Journal of Physical Chemistry Letters. 2011;2(20):2582–2586.
16. Liu P, Harder E, Berne B. On the calculation of diffusion coefficients in confined fluids and interfaces with an application to the liquid-vapor interface of water. The Journal of Physical Chemistry B. 2004;108(21):6595–6602.
17. Hinczewski M, Hansen Y, Dzubiella J, Netz R. How the diffusivity profile reduces the arbitrariness of protein folding free energies. The Journal of chemical physics. 2010;132(24):245103. pmid:20590217
18. Hess B, Kutzner C, Spoel D, Lindahl E. GROMACS 4: Algorithms for Highly Efficient, Load-Balanced, and Scalable Molecular Simulation. Journal of Computational Chemistry. 2008;4:435–447. pmid:26620784
19. Bussi G, Donadio D, Parrinello M. Canonical sampling through velocity rescaling. Journal of Chemical Physics. 2007;126:014101. pmid:17212484
20. Sedlmeier F, Hansen Y, Mengyu L, Horinek D, Netz R. Water Dynamics at Interfaces and Solutes: Disentangling Free Energy and Diffusivity Contributions. Journal of Statistical Physics. 2011;145(2):240–252.
21. Xing C, Faller R. Interactions of lipid bilayers with supports: a coarse-grained molecular simulation study. The Journal of Physical Chemistry B. 2008;112(23):7086–7094. pmid:18461982
22. Xing C, Faller R. Coarse-grained simulations of supported and unsupported lipid monolayers. Soft Matter. 2009;5(22):4526–4530.
23. Xing C, Faller R. Density imbalances and free energy of lipid transfer in supported lipid bilayers. The Journal of chemical physics. 2009;131(17):175104. pmid:19895045
24. Klauda J, Venable R, MacKerell A, Pastor R. Considerations for lipid force field development. Current topics in membranes. 2008;60:1–48.

25. Klauda J, Brooks B, Pastor R. Dynamical motions of lipids and a finite size effect in simulations of bilayers. The Journal of chemical physics. 2006;125(14):144710. pmid:17042634
26. Castro-Román F, Benz R, White S, Tobias D. Investigation of finite system-size effects in molecular dynamics simulations of lipid bilayers. The Journal of Physical Chemistry B. 2006;110(47):24157–24164. pmid:17125387
27. Lyubartsev A, Rabinovich A. Force field development for lipid membrane simulations. Biochimica et Biophysica Acta (BBA)-Biomembranes. 2016;. pmid:26766518
28. Camley B, Lerner M, Pastor R, Brown F. Strong influence of periodic boundary conditions on lateral diffusion in lipid bilayer membranes. The Journal of chemical physics. 2015;143(24):243113. pmid:26723598
29. Poger D, Caron B, Mark A. Validating lipid force fields against experimental data: Progress, challenges and perspectives. Biochimica et Biophysica Acta (BBA)-Biomembranes. 2016;1858(7):1556–1565. pmid:26850737
30. Falck E, Róg T, Karttunen M, Vattulainen I. Lateral diffusion in lipid membranes through collective flows. Journal of the American Chemical Society. 2008;130(1):44–45. pmid:18078342
31. Roark M, Feller S. Molecular dynamics simulation study of correlated motions in phospholipid bilayer membranes. The Journal of Physical Chemistry B. 2009;113(40):13229–13234. pmid:19754078
32. Camley B, Brown F. Diffusion of complex objects embedded in free and supported lipid bilayer membranes: role of shape anisotropy and leaflet structure. Soft Matter. 2013;9(19):4767–4779.
33. McGillivray D, Valincius G, Vanderah D, Febo-Ayala W, Woodward J, Heinrich F, et al. Molecular-scale structural and functional characterization of sparsely tethered bilayer lipid membranes. Biointerphases. 2007;2(1):21–33. pmid:20408633
34. Naumann C, Prucker O, Lehmann T, R ü J, Knoll W, Frank C. The polymer-supported phospholipid bilayer: Tethering as a new approach to substrate-membrane stabilization. Biomacromolecules. 2001;3:27–35. pmid:11866552

35. Wagner M, Tamm L. Tethered polymer-supported planar lipid bilayers for reconstitution of integral membrane proteins: silane-PEG-lipid as a cushion and covalent linker. Biophysical Journal. 2000;79:1400–1414. pmid:10969002
36. Santos F, Franzese G. Relations between the diffusion anomaly and cooperative rearranging regions in a hydrophobically nanoconfined water monolayer. Physical Review E. 2012;85(1):010602. pmid:22400504
37. Yamamoto E, Akimoto T, Yasui M, Yasuoka K. Origin of subdiffusion of water molecules on cell membrane surfaces. Nature, Scientific Reports. 2014;4:4720. pmid:24739933
38. Singh M, Kant R. Theory of Anomalous Dynamics of Electric Double Layer at Heterogeneous and Rough Electrodes. The Journal of Physical Chemistry C. 2014;118(10):5122–5133.
39. Róg T, Murzyn K, Pasenkiewicz-Gierula M. The dynamics of water at the phospholipid bilayer surface: a molecular dynamics simulation study. Chemical Physics Letters. 2002;352:323–327.
40. Flenner E, Das J, Rheinstädter M, Kosztin I. Subdiffusion and lateral diffusion coefficient of lipid atoms and molecules in phospholipid bilayers. Physical Review E. 2009;79:011907. pmid:19257069
41. Heinrich F, Ng T, Vanderah D, Shekhar P, Mihailescu M, Nanda H, et al. A new lipid anchor for sparsely tethered bilayer lipid membranes. Langmuir. 2009;25(7):4219–4229. pmid:19714901
42. Jiang F, Bouret Y, Kindt J. Molecular Dynamics Simulations of the Lipid Bilayer Edge. Biophys J. 2004;87:182–192. pmid:15240456
43. Zhelev D, Needham D. Tension-stabilized pores in giant vesicles: determination of pore size and pore line tension. Biochim Biophys Acta. 1993;1147:89–104. pmid:8466935
44. Wohlert J, Otter W, Edholm O, Briels W. Free energy of a trans-membrane pore calculated from atomistic molecular dynamics simulations. The Journal of chemical physics. 2006;124(15):154905. pmid:16674263
45. Jiang Y, Kindt J. Simulations of edge behaviour in a mixed bilayer: Flctuation analysis. J Chem Phys. 2007;126:045105–9. pmid:17286515

46. Joannis J, Jiang F, Kindt J. Coarse-Grained Model Simulations of Mixed-Lipid Systems: Composition and Line Tension of a Stabilized Bilayer Edge. Langmuir. 2006;22:998–1005. pmid:16430259
47. Asgari M, Biria A. Free energy of the edge of an open lipid bilayer based on the interactions of its constituent molecules. International journal of non-linear mechanics. 2015;76:135–143. pmid:26213414
48. Jähnig F. What is the surface tension of a lipid bilayer membrane? Biophysical journal. 1996;71(3):1348–1349. pmid:8874009
49. Goetz R, Lipowsky R. Computer simulations of bilayer membranes: self-assembly and interfacial tension. The Journal of chemical physics. 1998;108(17):7397–7409.
50. Marrink S, Mark A. Effect of undulations on surface tension in simulated bilayers. The Journal of Physical Chemistry B. 2001;105(26):6122–6127.
51. Feller S, Zhang Y, Pastor R. Computer simulation of liquid/liquid interfaces. II. Surface tension-area dependence of a bilayer and monolayer. The Journal of chemical physics. 1995;103(23):10267–10276.
52. Gauthier N, Masters T, Sheetz M. Mechanical feedback between membrane tension and dynamics. Trends in cell biology. 2012;22(10):527–535. pmid:22921414
53. Lieber A, Yehudai-Resheff S, Barnhart E, Theriot J, Keren K. Membrane tension in rapidly moving cells is determined by cytoskeletal forces. Current Biology. 2013;23(15):1409–1417. pmid:23831292
54. Simunovic M, Voth G. Membrane tension controls the assembly of curvature-generating proteins. Nature communications. 2015;6. pmid:26008710
55. Torrie G, Valleau J. Nonphysical sampling distributions in Monte Carlo free-energy estimation—Umbrella sampling. J Comput Phys. 1977;23:187–199.
56. Kumar S, Rosenberg J, Bouzida D, Swendsen R, Kollman P. The weighted histogram analysis method for free-energy calculations on biomolecules. 1. The method. J Comput Chem. 1992;13:1011–1021.

57. Bennett W, Sapay N, Tieleman D. Atomistic Simulations of Pore Formation and Closure in Lipid Bilayers. Biophys J. 2014;106:210–214. pmid:24411253

58. Dorairaj S, Allen T. On the thermodynamic stability of a charged arginine side chain in a transmembrane helix. Proceedings National Academy of Science USA. 2007;104:4943–4948. pmid:17360368

59. Bennett W, MacCallum J, Tieleman D. Thermodynamic Analysis of the Effect of Cholesterol on Dipalmitoylphosphatidylcholine Lipid Membranes. J Am Chem Soc. 2009;131:1972–1978. pmid:19146400

60. Bennett W, Tieleman D. Water defect and pore formation in atomistic and coarse-grained lipid membranes: pushing the limits of coarse graining. J Chem Theory Comput. 2011;7:2981–2988. pmid:26605486

61. Tieleman P, Marrink S. Lipids out of equilibrium: Energetics of desorption and pore mediated flip-flop. Journal of the American Chemical Society. 2006;128(38):12462–12467. pmid:16984196

62. Hu Y, Sinha S, Patel S. Investigating hydrophilic pores in model lipid bilayers using molecular simulations: correlating bilayer properties with pore-formation thermodynamics. Langmuir. 2015;31(24):6615–6631. pmid:25614183

63. Parisio G, Ferrarini A, Sperotto M. Model studies of lipid flip-flop in membranes. International Journal of Advances in Engineering Sciences and Applied Mathematics. 2016;p. 1–13.

64. Awasthi N, Hub J. Simulations of pore formation in lipid membranes: reaction coordinates, convergence, hysteresis, and finite-size effects. Journal of chemical theory and computation. 2016;. pmid:27254744

65. Polak A, Bonhenry D, Dehez F, Kramar P, Miklavčič D, Tarek M. On the electroporation thresholds of lipid bilayers: molecular dynamics simulation investigations. The Journal of membrane biology. 2013;246(11):843–850. pmid:23780415

66. Piggot T, Holdbrook D, Khalid S. Electroporation of the E. coli and S. aureus membranes: molecular dynamics simulations of complex bacterial membranes. The Journal of Physical Chemistry B. 2011;115(45):13381–13388. pmid:21970408

67. Jo S, Lim J, Klauda J, Im W. CHARMM-GUI membrane builder for

mixed bilayers and its application to yeast membranes. Biophysical journal. 2009;97(1):50–58. pmid:19580743

68. Raicu V, Popescu A. Integrated molecular and cellular biophysics. Springer; 2008.
69. Shinoda W, Mikami M, Baba T, Hato M. Dynamics of a highly branched lipid bilayer: a molecular dynamics study. Chemical physics letters. 2004;390(1):35–40.
70. Shinoda W, Mikami M, Baba T, Hato M. Molecular dynamics study on the effect of chain branching on the physical properties of lipid bilayers: structural stability. The Journal of Physical Chemistry B. 2003;107(50):14030–14035.
71. Hansen Y, Gekle S, Netz R. Anomalous Anisotropic Diffusion Dynamics of Hydration Water at Lipid Membranes. Physical Review Letters. 2013;111:118103. pmid:24074121
72. Pastushenko V, Chizmadzhev Y, Arakelyan V. Electric breakdown of bilayer lipid membranes: II. Calculation of the membrane lifetime in the steady-state diffusion approximation. Journal of Electroanalytical Chemistry and Interfacial Electrochemistry. 1979;104(0):53—62.
73. Barnett A, Weaver J. Electroporation: a unified, quantitative theory of reversible electrical breakdown and mechanical rupture in artificial planar bilayer membranes. Bioelectrochemistry and bioenergetics. 1991;25(2):163–182.
74. Freeman S, Wang M, Weaver J. Theory of electroporation of planar bilayer membranes: predictions of the aqueous area, change in capacitance, and pore-pore separation. Biophysical Journal. 1994;67(1):42–56. pmid:7919016
75. Glaser R, Leikin S, Chernomordik L, Pastushenko V, Sokirko A. Reversible electrical breakdown of lipid bilayers: formation and evolution of pores. Biochimica et Biophysica Acta (BBA)-Biomembranes. 1988;940(2):275–287. pmid:2453213
76. Neu J, Krassowska W. Asymptotic model of electroporation. Physical Review E. 1999;59:3471–3482.

77. Krassowska W, Filev P. Modeling electroporation in a single cell. Biophysical Journal. 2007;92(2):404—417. pmid:17056739
78. Smith K, Neu J, Krassowska W. Model of creation and evolution of stable electropores for DNA delivery. Biophysical Journal. 2004;86(5):2813—2826. pmid:15111399

CHAPTER 12

LIPID EXCHANGE MECHANISM OF THE CHOLESTERYL ESTER TRANSFER PROTEIN CLARIFIED BY ATOMISTIC AND COARSE-GRAINED SIMULATIONS

Artturi Koivuniemi[1], Timo Vuorela[1], Petri T. Kovanen[2], Ilpo Vattulainen[1,3,4], Marja T. Hyvönen[3,5]

[1] Department of Physics, Tampere University of Technology, Tampere, Finland

[2] Wihuri Research Institute, Helsinki, Finland

[3] Department of Applied Physics, Aalto University School of Science and Technology, Espoo, Finland

[4] MEMPHYS – Centre for Biomembrane Physics, University of Southern Denmark, Odense, Denmark

[5] Department of Physics, University of Oulu, Oulu, Finland

ABSTRACT

Cholesteryl ester transfer protein (CETP) transports cholesteryl esters, triglycerides, and phospholipids between different lipoprotein fractions in blood plasma. The inhibition of CETP has been shown to be a sound strategy to prevent and treat the development of coronary heart disease. We employed molecular dynamics simulations to unravel the mechanisms associated with the CETP-mediated lipid exchange. To this end we used both atomistic and coarse-grained models whose results were consistent with each other. We found CETP to bind to the surface of high density lipoprotein (HDL) -like lipid droplets through its charged and tryptophan residues. Upon binding, CETP rapidly (in about 10 ns) induced the formation of a small hydrophobic patch to the phospholipid surface of the droplet, opening a route from the core of the lipid droplet to the binding pocket of CETP. This was followed by a conformational change of helix X of CETP to an open state, in which we found the accessibility of cholesteryl esters to the C-terminal tunnel opening of CETP to increase. Furthermore, in the absence of helix X, cholesteryl esters rapidly diffused into CETP through the C-terminal opening. The results provide compelling evidence that helix X acts as a lid which conducts lipid exchange by alternating the open and closed states. The findings have potential for the design of novel molecular agents to inhibit the activity of CETP.

INTRODUCTION

Cholesteryl ester transfer protein (CETP) is a 476-residue-long glycoprotein which promotes the transfer of cholesteryl esters (CEs), triacylglycerols (TGs) and phospholipids (PLs) between the different lipoprotein fractions (high density lipoprotein (HDL), low density lipoprotein (LDL), and very low density lipoprotein (VLDL)) in human blood plasma. CETP is believed to mediate the transfer by a hetero-exchange mechanism in which CEs are carried from HDL to VLDL and LDL particles, and TGs are carried in the opposite direction from VLDL and LDL to HDL particles, resulting in CE depletion and TG enrichment of HDL [1]. Interestingly, CETP is structurally homologous to the phospholipid transfer protein (PLTP), the lipopolysaccharide binding protein (LBP), and the bactericidal/permeability-increasing protein (BPI) [1]. As all these proteins are able to bind phospholipids, similarity in their transportation mechanisms has been suggested. Importantly, however, CETP is the only protein able to transfer neutral lipids (cholesteryl esters and triglycerides) in human plasma [2].

The broad interest to understand CETP and its lipid trafficking properties stems from the fact that it has a potentially protective role in the development of cardiovascular diseases, in particular atherosclerosis, which are currently the main cause of death in Western countries, claiming ~17 million lives a year. The role of CETP in the development of atherosclerosis became evident when it was found that CETP deficiency and the inhibition of CETP lower LDL and increase HDL levels in human plasma [3]. High HDL levels have been clinically found to be inversely correlated with the development of atherosclerosis, since HDL particles are considered crucial components in the transport of cholesterol from atherosclerotic plaques back into the systemic circulation. Unfortunately, the clinical trial with the first oral anti-atherogenic drug candidate with a CETP-inhibitory activity, torcetrapib, was unsuccessful because of its potentially lethal side effects [4]. Treatment with torcetrapib increased blood pressure and circulating aldosterone levels and also altered serum electrolyte levels. However, subsequent studies indicated that these adverse effects of torcetrapib were unrelated to the inhibition of CETP and are not necessarily shared by the other members of the class of CETP inhibitors. Indeed, a recent clinical trial showed that another CETP inhibitor, anacetrapib, effectively raises HDL and has an acceptable side-effect profile in patients with coronary heart disease or risk factors for coronary heart disease [5]. Importantly, a recent meta-analysis of 92 studies involving 113,833 participants concluded that the CETP genotypes that have lower CETP activity are associated with a decreased coronary risk [6].

Considering the central role of CETP in the development of coronary atherosclerosis and its complications, we face an outstanding challenge to better understand the mechanisms associated with CETP functions. Recently, Qiu et al. resolved the X-ray structure of CETP showing that it carries CE molecules inside a long hydrophobic tunnel, whose ends are plugged by phospholipids (Figure 1) [7]. This kind of hydrophobic tunnel is unique among proteins, and it was speculated that CEs diffuse into and out from the tunnel through the two tunnel openings, which are closed by PLs during the transportation in aqueous surroundings. In addition, based on the X-ray structure it has been speculated that CETP is attached to lipoproteins via its concave surface where also the two hydrophobic tunnel openings reside [7]. Further, it has been proposed that the formation of CETP-lipoprotein complexes is modulated by pH, surface pressure, and the ionic interactions between CETP and phospholipids [8], [9]. Fluorescence quenching has been used to demonstrate that the interaction between tryptophan residues of CETP and PLs could be important in the attachment [10]. Regarding the lipid exchange mechanism of CETP, helix X has been suggested to play a role in lipid loading and unloading by acting as a lid at the C-terminal tunnel opening, being in the open state when the exchange of lipids takes place, and in the closed state when CETP detaches from the lipoprotein surface to become surrounded by aqueous medium [5]. Various mutational studies further suggest that helix X is possibly crucial in the transfer of CEs and TGs but not in the transfer of PLs [11], [12].

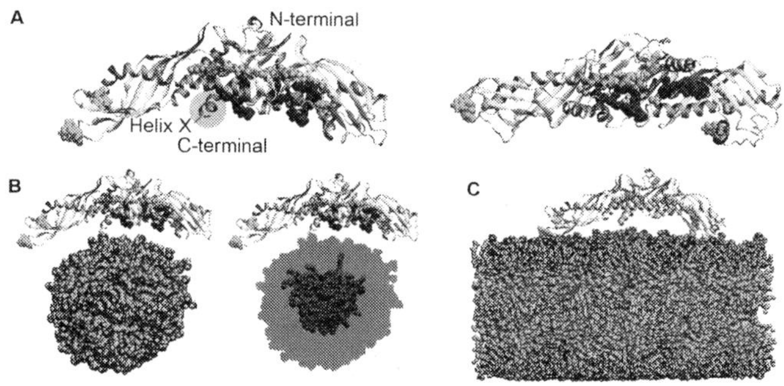

Figure 1: Structure of CETP and starting configurations for simulations. A) X-ray structure of CETP from the side (left) and bottom (right). Two DOPCs (grey and blue spheres) plug the tunnel openings that lead to the hydrophobic tunnel where two CETP-bound CEs (orange spheres) are located. Helix X is labelled and marked with a red sphere. B) The starting configuration for droplet simulations. C) The starting configuration for lipid trilayer simulation. POPCs and DOPCs are coloured as grey, CEs are orange, head group nitrogens are blue, and Trp residues green. Water molecules were removed from the snapshots for clarity.

The above findings and suggestions are appealing and insightful, but call for better understanding of the structure-function relationship and of the dynamics that drive CE, TG and PL transfer. In essence, atomic and molecular scale insight into the lipid exchange between CETP and lipoproteins is limited, which largely stems from exceptional difficulties to experimentally probe the related transient processes in the nanometer scale. In the current study, our objective is to complement experiments through atomistic and coarse-grained molecular dynamics simulations to investigate the binding of CETP to a small lipid droplet and a planar lipid trilayer, and to determine the initial stages of the lipid exchange mechanism. By doing so, we can follow the lipid exchange in atomic detail, shed light on its mechanism, consider the effect of lipoprotein curvature, and unravel the dynamics of the related processes. These mechanisms and phenomena are considered over a multitude of time scales by bridging atomistic and coarse-grained simulations, which are shown to provide consistent results. The present study paves the way for future simulations to elucidate interactions of anacetrapib with CETP and CETP-lipoprotein complexes, with an objective to unlock its inhibitory mechanism. Given the significant role of CETP in cardiovascular diseases, the broad interest of the topic is hoped to attract substantial interest to extend the present work.

RESULTS

Flexible structure of CETP helps it to bind to curved lipoproteins

We carried out three 100 ns atomistic simulations for fully hydrated systems containing CETP with different interior lipid compositions and a small pre-equilibrated HDL-sized lipid droplet composed of POPCs and CEs (A1, A2, A3; Figure 1; see Materials and Methods). In addition to these spherical droplets, CETP was simulated with a pre-equilibrated planar POPC-CE trilayer system (A4; Figure 1C) to study the effect of less curved lipoprotein particles, like VLDL and LDL, on the conformation of CETP.

Root mean square deviation (RMSD) profiles indicate that the structures do not deviate considerably from the X-ray structure (Figure 2B). The radius of gyration fluctuated between 3.2 and 3.5 nm, and its profiles together with snapshots from simulation trajectories show that the conformation of CETP is able to bend to bind to surfaces with different curvatures (Figure 2A). In the case of spherical A3 the curvature of CETP is clearly higher than in the planar A4 system, and it became apparent that the conformation of CETP is not able to rearrange sufficiently to fully match the planar surface. Nonetheless, our results imply that the structure of CETP is elastic and facilitates the binding of CETP to different lipoprotein surfaces with varying curvatures. Yet, due to its inherent curvature that closely matches the curvature of HDL, CETP prefers to bind to HDL-sized particles compared to larger VLDL-sized particles. Consequently, we propose that the free energy change associated with the binding of CETP to HDL is more favorable compared to the formation of a CETP-VLDL complex.

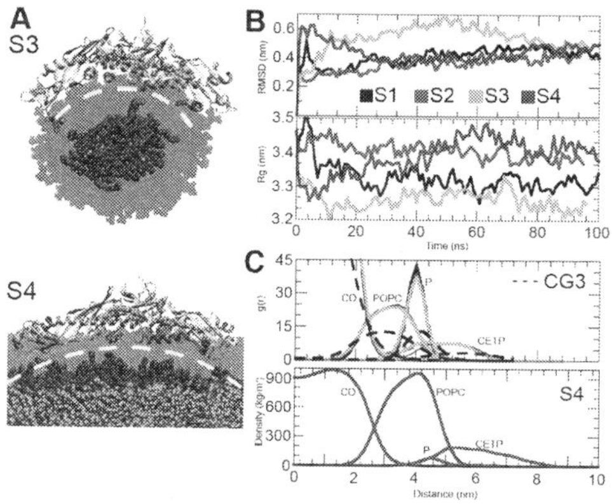

Figure 2: The binding of CETP to lipid surfaces with different curvatures. A) Snapshots from the end of atomistic simulations A3 and A4. POPCs are transparent and grey, and COs are orange. Water molecules were omitted for clarity. CETP is rendered using secondary structures and Trp residues are marked with green color. Dashed and yellow lines present the curvature of CETP. B) RMSD and radii of gyration profiles for CETP in droplet and trilayer simulations. C) Radial distribution functions and density profiles for the droplet and trilayer systems, respectively.

Radial distribution functions and density profiles indicate that CETP does not penetrate deeper than to the level of POPC phosphate groups (Figure 2C). Therefore, in all atomistic simulations the core CEs were observed not to interact directly with CETP, as instead they were found to reside only in the core. This suggests that the surface-core lipid ratio is important for the exchange of neutral lipids by CETP.

During the simulation A2, CETP-bound DOPCs did not diffuse into the lipid droplet. However, we found that during the simulation the hydrophobic tunnel of CETP collapsed, which strongly suggests that the structure of CETP is not stable without interior lipids. This finding is important regarding the lipid exchange process of CETP as it suggests that during the neutral lipid exchange, the hydrophobic cavity is not empty at any point. We return to this matter later.

Salt bridges and tryptophans stabilize CETP-lipoprotein complexes

In atomistic lipid droplet simulations, we calculated the number of salt bridges that formed between CETP and POPCs as a function of time, in order to characterize the key charged residues involved in the attachment of CETP. The number of salt bridges that formed between the positively charged lysine residues of CETP and the negatively charged phosphate (P) groups of POPCs stabilized to a level of 12–20 (Figure 3A). Salt bridging of lysines is much more efficient in A2 and A3 than in A1 (19–20 compared to 12, see Figure 3A). The number of salt-bridges between arginines and P groups was on average two or three. Additionally, we calculated the number of salt bridges formed by the negatively charged Asp and Glu residues and found that Asp residues were able to form 6–8 and Glu residues 2–4 salt bridges with the positively charged choline groups. Amino acids that form most of the salt bridges are shown in Figure 3, revealing that they are mainly located at the edge of the concave surface of CETP.

In the spirit of the earlier Trp quenching study [10], we inspected more carefully the behavior of Trps during binding. In all droplet simulations, Trp299 formed hydrogen bonds with POPCs (Figure 3B). Trp264 stayed buried inside the structure of the protein and Trp162 was able to interact with the water molecules. In A1 and A2, Trp105 and Trp106 were located facing the water phase, while in A3 the flap Ω_5 interacted with POPCs by anchoring Trps 105 and 106 to the carbonyl region of POPC surface, highlighted in Figure 2. In the trilayer simulation only two Trp residues (105 and 299) were able to interact with the POPC surface.

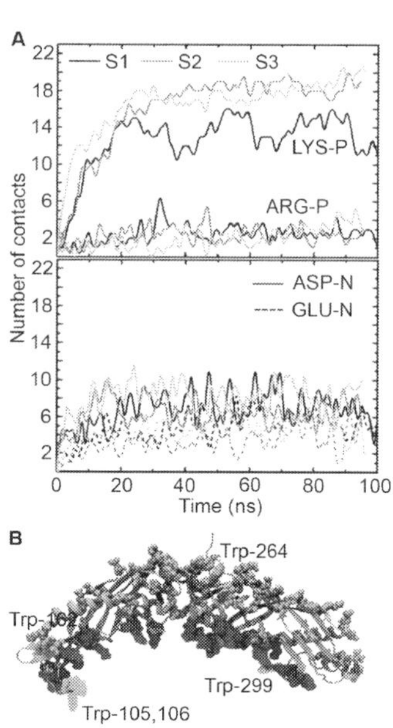

Figure 3: Electrostatic interactions between CETP and lipid droplet. A) Number of salt bridges formed between the charged residues of CETP and the head groups of POPCs as a function of time. The upper profile shows the number of contacts between the positively charged residues and P atoms of POPCs, and the lower profile shows the number of contacts between the negatively charged amino acids and N atoms of POPCs. B) Salt bridge-forming positively (red) and negatively (blue) charged amino acids marked to the structure of CETP. Trp residues are labeled and green.

Our results highlight the importance of electrostatic interactions between CETP and phospholipids in the formation of CETP-droplet complexes. The results provide compelling evidence that three Trp residues anchor CETP to lipid droplets, introducing additional stability to CETP-lipoprotein complexes where the curvature of CETP and a lipoprotein matches.

Coarse-grained simulations reveal that the ratio of surface and core lipids in lipid droplets is an important modulator of CETP activity

Interpretation of atomistic simulations requires care due to the limited time and length scales that are feasible through atomistic studies. For example, the diffusion of lipids in HDLs is slow compared to the time scales we have simulated and, thus, claims regarding the principal binding site and penetration depth of CETP must be carefully considered. In order to add liability to our atomistic simulations, we also carried out coarse-grained simulations, covering time scales beyond 2 μs.

CG simulations support and validate atomistic simulations by showing that the concave surface is the principal lipoprotein binding site of CETP. We did not observe any deviations from this conclusion during the three independent 2-microsecond simulations. Radial distribution functions shown in Figure 2 depict a similar distribution of molecules as in atomistic simulations.

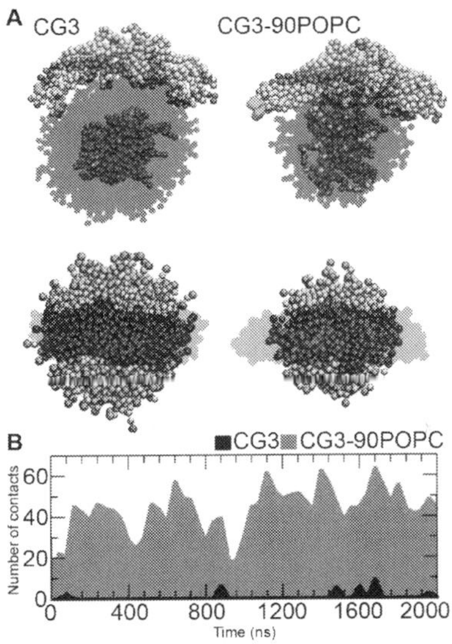

Figure 4: Interaction of CETP with core lipids.A) Snapshots from the coarse-grained simulations CG3 (left) and CG3-90POPC (right). The upper snapshots show side views and the lower ones top views of CETP bound to a lipid droplet. In the latter, the hydrophobic patch under CETP is clearly visible. The structure of CETP has been rendered using secondary structure information in upper snapshots (β-sheets are yellow, α-helixes violet and random coils gray) or as dark transparent phantom in lower snapshots. The green spheres are Trp residues. POPCs are transparent in the upper snapshots, while those in the bottom snapshots are visible as grey (the choline head groups are visible as blue). CEs are rendered with

orange spheres. Water molecules were omitted for clarity. B) Number of contacts between core CEs and CETP with different surface-core lipid ratios as a function of time.

However, intriguingly we found that POPCs which were in contact with the concave surface of CETP migrated away from the tunnel openings, forming a small hydrophobic patch under the concave surface (Figure 4A). In essence, CETP drives phospholipids to diffuse away from the slightly hydrophobic tunnel openings to its edges where most of the salt bridge-forming amino acids reside. We analyzed the spatial densities of the polar beads (GL1, GL2, and NC3, PO4, using the descriptions of the Martini model) of POPCs to clarify the patch formation more clearly. The spatial density map revealed the formation of a hydrophobic patch under the concave surface, and specifically in the region where the N- and C-terminal tunnel openings reside.

The time associated with the formation of the hydrophobic patch is difficult to estimate accurately, since the process fluctuates depending on the dynamics of the CETP-droplet complex. In practice we found the patch to emerge in roughly 10–40 ns, and it increased to a size of about 1 nm×3 nm in 100–500 nanoseconds, depending on the system studied. At longer times the patch fluctuated quite a lot but there was a trend showing a slow increase in size, suggesting that the total formation time may be of the order of microseconds.

To gain further support for patch formation, as predicted by CG simulations, we repeated the analysis with two additional CG simulations where we used PME for electrostatics with the non-polarizable Martini water model, and PME with the polarizable Martini water model [13]. With the polarizable water model the solubility of charged species to apolar media should be better described compared to the standard Martini model. In both additional CG simulations, hydrophobic patch formation was observed too.

Given that the patch emerged in CG simulations in tens of nanoseconds, and the atomistic simulations lasted for 100 ns, we returned to our atomistic simulation data to consider this aspect in atomic detail. We analyzed the spatial density profile of POPC head groups and observed similar hydrophobic patch formation. For example, in A3 we noticed that a small hydrophobic patch was formed under CETP already in ~20 ns, and the patch slowly grew in size and number as in ~80–100 ns there were two patches close to one another (data not shown). The fact that also 100 ns atomistic simulations show the hydrophobic patch formation confirms that the CETP-lipoprotein interaction is strong specifically under the concave surface and promotes the formation of a path between the droplet core and CETP.

The hydrophobic patch formation exposes the hydrophobic parts of the lipids to the concave surface where the hydrophobic tunnel openings are located. However, hardly any of the CEs were in contact with CETP, as can be seen from Figure 4. Thus, we reduced the number of POPCs from 180 to 90 (CG3-90POPC) and simulated the system again for 2 μs in order to see if the lower surface pressure of a lipid droplet would promote the solubility of CEs to the surface lipid monolayer and the interaction between CETP and CEs. Indeed, the contacts between core CEs and CETP increased. Clearly, the concave surface of CETP has some affinity for CEs, over random thermal fluctuations, as the hydrophobic patch under CETP guides core CEs to the concave surface. However, the surface pressure must be low enough for CEs to localize to the surface monolayer and CETP to bind to the surface. This implies that the ratio of surface and core lipids (surface pressure) and the formation of the small hydrophobic patch under CETP are important factors modulating the core lipid transfer activity of CETP.

Structure fluctuations show that the hinge region of helix X is highly mobile

Root mean square fluctuations (RMSFs) of the protein backbone were analyzed after 40 ns of atomistic lipid droplet simulations to find the regions, which wobble the most after CETP attached to the surface of the lipid droplet (Figure 5). This was done as follows. First, the RMSF of backbone atoms was computed by fitting the atomic positions to the reference structure (average structure of CETP after its binding to the lipid droplet surface) and then calculating the average distance deviation from the reference structure. The RMSFs of individual backbone atoms were then averaged per residue to determine the residual RMSF profile. Similar results were observed in all droplet simulations. The N- and C-terminal ends and loop regions (marked by omegas) of CETP showed high fluctuations, as expected. We also found that in the helix X region (residues 460–476) the conformational fluctuations peaked near the residue 462. This region has previously been proposed to be a potential hinge region of helix X with elevated B-factors [7]. In addition, it was found that the flaps Ω_1 and Ω_2 resulted in high fluctuations to the RMSF profile as was also proposed based on the B-factors of the X-ray structure of CETP [7]. In addition to the suggested high fluctuations, we found that another five regions of CETP were also highly fluctuating in each simulation. These regions were Ω_3 (residues 380–400), Ω_4(residues 40–50), Ω_5 (residues 90–110), Ω_6 (residues 150–170), and Ω_7 (residues 230–260). All regions are found in the loops, and hence high fluctuations can be expected. Previously it has been speculated that the hinge region could promote the needed flexibility to helix X that is important in the lipid exchange process [5]. To study further the role of helix X in lipid exchange, we did two additional atomistic simulations to probe its role in the lipid exchange process, see below.

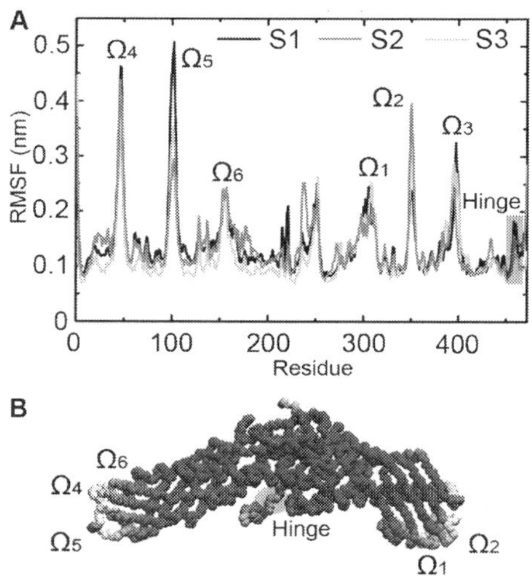

Figure 5: Dynamical properties of CETP. A) Root mean square fluctuations for atomistic droplet simulations. Loop regions are marked with omegas and the hinge region of helix X has been slightly darkened. B) Residual B-factors mapped to the backbone structure of CETP. Red color indicates the most rigid structures, whereas white and blue indicate the most flexible structural regions. The hinge region of helix X is marked with a transparent blue sphere.

Helix X regulates the accessibility of cholesteryl esters inside CETP

Earlier point and deletion mutations suggest that helix X is important in the transfer of core lipids, while it is not needed in phospholipid transfer [12]. Since we found that the hydrophobic patch was formed under the concave surface of CETP in both CG and atomistic simulations, we asked if the fully formed hydrophobic patch could induce changes to the conformation of helix X. To test this hypothesis, we did one additional atomistic simulation with 90 POPCs (that is, starting from the system A3-90POPC) where we expanded the hydrophobic patch under the concave structure by removing POPCs near the two tunnel openings of CETP, so that helix X was only able to interact with the hydrophobic parts of POPCs and CEs. Here, it is worth to mention that atomistic simulations are the only method of choice for this purpose, since this kind of conformational change can not take place in our CG simulations, where we used the elastic network model to keep the secondary structure of CETP stable [14]. We found that the conformation of helix X rearranged and became buried inside the hydrophobic cavity of CETP, where it interacted with CETP-bound CE (Figure 6). This conformational change generated a hydrophobic pathway from the droplet surface to the tunnel, increasing the accessibility of core CEs to the hydrophobic tunnel of CETP.

To further assess the regulatory role of helix X, we created a deletion mutant of CETP, in which the residues 462–476 (helix X) were removed from the structure, and we simulated this structure for 80 ns. Deletion mutation simulation revealed that three CEs readily diffused into CETP when helix X was completely removed from the structure (see the spatial density maps and the number of contacts plot in Figures 6B and 6C). This provides further support for the view that helix X acts as a lid at the C-terminal tunnel opening, and that its conformation regulates the accessibility of CEs to the hydrophobic tunnel.

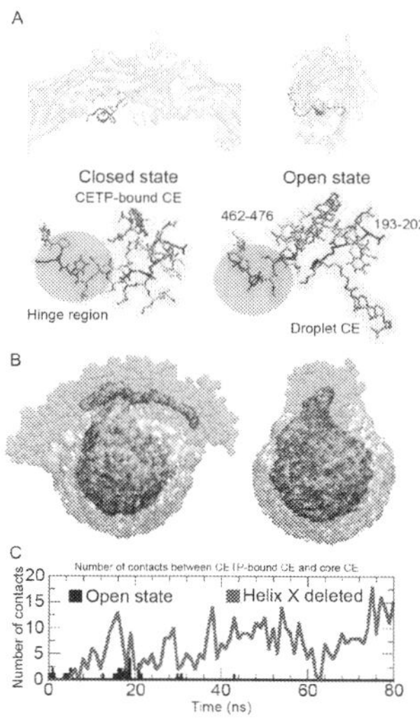

Figure 6: Hypothesis for the initial event of helix X assisted core lipid exchange. A) Two RMSD-fitted snapshots from the simulation A3-90POPC showing the rearrangement of helix X (darkened colour). The green conformation is for the open state and the blue one for the closed state. A more detailed structure of helix X and the role of the hinge region during the conformational change (red and transparent region) are shown in the lower snapshots. CEs are shown as orange sticks. The residues 462–476 and 193–202 of CETP have been rendered using sticks, and coloring is based on the polarity of residues. B) Spatial number density of POPCs (grey and transparent) and CEs (orange) during the simulation A3-90POPC. Core CEs diffuse into the hydrophobic tunnel of CETP (green spheres) without helix X. C) Number of contacts between core CEs and interior CE-473 when helix X is in the open state (black) or completely removed (red).

DISCUSSION

Previously, the role of electrostatic interactions in the formation of isolable CETP-lipoprotein complexes was demonstrated by Pattnaik et al., who showed that (in addition to CETP-HDL complexes) CETP was able to form isolable complexes with LDL and VLDL particles when negative surface charge was increased by phospholipase A_2 digestion or by acylation of phospholipid amino groups. They reached the conclusion that the phospholipid phosphate groups are the primary sites for the interaction of lipoproteins with CETP [8]. They also found that the formation of isolated CETP-HDL complexes was hindered by decreasing the pH, introducing positive divalent ions into the solution, or by digesting lipoproteins by phospholipase C. Moreover, Nishida et al. reported that the affinity of CETP for various lipoproteins is governed by a delicate balance of electrostatics and hydrophobic interactions [15]. The importance of electrostatic interaction in CETP binding has been shown also by several point mutation studies applied to the positively charged lysine residues of CETP [16]. Our results are in agreement with experiments, as in the present simulations most of the salt-bridges with the negatively charged phosphate groups of POPCs were formed by the Lys residues at the concave surface, when CETP fastened to the lipid surface. However, also Glu and Asp residues formed salt-bridges with the positively charged choline groups of POPCs, although the ratio of salt-bridges formed by the positively and negatively charged amino acids is approximately 1.8, indicating that mostly the positively charged amino acids contribute to the formation of CETP-lipoprotein complexes. Another important factor playing a role in the binding of CETP is Trp residues located in the flaps Ω_5 and Ω_1 that were found to become buried into the lipid matrix. Most likely Trp residues add more stability to the CETP-lipoprotein complexes by anchoring CETP to a lipoprotein surface. Interestingly, Desmuraux et al. made mutations to the structurally similar flap Ω_5 region (Trp-91, Phe-92 and Phe-93) of PLTP and showed that the phospholipid transfer activity of PLTP from liposomes to HDL particles decreased up to 60% [17]. This finding together with our results suggests that the flexible flap Ω_5 region of CETP and Trp residues therein are crucial in the binding of CETP to HDL particles, playing an important role in the CETP-mediated lipid transfer.

Penetration depth of CETP is an important factor in CETP-mediated lipid exchange, as it determines how efficiently the neutral core lipids are able to interact with CETP. Previous studies have shown that the exclusion pressure of CETP is lower than the exclusion pressure of other apolipoproteins, like apoA-I [9], [10]. Moreover, it has been argued that the weaker penetration of CETP to the emulsion particles compared to apoA-I makes the activation energy of the attachment and detachment of CETP lower, rendering the transportation process more efficient [9], [10].

Our atomistic and CG simulation results showed that CETP is not able to bury its amino acid residues deeper than to the level of the phosphate groups of POPCs. The above findings therefore imply that core lipids have to diffuse to or reside at the surface to enter CETP. Therefore, the amount of core lipids at the lipoprotein surface is an important factor modulating the activity of CETP, as has been suggested previously based on liposome studies [18], and it can be promoted by defects as is outlined below. The number of surface-located neutral lipids can be regulated by the lipid and apolipoprotein composition of lipoprotein particles.

Interestingly, we found that when CETP attaches to the surface by the aid of electrostatic interactions, the head groups of POPCs moved aside, providing access to the hydrophobic lipid region. In this manner, the two tunnel openings of the concave surface are exposed to the hydrophobic lipid matrix of the lipoprotein. The hydrophobic patch formation facilitates, by generating a defect to the surface monolayer, the diffusion of core lipids to the surface monolayer region located under CETP (Figure 6B). Thus, the localization of neutral lipids at the surface monolayer itself is not crucial to allow CETP to exchange neutral lipids between lipoproteins but the neutral lipids can enter CETP through the formation of the hydrophobic patch. Consequently, we envision that the activity of CETP could be inhibited by nonpolar drugs that are transferred into the hydrophobic tunnel of CETP through the hydrophobic core of lipoproteins. Further, we observed that the concave surface interacted directly with CEs that diffused more readily to the hydrophobic tunnel openings when the surface-core lipid ratio was decreased.

Finally, we found CEs to diffuse into the hydrophobic tunnel of CETP and interact with CETP-bound CE when the conformation of helix X was in the open state or completely removed. A previous mutational study argued that CETP is not able to transfer neutral lipids when helix X is removed from the structure [11], [12]. However, CETP is able to transfer phospholipids without helix X. Our results showed that the conformation of helix X rearranges, and helix X moves inside the hydrophobic tunnel of CETP where it can interact with CETP-bound CE. Given this, we suggest that there are two important functional properties of helix X that make the neutral lipid exchange possible. First, helix X is able to facilitate the neutral lipid exchange by opening the hydrophobic pathway from a lipoprotein surface to the hydrophobic tunnel of CETP. Second, helix X promotes the diffusion of neutral lipids from the hydrophobic tunnel to lipoproteins by filling the volume of CETP-bound neutral lipid when it diffuses out from CETP. Afterwards, another neutral lipid from the lipoprotein core or inside CETP could take the place of helix X after which the C-terminal tunnel opening closes again. We propose that helix X is needed to prevent the structure of CETP from collapsing as was registered in the simulation A1 without the CETP-bound lipids. The above reasons would explain why helix X is important in the neutral lipid exchange, but not in the exchange of phospholipids.

It is tempting to contemplate the possible roles of helix X in the inhibition of CETP. It has been reported that dalcetrapib, a novel CETP inhibitor, binds covalently to CETP by forming a disulphide bond with Cys-13, which is located inside the hydrophobic tunnel of CETP [19]. In addition, it has been suggested that the disulphide bond formation is a necessary requirement for the dalcetrapib-mediated CETP inhibition. However, that is not the case with torcetrapib or anacetrapib (another novel CETP inhibitors), both of which bind reversibly to CETP [19]. Yet, all inhibitors stabilize HDL-CETP complexes, which has been found to be the second major inhibitory mechanism of the neutral lipid transfer exerted by the synthetic CETP inhibitors [19]. Based on our simulations, we can hypothesize that helix X is locked to the open state when inhibitors are bound to CETP. The driving force for this could be the small size of an inhibitor that enforces helix X to be located inside the hydrophobic tunnel and, thus, prevents collapse of the lipid pocket. Another reason could be the more favorable interaction between helix X and the inhibitor inside the tunnel, which could conceivably force the conformation of helix X to the open state, or change the conformation distribution to favor the open state. Consequently, the detachment of CETP from the surface of HDL would be hindered since the helix X is not able to shield the hydrophobic tunnel opening of the lipid pocket when CETP is completely in the aqueous phase. In addition, the open state could prevent the binding of phospholipids to the C-terminal tunnel opening, which, based on the X-ray structure of CETP, is known to be occupied by phospholipids when CETP is not attached to a lipoprotein surface. A reduced ability of CETP to bind and transport phospholipids could further stabilize the HDL-CETP complex.

In summary, we have provided a detailed atomistic picture regarding the initial steps in the lipid exchange mechanism of CETP and, furthermore, we have offered a plausible mechanism for the exchange of neutral lipids mediated by CETP. Overall, our work paves the way for additional future studies to elucidate interactions of the available promising CETP inhibiting drugs, such as anacetrapib and dalcetrapib, with CETP and CETP-lipoprotein complexes. Our findings for the factors that affect the lipid exchange process can also be exploited in the design of novel molecular agents capable of inhibiting the activity of CETP, one possible strategy being the design of nonpolar drugs which can be transferred into the hydrophobic tunnel of CETP. Together with recent simulation models for both HDL and LDL [20], these ideas are a reasonable goal already at present.

MATERIALS AND METHODS

System setup

The coordinate file of CETP in the PDB format with an accession code 2OBD was acquired from the RCSB Protein Data Bank. In addition to the protein, the structure provides information of the lipids carried by CETP: there are two CEs located inside the long hydrophobic tunnel of CETP, and two dioleoylphosphatidylcholine (DOPC) lipids that cover the two endings of the hydrophobic tunnel. The charge state of CETP was chosen to represent the physiological pH that is 7.4. A detailed explanation of the protein structure is given elsewhere [7].

For atomic-scale simulations, three different setups were constructed by combining pre-equilibrated lipid droplets consisting of 180 palmitoyloleoyl-PC (POPC) and 35 CE molecules [21]. In each system, CETP was placed approximately at a distance of 1 nm from the surface of the lipid droplet (Figure 2). In the first simulated system (A1), the two DOPCs and CEs were removed from CETP. In the second simulation (A2), both DOPCs and CEs were included, while in the third simulation (A3) only CEs were kept inside CETP. We also simulated CETP with a planar trilayer system composed of 512 POPCs and 796 CEs (A4). The droplet systems were solvated with ~180,000 water molecules at a salt concentration of 0.2 M including counter ions, while the trilayer system included ~50,000 water molecules. Altogether, the systems included ~500,000 atoms. Finally, three additional atomistic systems were constructed to characterize the role of helix X in lipid exchange in more detail (see text). First, we studied the effect of the hydrophobic patch on the structure of helix X by removing half of the POPCs from A3 at 100 ns (A3-90POPC). Second, we also considered CETP through its helix X deletion mutant to probe the regulatory role of the helix. Third, we used A3-90POPC as a basis and removed some of the surface lipids to model the complete formation of a hydrophobic patch under the concave surface of CETP. The context of these simulations to the studied processes will become clear in the discussion below.

In addition to atomistic simulations, we carried out four coarse-grained (CG) simulations. First, the system A3 was directly coarse grained (in the text, we refer to this simulation as CG3) by using a script that is available at the homepage of the Martini force field. Second, 90 POPCs were removed from CG3, ending up in the system denoted as CG3-90POPC. The systems CG3 and CG3-90POPC were simulated under standard Martini model (see below) conditions with regard to electrostatics (using truncation of electrostatic interactions) and the water model that is non-polarizable. To clarify the influence of long-range interactions and the water model, we simulated two additional systems. That is, in the third case we focused on the system CG3 which was simulated with full electrostatics using the particle mesh Ewald (PME) method [22]. Finally, in the fourth coarse-grained model, we simulated the system CG3 using both PME and the polarizable Martini water model [13].

Simulation parameters and force field

The GROMACS simulation package with version 4.0 was used in the simulations [23]. In atomistic studies, we used the Nose-Hoover thermostat [24], [25] with a coupling constant of 1.0 ps to set the temperature to 330 K in which the particle core is certainly in liquid state. The pressure was set to 1 bar using the Parrinello-Rahman barostat [26] with isotropic pressure coupling and a coupling constant of 0.1 ps. The van der Waals interactions were chosen to have a cutoff at 1.0 nm. Electrostatic interactions were evaluated by the particle mesh Ewald technique with a real space cut-off of 1.0 nm [22]. Water molecules were described using the SPC water model. All non-water bonds were constrained using the LINCS algorithm [27] and the SETTLE algorithm [28] was used to constrain water molecules, allowing the use of a time step of 2 fs in the integration of equations of motion. Berger parameters [29] were used for lipids, while the GROMOS53A6 force-field [30] was employed for the protein. Combination rules were introduced for the interactions between lipids and the protein. The four leading atomistic systems (A1–A4) were simulated for 100 ns, and the last two ones that focused on helix X for 80 ns. The total simulation time of atomistic simulations was 0.56 μs.

CG simulations were also carried out with GROMACS, using the Martini force field with an extension to proteins [31], [32]. The ElNeDyn elastic network model was used to keep the structure of CETP stable [14]. The Berendsen thermostat and barostat were used with time constants of 1.0 ps. Temperature was set to 320 K and isotropic pressure coupling was used with pressure set to 1 bar. Cut-off distance for electrostatic interactions was set to 1.2 nm. For Lennard-Jones interactions we used a cut-off of 1.2 nm, and Lennard-Jones interactions were shifted to zero from 0.9 nm. Time step was 25 fs. The simulation time of each CG system was beyond 2 μs, and the times reported here are given in units of the effective Martini time. All rendered figures were done by VMD [33].

ACKNOWLEDGMENTS

We would like to thank CSC – The IT Center for Science for providing supercomputer facilities.

AUTHOR CONTRIBUTIONS

Conceived and designed the experiments: AK TV PK IV MH. Performed the experiments: AK TV. Analyzed the data: AK. Contributed reagents/materials/analysis tools: AK TV. Wrote the paper: AK TV PK IV MH.

REFERENCES

1. Tall AR (1993) Plasma cholesteryl ester transfer protein. J Lipid Res 34: 1255–1274.AR Tall1993Plasma cholesteryl ester transfer protein.J Lipid Res3412551274

2. Hesler CB, Tall AR, Swenson TL, Weech PK, Marcel YL, et al. (1988) Monoclonal antibodies to the M_r 74,000 cholesteryl ester transfer protein neutralize all of the cholesteryl ester and triglyceride transfer activities in human plasma. J Biol Chem 263: 5020–5023.CB HeslerAR TallTL SwensonPK WeechYL Marcel1988Monoclonal antibodies to the M_r 74,000 cholesteryl ester transfer protein neutralize all of the cholesteryl ester and triglyceride transfer activities in human plasma.J Biol Chem26350205023
3. Polk D, Shah PK (2008) Cholesterol ester transfer protein (CETP) and atherosclerosis. Drug Discov Today Ther Strateg 4: 137–145.D. PolkPK Shah2008Cholesterol ester transfer protein (CETP) and atherosclerosis.Drug Discov Today Ther Strateg4137145
4. Tall AR, Yvan-Charvet L, Wang N (2007) The failure of torcetratib: Was it the molecule or the mechanism. Arterioscler Thromb Vasc Biol 27: 257–260.AR TallL. Yvan-CharvetN. Wang2007The failure of torcetratib: Was it the molecule or the mechanism.Arterioscler Thromb Vasc Biol27257260
5. Cannon CP, Shah S, Dansky HM, Davidson M, Brinton EA, et al. (2010) Safety of anacetrapib in patients with or at high risk for coronary heart disease. N Engl J Med 363: 2406–2415.CP CannonS. ShahHM DanskyM. DavidsonEA Brinton2010Safety of anacetrapib in patients with or at high risk for coronary heart disease.N Engl J Med36324062415
6. Vasan RS, Pencina MJ, Robins SJ, Zachariah JP, Kaur G, et al. (2009) Association of circulating cholesteryl ester transfer protein activity with incidence of cardiovascular disease in the community. Circulation 120: 2414–2420.RS VasanMJ PencinaSJ RobinsJP ZachariahG. Kaur2009Association of circulating cholesteryl ester transfer protein activity with incidence of cardiovascular disease in the community.Circulation12024142420
7. Qiu X, Mistry A, Ammirati MJ, Chrunyk BA, Clark RW, et al. (2007) Crystal structure of cholesteryl ester transfer protein reveals a long tunnel and four bound lipid molecules. Nat Struct Mol Biol 14: 106–113.X. QiuA. MistryMJ AmmiratiBA ChrunykRW Clark2007Crystal structure of cholesteryl ester transfer protein reveals a long tunnel and four bound lipid molecules.Nat Struct Mol Biol14106113

8. Pattnaik NM, Zilversmit DB (1979) Interaction of cholesteryl ester exhange protein with human plasma lipoproteins and phoshpholipid vesicles. J Biol Chem 254: 2782–2786.NM PattnaikDB Zilversmit1979Interaction of cholesteryl ester exhange protein with human plasma lipoproteins and phoshpholipid vesicles.J Biol Chem25427822786
9. Weinberg RB, Cook VR, Jones JB, Kussie P, Tall AR (1994) Interfacial properties of recombinant human cholesterol ester transfer protein. J Biol Chem 269: 29588–29591.RB WeinbergVR CookJB JonesP. KussieAR Tall1994Interfacial properties of recombinant human cholesterol ester transfer protein.J Biol Chem2692958829591
10. Rajaram OV, Sawyer WH (1996) Penetration of an emulsion surface by cholesteryl ester transfer protein. Eur Biophys J 25: 31–36.OV RajaramWH Sawyer1996Penetration of an emulsion surface by cholesteryl ester transfer protein.Eur Biophys J253136
11. Wang S, Wang X, Deng L, Rassart E, Milne RW, et al. (1993) Point mutagenesis of carboxyl-terminal amino acids of cholesteryl ester transfer protein. J Biol Chem 268: 1955–1959.S. WangX. WangL. DengE. RassartRW Milne1993Point mutagenesis of carboxyl-terminal amino acids of cholesteryl ester transfer protein.J Biol Chem26819551959
12. Wang S, Deng L, Milne RW, Tall AR (1992) Identification of a sequence within the C-terminal 26 amino acids of cholesteryl ester transfer protein responsible for binding a neutralizing monoclonal antibody and necessary for neutral lipid transfer activity. J Biol Chem 267: 17487–17490.S. WangL. DengRW MilneAR Tall1992Identification of a sequence within the C-terminal 26 amino acids of cholesteryl ester transfer protein responsible for binding a neutralizing monoclonal antibody and necessary for neutral lipid transfer activity.J Biol Chem2671748717490
13. Yesylevskyy SO, Schafer LV, Sengupta D, Marrink SJ (2010) Polarizable water model for the coarse-grained MARTINI force field. PLoS Comput Biol 6: e1000810.SO YesylevskyyLV SchaferD. SenguptaSJ Marrink2010Polarizable water model for the coarse-grained MARTINI force field.PLoS Comput Biol6e100081010.1371/journal.pcbi.1000810. 10.1371/journal.pcbi.1000810.

14. Periole X, Cavalli M, Marrink S, Ceruso MA (2009) Combining an elastic network with a coarse-grained molecular force field: Structure, dynamics, and intermolecular recognition. J Chem Theory Comput 5: 2531–2543.X. PerioleM. CavalliS. MarrinkMA Ceruso2009Combining an elastic network with a coarse-grained molecular force field: Structure, dynamics, and intermolecular recognition.J Chem Theory Comput525312543
15. Nishida HI, Arai H, Nishida T (1993) Cholesterol ester transfer mediated by lipid transfer protein as influenced by changes in the charge characteristics of plasma lipoproteins. J Biol Chem 268: 16352–16360.HI NishidaH. AraiT. Nishida1993Cholesterol ester transfer mediated by lipid transfer protein as influenced by changes in the charge characteristics of plasma lipoproteins.J Biol Chem2681635216360
16. Jiang X, Bruce C, Cocke T, Wang S, Boguski M, et al. (1995) Point mutagenesis of positively charged amino acids of cholesteryl ester transfer protein: Conserved residues within the lipid transfer/ lipopolysaccharide binding protein gene family essential for function. Biochemistry 7258–7263.X. JiangC. BruceT. CockeS. WangM. Boguski1995Point mutagenesis of positively charged amino acids of cholesteryl ester transfer protein: Conserved residues within the lipid transfer/lipopolysaccharide binding protein gene family essential for function.Biochemistry72587263
17. Desrumaux C, Labeur c, Verhee A, Tavernier J, Vandekerckhove J, et al. (2001) A hydrophobic cluster at the surface of the human plasma phospholipid transfer protein is critical for activity on high density lipoproteins. J Biol Chem 276: 5908–5915.C. Desrumauxc. LabeurA. VerheeJ. TavernierJ. Vandekerckhove2001A hydrophobic cluster at the surface of the human plasma phospholipid transfer protein is critical for activity on high density lipoproteins.J Biol Chem27659085915
18. Morton RE, Greene DJ (2003) The surface cholesteryl ester content of donor and acceptor particles regulates CETP: A liposome-based approach to assess the substrate properties of lipoproteins. J Lipid Res 44: 1364–1372.RE MortonDJ Greene2003The surface cholesteryl ester content of donor and acceptor particles regulates CETP: A liposome-based approach to assess the substrate properties

of lipoproteins.J Lipid Res4413641372

19. Ranalletta M, Bierilo KK, Chen Y, Milot D, Chen Q, et al. (2010) Biochemical characterization of cholesteryl ester transfer protein inhibitors. J Lipid Res 51: 2739–2752.M. RanallettaKK BieriloY. ChenD. MilotQ. Chen2010Biochemical characterization of cholesteryl ester transfer protein inhibitors.J Lipid Res5127392752
20. Vuorela T, Catte A, Niemela PS, Hall A, Hyvonen MT, et al. (2010) Role of lipids in spheroidal high density lipoproteins. PLoS Comput Biol 6: e1000964.T. VuorelaA. CattePS NiemelaA. HallMT Hyvonen2010Role of lipids in spheroidal high density lipoproteins. PLoS Comput Biol6e1000964
21. Koivuniemi A, Heikelä M, Kovanen PT, Vattulainen I, Hyvönen MT (2009) Atomistic simulations of phosphatidylcholines and cholesteryl esters in high-density lipoprotein-sized lipid droplet and trilayer: Clues to cholesteryl ester transport and storage. Biophys J 96: 4099–4108.A. KoivuniemiM. HeikeläPT KovanenI. VattulainenMT Hyvönen2009Atomistic simulations of phosphatidylcholines and cholesteryl esters in high-density lipoprotein-sized lipid droplet and trilayer: Clues to cholesteryl ester transport and storage.Biophys J9640994108
22. Essmann U, Perera L, Berkowitz ML, Darden T, Lee H, et al. (1995) A smooth particle mesh ewald method. J Chem Phys 103: 8577–8593.U. EssmannL. PereraML BerkowitzT. DardenH. Lee1995A smooth particle mesh ewald method.J Chem Phys10385778593
23. Hess B, Kutzner C, van der Spoel D, Lindahl E (2008) GROMACS 4: Algorithms for highly efficient, load-balanced, and scalable molecular simulation. J Chem Theory Comput 4: 435–447.B. HessC. KutznerD. van der SpoelE. Lindahl2008GROMACS 4: Algorithms for highly efficient, load-balanced, and scalable molecular simulation.J Chem Theory Comput4435447
24. Nose S, Klein ML (1983) Constant pressure dynamics for molecular systems. Mol Phys 52: 1055–1076.S. NoseML Klein1983Constant pressure dynamics for molecular systems.Mol Phys5210551076
25. Hoover WG (1985) Canonical dynamics: Equilibrium phase-space distributions. Phys Rev A 31: 1695–1697.WG Hoover1985Canonical dynamics: Equilibrium phase-space distributions.Phys Rev

A3116951697

26. Shinoda W, Fukada T, Okazaki S, Okada I (1995) Molecular dynamics simulation of the dipalmitoylphosphatidylcholine (DPPC) lipid bilayer in the fluid phase using the nosé-parrinello-rahman NPT ensemble. Chem Phys Lett 232: 308–322.W. ShinodaT. FukadaS. OkazakiI. Okada1995Molecular dynamics simulation of the dipalmitoylphosphatidylcholine (DPPC) lipid bilayer in the fluid phase using the nosé-parrinello-rahman NPT ensemble.Chem Phys Lett232308322
27. Hess B, Bekker H, Berendsen HJC, Fraaije JGEM (1997) LINCS: A linear constraint solver for molecular simulations. J Comp Chem 18: 1463–1472.B. HessH. BekkerHJC BerendsenJGEM Fraaije1997LINCS: A linear constraint solver for molecular simulations.J Comp Chem1814631472
28. Miyamoto S, Kollman PA (1992) Settle: An analytical version of the SHAKE and RATTLE algorithm for rigid water models. J Comp Chem 13: 952–962.S. MiyamotoPA Kollman1992Settle: An analytical version of the SHAKE and RATTLE algorithm for rigid water models.J Comp Chem13952962
29. Berger O, Edholm O, Jähnig F (1997) Molecular dynamics simulations of a fluid bilayer of dipalmitoylphosphatidylcholine at full hydration, constant pressure, and constant temperature. Biophys J 72: 2002–2013.O. BergerO. EdholmF. Jähnig1997Molecular dynamics simulations of a fluid bilayer of dipalmitoylphosphatidylcholine at full hydration, constant pressure, and constant temperature.Biophys J7220022013
30. Schuler LD, Daura X, van Gunsteren W (2001) An improved GROMOS96 force field for aliphatic hydrocarbons in the condensed phase. J Comp Chem 22: 1205–1218.LD SchulerX. DauraW. van Gunsteren2001An improved GROMOS96 force field for aliphatic hydrocarbons in the condensed phase.J Comp Chem2212051218
31. Marrink SJ, de Vries AH, Mark AE (2004) Coarse grained model for semi-quantitative lipid simulations. J Phys Chem B 108: 750–760.SJ MarrinkAH de VriesAE Mark2004Coarse grained model for semi-quantitative lipid simulations.J Phys Chem B108750760

32. Monticelli L, Kandasamy SK, Periole X, Larson RG, Tieleman DP, et al. (2008) The MARTINI coarse grained forcefield:Extension to proteins. J Chem Theory Comput 4: 819–834.L. MonticelliSK KandasamyX. PerioleRG LarsonDP Tieleman2008The MARTINI coarse grained forcefield:Extension to proteins.J Chem Theory Comput4819834
33. Humprey W, Dalke A, Schulten K (1996) VMD - visual molecular dynamics. J Molec Graphics 14: 33–38.W. HumpreyA. DalkeK. Schulten1996VMD - visual molecular dynamics.J Molec Graphics143338

CITATIONS

CHAPTER 1

Ken Takahashi, Takayuki Oda and Keiji Naruse, Coarse-grained molecular dynamics simulations of biomolecules, DOI: 10.3934/biophy.2014.1.1.

CHAPTER 2

Junfeng Gu, Fang Bai, Honglin Li and Xicheng Wang, A Generic Force Field for Protein Coarse-Grained Molecular Dynamics Simulation, doi:10.3390/ijms131114451.

CHAPTER 3

Ronald D. Hills Jr. and Charles L. Brooks, Insights from Coarse-Grained Gō Models for Protein Folding and Dynamics, doi: 10.3390/ijms10030889.

CHAPTER 4

Jacek Wabik, Mateusz Kurcinski, Andrzej Kolinski **and** Rino Ragno, Coarse-Grained Modeling of Peptide Docking Associated with Large Conformation Transitions of the Binding Protein: Troponin I Fragment–Troponin C System, doi:10.3390/molecules200610763.

CHAPTER 5

Peter Spijker, Bram van Hoof, Michel Debertrand, Albert J. Markvoort, Nagarajan Vaidehi **and Peter A.J.** Hilbers, Coarse Grained Molecular Dynamics Simulations of Transmembrane Protein-Lipid Systems, doi: 10.3390/ijms11062393.

CHAPTER 6

Hall BA, Armitage JP, Sansom MSP (2011) Transmembrane Helix Dynamics of Bacterial Chemoreceptors Supports a Piston Model of Signalling. PLoS Comput Biol 7(10): e1002204. https://doi.org/10.1371/journal.pcbi.1002204.

CHAPTER 7

Nikolay Korolev, Di Luo, Alexander P. Lyubartsev and Lars Nordenskiöld, A Coarse-Grained DNA Model Parameterized from Atomistic Simulations by Inverse Monte Carlo, doi: 10.3390/polym6061655.

CHAPTER 8

Yichen He, Yazhuo Shang, Yu Liu, Shuangliang Zhao and Honglai Liu, Melting dynamics of short dsDNA chains in saline solutions, https://doi.org/10.1186/s40064-015-1581-7.

CHAPTER 9

Aarón Ayllón Benítez, José Ginés Hernández Cifre, Francisco Guillermo Díaz Baños and José García de la Torre, Prediction of solution properties and dynamics of RNAs by means of Brownian dynamics simulation of coarse-grained models: Ribosomal 5S RNA and phenylalanine transfer RNA, DOI 10.1186/s13628-015-0025-7.

CHAPTER 10

Liu L, Chen S-J (2012) Coarse-Grained Prediction of RNA Loop Structures. PLoS ONE 7(11): e48460. https://doi.org/10.1371/journal.pone.0048460.

CHAPTER 11

Hoiles W, Gupta R, Cornell B, Cranfield C, Krishnamurthy V (2016) The Effect of Tethers on Artificial Cell Membranes: A Coarse-Grained Molecular Dynamics Study. PLoS ONE 11(10): e0162790. https://doi.org/10.1371/journal.pone.0162790.

CHAPTER 12

Koivuniemi A, Vuorela T, Kovanen PT, Vattulainen I, Hyvönen MT (2012) Lipid Exchange Mechanism of the Cholesteryl Ester Transfer Protein Clarified by Atomistic and Coarse-grained Simulations. PLoS Comput Biol 8(1): e1002299. https://doi.org/10.1371/journal.pcbi.1002299.

INDEX